Communications in Computer and Information Science 550

Commenced Publication in 2007
Founding and Former Series Editors:
Alfredo Cuzzocrea, Dominik Ślęzak, and Xiaokang Yang

More information about this series at http://www.springer.com/series/7899

Sebastiano Battiato · Sabine Coquillart
Julien Pettré · Robert S. Laramee
Andreas Kerren · José Braz (Eds.)

Computer Vision, Imaging and Computer Graphics – Theory and Applications

International Joint Conference, VISIGRAPP 2014
Lisbon, Portugal, January 5–8, 2014
Revised Selected Papers

 Springer

Editors

Sebastiano Battiato
Dipartimento di Matematica e Informatica
Università di Catania
Catania, Catania
Italy

Sabine Coquillart
Inria/ZIRST
Saint Ismier
France

Julien Pettré
INRIA-Rennes/MimeTIC Tean
Rennes cedex
France

Robert S. Laramee
Swansea University
Swansea
UK

Andreas Kerren
Computer Science
Linnaeus University
Växjö
Sweden

José Braz
do IPS
Escola Superior de Tecnologia
Setúbal
Portugal

ISSN 1865-0929 ISSN 1865-0937 (electronic)
Communications in Computer and Information Science
ISBN 978-3-319-25116-5 ISBN 978-3-319-25117-2 (eBook)
DOI 10.1007/978-3-319-25117-2

Library of Congress Control Number: 2015957120

This Springer imprint is published by SpringerNature
The registered company is Springer International Publishing AG Switzerland

Preface

This book includes the extended versions of selected papers from VISIGRAPP 2014, the International Joint Conference on Computer Vision, Imaging and Computer Graphics Theory and Applications, which was held in Lisbon, Portugal, during January 5–8, 2014. The conference was organized by the Institute for Systems and Technologies of Information, Control and Communication (INSTICC), in cooperation with Eurographics, technically co-sponsored by the IEEE Computer Society, IEEE VGMT, and IEEE TCMC.

VISIGRAPP comprises three conferences, namely, the International Conference on Computer Vision Theory and Applications (VISAPP), the International Conference on Computer Graphics Theory and Applications (GRAPP), and the International Conference on Information Visualization Theory and Applications (IVAPP).

VISIGRAPP received 543 paper submissions from more than 50 countries. After a rigorous double-blind evaluation, only 20 % of the papers were accepted and published as full papers. These numbers show that our conference is aiming for the highest scientific standards, and that it can now be considered a well-established venue for researchers in the broad fields of computer vision, image analysis, computer graphics, and information visualization. From the set of full papers, 22 were selected for inclusion in this book. The selection process was based on quantitative and qualitative evaluation results provided by the Program Committee reviewers as well as the feedback on paper presentations provided by the session chairs during the conference. After selection, the accepted papers were further revised and extended by the authors. Our gratitude goes to all contributors and reviewers, without whom this book would not have been possible. Apart from the full papers, 26 % of the papers were accepted for short presentations and 24 % accepted for poster presentations. However, these works were not considered for the present book. We do not expect that each individual reader is equally interested in all 22 of the selected VISIGRAPP papers. However, the diversity of these papers makes it very likely that all readers can find something of interest in this selection.

As VISAPP 2014 constitutes the largest part of VISIGRAPP with 382 submissions, we decided to select and integrate 13 extended full papers aiming to cover different aspects and areas related to computer vision, such as image formation and pre-processing, image and video analysis and understanding, motion tracking, stereo vision, as well as diverse computer vision applications and services. Here, we would like to mention that when we selected the papers from VISAPP for this book, our intention was to cover and highlight research from different areas and subareas related to computer vision. These papers were mainly competing with other VISAPP papers having similar content, and therefore, we want to explicitly acknowledge that other high-quality papers accepted at the conference could have been integrated in this book if we had enough space for them.

Concerning GRAPP 2014, 85 papers were submitted, and we decided to include five extended full papers in this book. We tried to cover the main areas of computer graphics to bring the content of the book in line with the research addressed at the conference.

The four selected IVAPP 2014 papers are not only excellent representatives of the field of information visualization, but also form a quite balanced representation of the field itself. Above all, they are almost as diverse and exciting as the field of information visualization.

VISIGRAPP 2014 also included four invited keynote lectures, presented by internationally renowned researchers, whom we would like to thank for their contribution to reinforcing the overall quality of the conference. They were (in alphabetical order): Miroslaw Z. Bober (University of Surrey, UK), Stephan Diehl (University of Trier, Germany), Andreas Holzinger (Medical University Graz, Austria), and Gudrun Klinker (TU München, Germany).

We wish to thank all those who supported VISIGRAPP and helped to organize the conference. On behalf of the conference Organizing Committee, we would like to especially thank the authors, whose work was the essential part of the conference and contributed to a very successful event. We would also like to thank the members of the Program Committee, whose expertise and diligence were instrumental in ensuring the quality of the final contributions. We also wish to thank all the members of the Organizing Committee whose work and commitment was invaluable. Last but not least, we would like to thank Springer for their collaboration and help in getting this book to print.

April 2015   Sebastiano Battiato
                                                                      Sabine Coquillart
                                                                          Julien Pettré
                                                                   Robert S. Laramee
                                                                      Andreas Kerren
                                                                             José Braz

Organization

Conference Chair

José Braz   Escola Superior de Tecnologia de Setúbal, Portugal

Program Co-chairs

GRAPP

Sabine Coquillart   Inria, France
Julien Pettré   Inria Rennes - Bretagne Atlantique, France

IVAPP

Robert S. Laramee   Swansea University, UK
Andreas Kerren   Linnaeus University, Sweden

VISAPP

Sebastiano Battiato   University of Catania, Italy

Organizing Committee

Bruno Encarnação   INSTICC, Portugal
Helder Coelhas   INSTICC, Portugal
Ana Guerreiro   INSTICC, Portugal
André Lista   INSTICC, Portugal
Filipe Mariano   INSTICC, Portugal
Andreia Moita   INSTICC, Portugal
Raquel Pedrosa   INSTICC, Portugal
Vitor Pedrosa   INSTICC, Portugal
Cláudia Pinto   INSTICC, Portugal
Cátia Pires   INSTICC, Portugal
Susana Ribeiro   INSTICC, Portugal
Rui Rodrigues   INSTICC, Portugal
Sara Santiago   INSTICC, Portugal
André Santos   INSTICC, Portugal
Fábio Santos   INSTICC, Portugal
Mara Silva   INSTICC, Portugal
José Varela   INSTICC, Portugal
Pedro Varela   INSTICC, Portugal

GRAPP Program Committee

Marco Agus	CRS4, Italy
Tremeau Alain	University of Saint Etienne, France
Marco Attene	National Research Council (CNR), Italy
Lilian Aveneau	University of Poitiers, France
Gérard Bailly	CNRS, France
Francesco Banterle	Visual Computing Lab, Italy
Bernd Bickel	Disney Research Zurich, Switzerland
Jiri Bittner	Czech Technical University in Prague, Czech Republic
Kristopher J. Blom	University of Barcelona, Spain
Manfred Bogen	Fraunhofer IAIS, Germany
David Bommes	Inria Sophia Antipolis, France
Stephen Brooks	Dalhousie University, Canada
Stefan Bruckner	University of Bergen, Norway
Pedro Cano	University of Granada, Spain
Maria Beatriz Carmo	Universidade de Lisboa, Portugal
L. G. Casado	University of Almeria, Spain
Eva Cerezo	University of Zaragoza, Spain
Teresa Chambel	Lasige, University of Lisbon, Portugal
Marc Christie	IRISA/Inria Rennes, France
Ana Paula Cláudio	Universidade de Lisboa, Portugal
Nicolas Courty	Université de Bretagne-Sud, France
Carsten Dachsbacher	Karlsruhe Institute of Technology, Germany
Kurt Debattista	University of Warick, UK
John Dingliana	Trinity College Dublin, Ireland
Jean-Michel Dischler	Université de Strasbourg, France
Thierry Duval	Irisa - Inria Rennes Bretagne Atlantique - Université de Rennes 1, France
Ramsay Dyer	University of Groningen, The Netherlands
Cathy Ennis	Utrecht University, The Netherlands
Marius Erdt	Fraunhofer IDM@NTU, Singapore
Petros Faloutsos	York University, Canada
Jie-Qing Feng	Zhejiang University, China
Luiz Henrique de Figueiredo	Impa, Brazil
Pablo Figueroa	Universidad de los Andes, Colombia
Fabian Di Fiore	Hasselt University, Belgium
Cedric Fleury	Université de Paris-Sud, France
Ioannis Fudos	University of Ioannina, Greece
Alejandro García-Alonso	University of the Basque Country, Spain
Djamchid Ghazanfarpour	University of Limoges, France
Enrico Gobbetti	CRS4, Italy
Stephane Gobron	HES-SO/HE-Arc/ISIC, Switzerland
Laurent Grisoni	LIFL, France
James Hahn	George Washington University, USA

Steve Pettifer	The University of Manchester, UK
Denis Pitzalis	UNESCO, France
Nicolas Pronost	Utrecht University, The Netherlands
Anna Puig	University of Barcelona, Spain
Inmaculada Remolar	Universitat Jaume I, Spain
Mickael Ribardière	University of Poitiers, XLIM, France
Tobias Ritschel	MPI Saarbruckücken, Germany
Inmaculada Rodríguez	University of Barcelona, Spain
Timo Ropinski	Linköping University, Sweden
Isaac Rudomin	BSC, Spain
Christian Sandor	University of South Austria, Australia
Luis Paulo Santos	Universidade do Minho, Portugal
Mateu Sbert	Universitat de Girona, Spain
Rafael J. Segura	Univesidad de Jaen, Spain
Etienne de Sevin	Masa Group, France
A. Augusto Sousa	FEUP/INESC Porto, Portugal
Frank Steinicke	Immersive Media Group, Germany
Ching-Liang Su	Da Yeh University, India
Susanne K. Suter	University of Florida, USA
Marco Tarini	Università degli Studio dell'Insubria, Italy
D. Tegolo	Università di Palermo, Italy
Matthias Teschner	University of Freiburg, Germany
Daniel Thalmann	Nanyang Technological University, Singapore
Juan Carlos Torres	Universidad de Granada, Spain
Torsten Ullrich	Fraunhofer Austria Research, Austria
Anna Ursyn	University of Northern Colorado, USA
Pere-Pau Vázquez	Universitat Politècnica De Catalunya, Spain
Luiz Velho	IMPA - Instituto de Matematica Pura e Aplicada, Brazil
Hannes Vilhjalmsson	Reykjavík University, Iceland
Andreas Weber	University of Bonn, Germany
Daniel Weiskopf	Universität Stuttgart, Germany
Burkhard Wuensche	University of Auckland, New Zealand
Lihua You	Bournemouth University, UK
Jian J. Zhang	Bournemouth University, UK

GRAPP Additional Reviewers

Theodoros Athanasiadis	University of Ioannina, Greece
Moritz Bächer	Disney Research Zurich, Switzerland
Tomas Barak	Czech Technical University in Prague, Czech Republic
Anton Bardera	University of Girona, Spain
Thomas Bashford-Rogers	University of Warwick, UK
Amit Bermano	ETH Zurich/Disney Research Zurich, Switzerland
Imma Boada	University of Girona, Spain
Annelies Braffort	CNRS, France
Gabriel Cirio	URJC Madrid, France

Neven El Sayed	University of South Australia, Australia
Sylvie Gibet	IRISA, France
Daniel Holden	Edinburgh University, UK
Laure Leroy	Université Paris 8, France
Mattia Natali	University of Beregn, Norway
Roberto Ribeiro	Uminho, Portugal
Davide Sobrero	CNR-IMATI, Italy
Veronika Solteszova	Christian Michelsen Research, Norway
Fabian Carlos Tommasini	FaMAF, Argentina
Andreas Vasilakis	University of Ioannina, Greece
He Wang	University of Edinburgh, UK
Xi Zhao	Edinburgh University, UK

IVAPP Program Committee

Wolfgang Aigner	Vienna University of Technology, Austria
Lisa Sobierajski Avila	Kitware Inc., USA
George Baciu	The Hong Kong Polytechnic University, Hong Kong, SAR China
Rita Borgo	Swansea University, UK
David Borland	University of North Carolina at Chapel Hill, USA
Anne Boyer	Loria - Inria Lorraine, France
Massimo Brescia	Istituto Nazionale di AstroFisica, Italy
Ross Brown	Queensland University of Technology, Brisbane, Australia
Maria Beatriz Carmo	Universidade de Lisboa, Portugal
Remco Chang	Tufts University, USA
Guoning Chen	University of Houston, USA
László Czúni	University of Pannonia, Hungary
Christoph Dalitz	Niederrhein University of Applied Sciences, Germany
Robertas Damasevicius	Kaunas University of Technology, Lithuania
Mihaela Dinsoreanu	Technical University of Cluj-Napoca, Romania
Csaba Domokos	National University of Singapore, Singapore
Georgios Dounias	University of the Aegean, Greece
Osman Hassab Elgawi	University of Birmingham, UK
Chi-Wing Fu	Nanyang Technological University, Singapore
Zhao Geng	Swansea University, UK
Mohammad Ghoniem	Centre de Recherche Public Gabriel Lippmann, Luxembourg
Wooi-Boon Goh	Nanyang Technological University, Singapore
David Gotz	IBM Research, USA
Martin Graham	Edinburgh Napier University, UK
Seokhee Hong	University of Sydney, Australia
Weidong Huang	CSIRO ICT Centre, Australia
Seiya Imoto	University of Tokyo, Japan
Mark W. Jones	Swansea University, UK

Rui José   University of Minho, Portugal
Johannes Kehrer   Vienna University of Technology, Austria
Jessie Kennedy   Edinburgh Napier University, UK
Andreas Kerren   Linnaeus University, Sweden
Martin Kraus   Aalborg University, Denmark
Simone Kriglstein   SBA Research, Austria
Denis Lalanne   University of Fribourg, Switzerland
Lars Linsen   Jacobs University, Bremen, Germany
Giuseppe Liotta   University of Perugia, Italy
Shixia Liu   Microsoft Research Asia, China
Ross Maciejewski   Arizona State University, USA
Luis Gustavo Nonato   Universidade de Sao Paulo, Brazil
Steffen Oeltze   University of Magdeburg, Germany
Benoît Otjacques   Centre de Recherche Public - Gabriel Lippmann,
                            Luxembourg
Alex Pang   University of California, Santa Cruz, USA
Torsten Reiners   Curtin University, Australia
Philip J. Rhodes   University of Mississippi, USA
Adrian Rusu   Rowan University, USA
Filip Sadlo   VISUS, University of Stuttgart, Germany
Angel Sappa   Computer Vision Center, Spain
Heidrun Schumann   University of Rostock, Germany
Marc Streit   Johannes Kepler Universität Linz, Austria
Yasufumi Takama   Tokyo Metropolitan University, Japan
Ying Tan   Peking University, China
Sidharth Thakur   Renaissance Computing Institute (RENCI), USA
Slobodan Vucetic   Temple University, USA
Chaoli Wang   Michigan Technological University, USA
Daniel Weiskopf   Universität Stuttgart, Germany
Kai Xu   Middlesex University, UK
Hsu-Chun Yen   National Taiwan University, Taiwan
Hongfeng Yu   University of Nebraska - Lincoln, USA
Xiaoru Yuan   Peking University, China
Blaz Zupan   University of Ljubljana, Slovenia

IVAPP Additional Reviewers

Bilal Alsallakh   Vienna University of Technology, Austria
Bertjan Broeksema   Centre de Recherche Public Gabriel Lippmann,
                            Luxembourg
Joao Comba   UFRGS, Brazil
Yoanne Didry   Centre de Recherche Public Gabriel Lippmann, France
Hanqi Guo   School of EECS, China
Kostiantyn Kucher   ISOVIS group, Linnaeus University, Sweden
Rosane Minghim   Universidade de Sao Paulo, São Carlos, Brazil, Brazil
Arlind Nocaj   University of Konstanz, Germany

Olivier Parisot   Centre de Recherche Public Gabriel Lippmann,
                                 Luxembourg
Johanna Schmidt   TU Wien, Austria
Guilherme Telles   University of Campinas, Brazil
Zuchao Wang   Peking University, China
Björn Zimmer   Linnaeus University, Sweden

VISAPP Program Committee

Amr Abdel-Dayem   Laurentian University, Canada
Tremeau Alain   University of Saint Etienne, France
Vicente Alarcon-Aquino   Universidad de las Americas Puebla, Mexico
Djamila Aouada   University of Luxembourg, Luxembourg
Jamal Atif   Université Paris-Sud 11, France
Xiao Bai   Beihang University, China
Lamberto Ballan   Università degli Studi di Firenze, Italy
Hichem Bannour   Atomic Energy Commission (CEA), France
Arrate Muñoz Barrutia   University of Navarra, Spain
Giuseppe Baruffa   University of Perugia, Italy
Mohamed Batouche   University Constantine 2, Algeria
Sebastiano Battiato   University of Catania, Italy
Saeid Belkasim   Georgia State University, USA
Fabio Bellavia   University of Palermo, Italy
Neil Bergmann   University of Queensland, Australia
Thilo Borgmann   TU Berlin, Germany
Adrian Bors   University of York, UK
Giosue Lo Bosco   University of Palermo, Italy
Djamal Boukerroui   Université de Technologie de Compiègne, France
Roland Bremond   Institut Français des Sciences et Technologies des
                                 Transports, de l'aménagement et des Réseaux
                                 (IFSTTAR), France
Marius Brezovan   University of Craiova, Romania
Egon L. van den Broek   University of Twente/Radboud UMC Nijmegen,
                                 The Netherlands
Ross Brown   Queensland University of Technology, Brisbane,
                                 Australia
Alfred Bruckstein   Technion, Israel
Arcangelo R. Bruna   STMicroelectronics, Italy
Xianbin Cao   Beihang University, China
Alice Caplier   GIPSA-lab, France
Franco Alberto Cardillo   Consiglio Nazionale delle Ricerche, Italy
Pedro Latorre Carmona   Universidad Jaume I, Spain
M. Emre Celebi   Louisiana State University in Shreveport, USA
Krzysztof Cetnarowicz   AGH - University of Science and Technology, Poland
Chee Seng Chan   University of Malaya, Malaysia
Vinod Chandran   Queensland University of Technology, Australia

Chin-Chen Chang	Feng Chia University, Taiwan
Hang Chang	Lawrence Berkeley National Lab, USA
Chung Hao Chen	Old Dominion University, USA
Samuel Cheng	University of Oklahoma, USA
Laurent Cohen	Université Paris Dauphine, France
David Connah	University of Bradford, UK
Donatello Conte	Université François Rabelais Tours, France
Guido de Croon	Delft University of Technology, The Netherlands
Fabio Cuzzolin	Oxford Brookes University, UK
Dima Damen	University of Bristol, UK
Roy Davies	Royal Holloway, University of London, UK
Larry Davis	University of Maryland College Park, USA
Emmanuel Dellandréa	Ecole Centrale de Lyon, France
Matteo Dellepiane	ISTI - CNR, Italy
David Demirdjian	Vecna, USA
Thomas M. Deserno	Aachen University of Technology (RWTH), Germany
Michel Devy	LAAS-CNRS, France
Sotirios Diamantas	University of Nebraska, USA
Yago Diez	University of Girona, Spain
Jana Dittmann	Otto-von-Guericke-Universität Magdeburg, Germany
Aijuan Dong	Hood College, USA
Jean-Luc Dugelay	Eurécom, Sophia Antipolis, France
Aysegul Dundar	Purdue University, USA
Mahmoud El-Sakka	The University of Western Ontario, Canada
Sergio Escalera	Computer Vision Center, Universitat de Barcelona, Spain
Grigori Evreinov	University of Tampere, Finland
Zhigang Fan	Xerox Corp., USA
Giovanni Maria Farinella	Università di Catania, Italy
Sanaa El Fkihi	Ensias, Université Mohammed V Souissi, Rabat, Morocco
David Fofi	Le2i, France
Tyler Folsom	QUEST Integrated Inc., USA
Gian Luca Foresti	Unversity of Udine, Italy
Andrea Fossati	ETH Zurich, Switzerland
Mohamed Fouad	MTC, Egypt
Roberto Fraile	University of Leeds, UK
John Qiang Gan	University of Essex, UK
Juan David Garcia	Universidad Nacional de Colombia, Colombia
Miguel A. Garcia-Ruiz	Algoma University, Canada
José Gaspar	ISR, IST/UTL, Portugal
Antonios Gasteratos	Democritus University of Thrace, Greece
Basilios Gatos	National Center for Scientific Research Demokritos, Greece
Luiz Goncalves	Federal University of Rio Grande do Norte, Brazil
Amr Goneid	The American University in Cairo, Egypt

Jordi Gonzàlez — Computer Vision Center, Spain
Manuel González-Hidalgo — Balearic Islands University, Spain
Bernard Gosselin — University of Mons, Belgium
Nikos Grammalidis — Centre of Research and Technology Hellas, Greece
Manuel Grana — University of the Basque Country, Spain
Christos Grecos — University of West of Scotland, UK
Jean-Yves Guillemaut — University of Surrey, UK
Andras Hajdu — Univerity of Debrecen, Hungary
Daniel Harari — Weizmann Institute of Science, Israel
Søren Hauberg — Max Planck Institute for Intelligent Systems, Germany
Kaoru Hirota — Tokyo Institute of Technology, Japan
Timothy Hospedales — Queen Mary University of London, UK
Hsi-Chin Hsin — National United University, Taiwan
Céline Hudelot — Ecole Centrale de Paris, France
Sae Hwang — University of Illinois at Springfield, USA
Daniela Iacoviello — Sapienza Università di Roma, Italy
Jiri Jan — University of Technology Brno, Czech Republic
Tatiana Jaworska — Polish Academy of Sciences, Poland
Ashoka Jayawardena — University of New England, Australia
Jiayan Jiang — Facebook, USA
Xiaoyi Jiang — University of Münster, Germany
Zhong Jin — Nanjing University of Science and Technology, China
Leo Joskowicz — The Hebrew University of Jerusalem, Israel
Martin Kampel — Vienna University of Technology, Austria
Mohan Kankanhalli — National University of Singapore, Singapore
Etienne Kerre — Ghent University, Belgium
Anastasios Kesidis — National Center For Scientific Research, Greece
Sehwan Kim — WorldViz LLC, USA
Nahum Kiryati — Tel Aviv University, Israel
Mario Köppen — Kyushu Institute of Technology, Japan
Andreas Koschan — University of Tennessee, USA
Constantine Kotropoulos — Aristotle University of Thessaloniki, Greece
Arjan Kuijper — Fraunhofer Institute for Computer Graphics Research, Darmstadt, Germany
Paul Kwan — University of New England, Australia
Agata Lapedriza — Universitat Oberta de Catalunya, Spain
Sébastien Lefèvre — Université de Bretagne Sud, France
Stan Z. Li — Chinese Academy of Sciences, China
Huei-Yung Lin — National Chung Cheng University, Taiwan
Xiuwen Liu — Florida State University, USA
Kar Seng Loke — Monash University, Malaysia
Nicolas Loménie — Université Paris Descartes, France
Angeles López — Universitat Jaume I, Spain
Jinhu Lu — Chinese Academy of Sciences, China
Rastislav Lukac — Foveon, Inc., USA
Lindsay MacDonald — University College London, UK

Muriel Visani	Université de La Rochelle, France
Frank Wallhoff	Jade University of Applied Science, Germany
Yu Wang	Auxogyn, Inc., USA
Zuoguan Wang	3M Company, USA
Joost van de Weijer	Autonomous University of Barcelona, Spain
Quan Wen	University of Electronic Science and Technology of China, China
Christian Wöhler	TU Dortmund University, Germany
Denis Fernando Wolf	University of Sao Paulo, Brazil
Stefan Wörz	University of Heidelberg, Germany
Pingkun Yan	Chinese Academy of Sciences, China
Guoan Yang	Xian Jiaotong University, China
Jucheng Yang	Tianjin University of Science and Technology, China
Lara Younes	Université de Reims Champagne-Ardenne, France
Yizhou Yu	University of Illinois, USA
Huiyu Zhou	Queen's University Belfast, UK
Yun Zhu	UCSD, USA
Li Zhuo	Beijing University of Technology, China
Peter Zolliker	Empa, Swiss Federal Laboratories for Materials Science and Technology, Switzerland
Ju Jia (Jeffrey) Zou	University of Western Sydney, Australia

VISAPP Additional Reviewers

Hassan Afzal	University of Luxembourg, Luxembourg
Laurent Caraffa	IFSTTAR, France
Mickaël Coustaty	L3i labs - University of La Rochelle, France
Girum Demisse	SnT, Ethiopia
Ricardo Galego	Instituto Superior Técnico/ISR, Portugal
Fausto Galvan	Università degli Studi di Udine, Italy
Shibo Gao	Northwestern Polytechnical University, China
Arnaud Gonguet	Alcatel-Lucent Bell Labs France, France
Sio-Song Ieng	IFSTTAR, France
Kassem Al Ismaeil	University of Luxembourg, Luxembourg
Stephan Jonas	Uniklinik RWTH Aachen, Germany
Zicheng Liao	University of Illinois at Urbana-Champaign, USA
Yugang Liu	National University of Singapore, Singapore
Giuseppe Loianno	University of Pennsylvania, USA
Rafik Mebarki	PRISMA Lab, Uniersity of Naples Federico II, Italy
Marco Moltisanti	Università Degli Studi di Catania, Italy
Akshay Pai	University of Copenhagen, Denmark
Giovanni Puglisi	University of Catania, Italy
Daniele Ravì	University of Catania, Italy
Michael Sapienza	Oxford Brookes University, UK
Sunando Sengupta	Oxford Brookes University, UK
Oleg Starostenko	UDLA, Mexico

Matteo Taiana	ISR/IST Lisboa, Portugal
Tiberio Uricchio	Università degli Studi di Firenze, Italy
Baoyuan Wang	Microsoft Research Asia, China
Quan Wang	Rensselaer Polytechnic Institute, USA
Ruobing Wu	The University of Hong Kong, Hong Kong
Yanjun Zhao	Georgia State University, USA
Olga Zoidi	Aristotle University of Thessaloniki, Greece

Invited Speakers

Miroslaw Z. Bober	University of Surrey, UK
Stephan Diehl	University of Trier, Germany
Andreas Holzinger	Medical University Graz, Austria
Gudrun Klinker	TU München, Germany

Contents

Computer Vision Theory and Applications

Invited Paper

Past, Present, and Future
of and in Software Visualization

Stephan Diehl[(✉)]

Computer Science Department, University of Trier, Trier, Germany
diehl@uni-trier.de

Abstract. In a selective retrospective of the history of software visualization we discuss examples of applying visualization techniques to analyze the past and present state of software. Based on this retrospective, we make various suggestions for future research. In particular, we argue that the prediction of future aspects of a software system is an important task, but that software visualization research has only scratched the surface of it and that speculative visualization will be one of the major future challenges.

Keywords: Software visualization · Visualization · Software engineering

1 Introduction

Software is used for all kinds of applications and this is in particular true for software that produces visualizations. In this paper we give a very brief and selective historical overview of the use of software for visualizing software itself[1]. As the title of this paper indicates, we discuss the past, present and future *in and of* software visualization. Thus, our goal is twofold: on the one hand we discuss topics and interesting approaches in the history of software visualization as well as current and possible future trends, on the other hand we also look at how visualization is used to show the past, the current state, and the possible future of software. Finally, we hope that this paper will inspire young researchers in the field and give them some ideas for future research.

Up to the late 90ties, research in software visualization focussed on the visualization of algorithms and programs [26]. This narrow view has changed since. Researchers began to visualize all kinds of information and artifacts related to software and its development process not only including technical artifacts like source code, data structures, as well as program state and code at runtime, but also requirements, design documents, bug reports, and software changes [11]. These artifacts actually provide information about three aspects of software: its structure, behavior and evolution. Thus, the rest of this paper will be organized along the historical (past, present, future) and software (structure, behavior, evolution) dimensions.

---

[1] For copyright reasons, we use examples from our own work to illustrate various approaches.

© Springer International Publishing Switzerland 2015
S. Battiato et al. (Eds.): VISIGRAPP 2014, CCIS 550, pp. 3–11, 2015.
DOI: 10.1007/978-3-319-25117-2_1

2   Visualizing the Structure of Software

In the early days of software visualization researchers were mostly interested in visualizing the structure of software and they developed many different graphical notations. Although, these notations were mainly intended for the design of software and diagrams were mostly drawn by hand, automatic generation of diagrams from existing source code was a topic as well. For example, in 1958 Scott presented a tool that computed control-flow diagrams and printed them with a line printer [25]. Many other notations were proposed later on including Jackson diagrams [14], Nassi-Shneiderman diagrams [22] and control-structure diagrams [9].

In the last fifteen years researchers every now and then developed tools that use 3D to visualize the structure of software (e.g. IMSoVision [21], Software Landscapes [3], city metaphor [23], UML City [19], and Code City [29]). From a visualization point of view these are either three-dimensional node-link diagrams or variants of information pyramids [1]. All of them use the size of boxes or buildings to indicate some software metrics.

Kuhn et al. [18] visualize software as a map of islands and integrated their visualization into the Eclipse IDE. The participants of their study "found the map most useful to explore search results", because it gave them a good estimate of the quantity and dispersion of these results.

Another recent trend is not only to visualize structure, but also to compare different structures, e.g. different software architectures. Figure 1 shows an example from our own work [5], a tool to compare two different hierarchical module decompositions, in other words the package structure, of the same software system (here: JFTP) as well as dependencies (here: static code dependencies and co-change coupling) within these structures. The horizontal and vertical icicle plots show the two considered structures. One decomposition is the original one, the other one was automatically computed using software clustering. The grey background color in the matrix indicates to what extent a module in the original decomposition is similar (with respect to the files it contains) to a module in the computed decomposition. In the shown figure, these cells are quite close to the diagonal, indicating a strong similarity of the two structures. The tool enabled us to better understand the influence of different concepts of coupling on software clustering [4].

3   Visualizing the Behavior of Software

Allegedly, the first animation of an algorithm was produced by Knowlton [17] in 1966 for demonstrating the features of the programming language L6 he had developed. It took another fifteen years before Baecker's video "Sorting out Sorting" [2] stirred new interest in algorithm animation. The most striking part of that video is the sorting race where the animations of nine different sorting algorithms are shown simultaneously on the screen. While quicksort finishes after 30 s on the given data set, for most of the other sorting algorithms there is almost

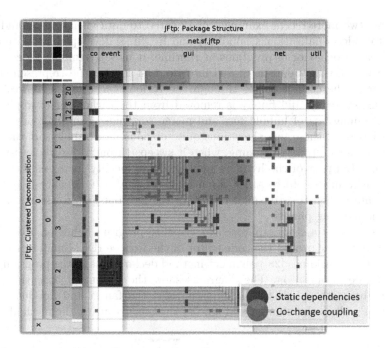

Fig. 1. Matrix-based visualization technique to compare different module decompositions and dependency graphs.

no visible progress, i.e. their data sets look as unsorted as at the beginning. The production of the video required not only a lot of programming, but also expertise in video recording and composing. So, researchers started to think about tool support for the development of algorithm animations. In 1985 Brown and Najork presented the Balsa tool which featured interesting events and multiple views which was followed by a series of other algorithm animation tools developed by Brown: Balsa II (1988), Zeus (1992), Anim3D, Zeus3D (1993), CAT (1996), JCAT (1997). In the 90ties many researchers started to develop their own tools often targeting particular kinds of algorithms, e.g. graph algorithms or geometrical algorithms. Tools developed during that period include Tango, Polka, Samba, Polka3D, Animal, CATAI, Daphne, Ganimal, Gasp, GeoWin, Jawaa, Jeliot, and Leonardo. Researchers did not only develop the tools, but also try to evaluate their usefulness. In a meta study of 24 studies on algorithm animation Hundhausen et al. [13] concluded: "Thus, according to our analysis, how students use AV technology, rather than what students see, appears to have the greatest impact on educational effectiveness." They also found that electronic learning material was more effective than lectures, and that algorithm animations were most successful, if the learner had to actively work on prediction or programming tasks.

While algorithm animation is mainly used for educational purposes, other approaches to visualize program behavior are meant to help developers. Next, we discuss two more recent examples, one related to data visualization and one related to code visualization. Choudhury and Rosen [8] developed a tool to analyze the cache behavior of programs at runtime. It uses a dynamic radial visualization to show how data is moved into the first level cache (represented by the center of the circle), passes through the various cache levels (represented by concentric rings), and finally is removed from the cache. For example, the resulting animations of bubble sort and merge sort make the difference of the cache behavior of these two algorithms quite obvious. For bubble sort the data is not cache local at the beginning, when it sweeps through the whole list, but as the unsorted part of the list gets smaller, cache locality increases, i.e. there are less cache misses. In contrast, for merge sort the cache locality is good at the beginning when it sorts smaller parts of the list, but gets worse at the end during the merging phases.

Figure 2 shows a static approach to visualize program behavior. Here, performance data is shown in the code editor of the Eclipse IDE. Small word-sized diagrams (c.f. sparklines [28]) augment method declarations in the program text with runtime information. The diagrams show the estimated execution time spent in the method and the methods it calls, the fan-in and fan-out of the method, as well as the number and kind of threads executing the method.

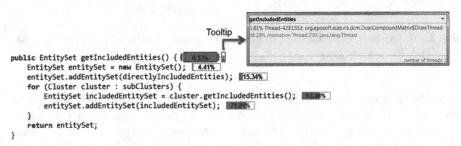

Fig. 2. Visually augmenting method declarations with performance data.

4   Visualizing the Evolution of Software

As Frederick Brooks [16] put it: "All successful software gets changed. [...] In short, the software product is embedded in a cultural matrix of applications, users, laws, and machine vehicles. These all change continually, and their changes inexorably force change upon the software product." Configuration management tools allow to archive changes to different kinds of artifcats of a software system and thus provide the data for later analysis of its evolution.

One of the early and probably the most influential visualizations of software evolution was SeeSoft [12]. It shrinks each line of the source code down to the size of a single pixel. The color of a pixel represents some metric value of the related line of code, e.g. the age of its last change. The user can travel through time using a slider to select versions of the files and the visualization is synchronously updated.

As discussed above, 3D representations have been used to show the architecture of a system. Steinbrückner and Lewerentz came up with a 3D visualization, called Software Cities [27] that shows the structure and its evolution in a static 3D image. In analogy to the evolution of cities in the real world, software cities grow down the hill. The original classes are represented by buildings on top of the hill, new classes are put on lower levels downhill.

As the interest in mining software repositories increased, in the last 15 years, researchers also began to develop methods to visualize information stored in these repositories. For example, several approaches for visualizing co-change information, i.e. the fact that several artifacts (files, classes, method, etc.) have been changed at the same time or as part of the same transaction, have been proposed including pixelmaps [30], ego-centric network visualization (Evolution Radar [10]), and graph clustering (CCVisu [6]).

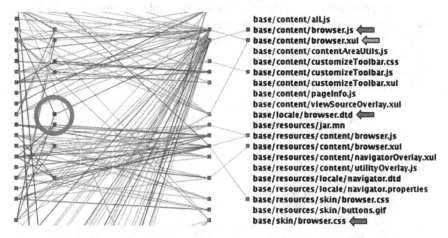

Fig. 3. EpoSee's Parallel coordindates visualization of sequence rules mined from the Mozilla Firefox repository.

When applying established data mining techniques (association and sequence rule mining) to learn from past software changes, we found that we had to use low thresholds leading to too many rules to inspect manually. Thus, we developed a visual data mining tool [7], see Fig. 3, that helps to find patterns and outliers in the rules mined from software repositories. The insights gained with our tool actually lead to the development of a recommender system for programming [31] which leverages the mined rules.

5   The Future in and of Software Visualization

To conclude this paper, we want to provide some suggestions for future research.

Learn from the past and predict the future. While there has been a lot of follow-up research on recommender systems and mining of software repositories, I am not aware of any approach to systematically use the information

to explore possible futures of a software system. In such a speculative approach the uncertainty would increase the farther one moves into the future.
Thus, the level of uncertainty should be visualized. Furthermore, information about alternative futures could be shown as context information using
focus-context techniques.

Visual standards. While researchers continue to produce novel visualizations
of the structure, behavior and evolution of software, there seem to be almost
no visual standards. Here, we need to learn from medicine: while it takes
medical doctors a long time to learn to read X-ray images, if finally pays off
and they become effective in making diagnoses. Empirical studies may help
to uncover emerging visual standards, areas were standards are needed and
ways to foster their use.

Beyond animation for showing structural change. Many software visualization approaches rely on graphs. For small graphs, changes can be shown
using graph animation. For larger graphs, these animations are confusing.
Either one has to develop automatic focusing techniques [24] to reduce the
size of the graphs, or, alternatively, one has to find static visualizations of
changes.

Algorithm explanation. Most work in algorithm animation addresses the visualization of the execution of programs with concrete input data. The visual
execution of programs with abstract input data, i.e. representations which
only reflect the relevant properties, has hardly ever been investigated [15].

Fig. 4. Collaborative and multi-modal interfaces: The image shows two collaborative
modeling tools (CREWW and CREWSpace) that leverage Wii remotes and smartphones as input devices for collaborative modelling.

3D visualization. Most of todays software visualization tools do not even exploit the graphics power of an average PC, laptop or even smartphone. Most existing 3D visualizations are basically three-dimensional graph representations which come with serious occlusion and navigation problems. In modern computer games narrative elements help the user to find her way through the virtual worlds.

Collaborative visualization and multi-modal interfaces. Novel display and interaction devices provide new opportunities for distributed as well as co-located collaboration. This collaboration can be both synchronous or asynchronous. For example, we have recently developed tools [20] to leverage gaming devices and smartphones in co-located modelling sessions in software engineering (see Fig. 4) using shared and personal workspaces. In a similar way, one could use these devices to collaboratively select and filter data sets, as well as to generate and navigate through visualizations. Moreover, following a divide and conquer approach visualizations could be created in a personal workspace, and then composed in a shared space.

End-user visualization. Today, software visualization tools are mainly used in education or by software developers. Some kinds of visualizations may be helpful for end-users who want to get some information about an application or smartphone app that they are using. For example, they may want to know what components are required for an application, what the licenses of these components are, and what resources they access.

Visualization and trust. Can visualization increase users' trust that applications do not leak condential information given that some kind of analysis and certification is able to provide reliable information. For example, before using a Web service a user might want to make sure that her credit card information is only passed through trusted intermediate Web services.

References

1. Andrews, K., Wolte, J., Pichler, M.: Information pyramids: a new approach to visualising large hierarchies. In: Proceedings of IEEE Visualization 1997 (1997)
2. Baecker, R.: Sorting Out Sorting. 30 minute color film (developed with assistance of Dave Sherman, distributed by Morgan Kaufmann, University of Toronto) (1981)
3. Balzer, M., Noack, A., Deussen, O., Lewerentz, C.: Software landscapes: visualizing the structure of large software systems. In: Proceedings of Joint EUROGRAPHICS - IEEE TCVG Symposium on Visualization, VisSym 2004, Konstanz, Germany, pp. 261–266 (2004)
4. Beck, F., Diehl, S.: On the impact of software evolution on software clustering. Empirical Softw. Eng. **18**(5), 970–1004 (2013)
5. Beck, F., Diehl, S.: Visual comparison of software architectures. Inf. Vis. **12**(2), 178–199 (2013)
6. Beyer, D.: Co-change visualization applied to PostgreSQL and ArgoUML: (MSR challenge report). In: Proceedings of the 2006 International Workshop on Mining Software Repositories, MSR, Shanghai, China, pp. 165–166 (2006)
7. Burch, M., Diehl, S., Weißgerber, P.: Visual data mining in software archives. In: Proceedings of ACM Symposium on Software Visualization (SOFTVIS05), St. Louis, MO. ACM Press, New York, NY, May 2005

8. Choudhury, A.N.M.I., Rosen, P.: Abstract visualization of runtime memory behavior. In: Proceedings of the 6th IEEE International Workshop on Visualizing Software for Understanding and Analysis, VISSOFT 2011, Williamsburg, VA, USA, pp. 1–8 (2011)

9. Cross, J.H., Hendrix, T.D., Maghsoodloo, S.: The control structure diagram: an overview and initial evaluation. Empirical Softw. Eng. **3**, 131–158 (1998)

10. D'Ambros, M., Lanza, M., Lungu, M.: Visualizing co-change information with the evolution radar. IEEE Trans. Softw. Eng. **35**(5), 720–735 (2009)

11. Diehl, S.: Software Visualization: Visualizing the Structure, Behaviour, and Evolution of Software. Springer, Heidelberg (2007)

12. Eick, S.G., Steffen Jr, J.L., Sumner Jr, E.E.: Seesoft TM – a tool for visualizing line oriented software statistics. IEEE Trans. Softw. Eng. **18**(11), 957–968 (1992)

13. Hundhausen, C.D., Douglas, S.A., Stasko, J.T.: A meta-study of algorithm visualization effectiveness. J. Vis. Lang. Comput. **13**(3), 259–290 (2002)

14. Jackson, M.: Principles of Program Design. Academic Press, Waltham (1975)

15. Johannes, D., Seidel, R., Wilhelm, R.: Algorithm animation using shape analysis: visualising abstract executions. In: Proceedings of ACM Symposium on Software Visualization(SOFTVIS 2005), St. Louis, MI, pp. 17–26. ACM Press, New York, NY (2005)

16. Brooks Jr, F.P.: No silver bullet - essence and accidents of software engineering. IEEE Comput. **20**(4), 10–19 (1987)

17. Knowlton, K.: Bell Telephone Laboratories Low-Level Linked List Language. 16-min black and white film. Bell Telephone Laboratories, MurrayHill (1966)

18. Kuhn, A., Erni, D., Nierstrasz, O.: Embedding spatial software visualization in the IDE: an exploratory study. In: Proceedings of the ACM 2010 Symposium on Software Visualization, Salt Lake City, UT, USA, pp. 113–122 (2010)

19. Lange, C.F.J., Chaudron, M.R.V.: Interactive views to improve the comprehension of UML models - an experimental validation. In: Proceedings of 15th International Conference on Program Comprehension (ICPC 2007), Banff, Alberta, Canada, pp. 221–230 (2007)

20. Lutz, R., Schäfer, S., Diehl, S.: Are smartphones better than CRC cards? In: Symposium on Applied Computing, SAC 2014, Gyeongju, Republic of Korea, pp. 987–994. ACM (2014)

21. Maletic, J.I., Leigh, J., Marcus, A., Dunlap, G.: Visualizing object-oriented software in virtual reality. In: Proceedings of Ninth International Workshop on Program Comprehension(IWPC 2001), Toronto, Canada, pp. 49–54. IEEE Computer Society Press, Washington, DC(2001)

22. Nassi, I., Shneiderman, B.: Flowchart techniques for structured programming. SIGPLAN Not. **8**(8), 12–26 (1973)

23. Panas, T., Epperly, T., Quinlan, D.J., Sbjrnsen, A., Vuduc, R.W.: Communicating software architecture using a unified single-view visualization. In: Proceedings of 2th International Conference on Engineering of Complex Computer Systems (ICECCS 2007), Auckland, New Zealand, pp. 217–228 (2007)

24. Reitz, F., Pohl, M., Diehl, S.: Focused animation of dynamic compound graphs. In: 13th International Conference on Information Visualisation, IV 2009, Barcelona, Spain, pp. 679–684. IEEE Computer Society (2009)

25. Scott, A.E.: Automatic preparation of flow chart listings. J. ACM Int. Bus. Mach. Corporation **5**(1), 57–66 (1958)

26. Stasko, J.T., Domingue, J., Brown, M.H., Price, B.A.: Software Visualization - Programming as a Multimedia Experience. MIT Press, Cambridge (1998)

27. Steinbrückner, F., Lewerentz, C.: Understanding software evolution with software cities. Inf. Vis. **12**(2), 200–216 (2013)
28. Tufte, E.R.: Beautiful Evidence. Graphis Press, New York (2006)
29. Wettel, R., Lanza, M., Robbes, R.: Software systems as cities: a controlled experiment. In: Proceedings of the 33rd International Conference on Software Engineering, ICSE 2011, Waikiki, Honolulu, HI, USA, pp. 551–560 (2011)
30. Zimmermann, T., Diehl, S., Zeller, A.: How history justifies system architecture (or not). In: Proceedings of 6th International Workshop on Principles of Software Evolution (IWPSE 2003), Helsinki, Finland, pp. 73–83 (2003)
31. Zimmermann, T., Weißgerber, P., Diehl, S., Zeller, A.: Mining version histories to guide software changes. IEEE Trans. Softw. Eng. **31**(6), 429–445 (2005)

Computer Graphics Theory and Applications

Distribution Interpolation of the Radon Transforms for Shape Transformation of Gray-Scale Images and Volumes

Márton József Tóth and Balázs Csébfalvi[(✉)]

Budapest University of Technology and Economics,
Magyar tudósok krt. 2, Budapest, Hungary
{tmarton,cseb}@iit.bme.hu
http://cg.iit.bme.hu/

Abstract. In this paper, we extend 1D distribution interpolation to 2D and 3D by using the Radon transform. Our algorithm is fundamentally different from previous shape transformation techniques, since it considers the objects to be interpolated as density distributions rather than level sets of density functions. First, we perform distribution interpolation on the precalculated Radon transforms of two different density functions, and then an intermediate density function is obtained by a consistent inverse Radon transform. This approach guarantees a smooth transition along all the directions the Radon transform is calculated for. Unlike the previous methods, our technique is able to interpolate between features that do not even overlap and it does not require a one dimension higher object representation. We will demonstrate that these advantageous properties can be well exploited for 3D modeling and metamorphosis.

Keywords: Shape-based interpolation · Image/volume morphing · Distribution interpolation · Radon transform

1 Introduction

Shape-based interpolation is mainly used for (1) modeling or reconstruction of 3D objects from 2D cross sections [12,13,21,24,26] and (2) morphing [8,17,25]. The major application fields of these techniques are Computer Aided Design (CAD), movie industry, and medical image processing and visualization. In CAD systems, 3D geometrical models can be built from contours defined in cross-sectional slices [18,24]. Surfaces that fit onto the contours are obtained by using a contour-interpolation method between the subsequent slices. In the movie industry, shape transformation is used for making special effects, such as morphing characters. In 3D medical imaging, it is usual that the resolution of a volumetric data set is lower along the z axis than along the x and y axes. Therefore, a shape-based interpolation technique is applied to produce intermediate slices to obtain an isotropic volume representation [12,13,21,24]. The most popular way

© Springer International Publishing Switzerland 2015
S. Battiato et al. (Eds.): VISIGRAPP 2014, CCIS 550, pp. 15–28, 2015.
DOI: 10.1007/978-3-319-25117-2_2

of automatic shape transformation is based on an Implicit Representation (IR) of 2D or 3D shapes [5,14]. An intermediate shape is simply produced as a level set of a function that is calculated by interpolating between the functions that implicitly define the initial and final shapes [13,21]. This approach is easy to implement and robust in a sense that topologically different shapes can be interpolated without searching for pairs of corresponding points. Nevertheless, in this paper, we show that the previous IR methods are able to make a smooth transition between two features only if they are overlapping, otherwise the features get disconnected. Furthermore, shape-based interpolation of gray-scale images (in other words, density functions) requires a one dimension higher representation than a shape-based interpolation of object boundaries [12]. To remedy these problems, we propose a fundamentally different approach for shape-based interpolation. Our major goal is to guarantee a continuous transition between the lower-dimensional projections of density functions to be interpolated. Therefore, we precalculate the Radon transforms [10] of the density functions, which represent the lower-dimensional projections from different angles, and apply a distribution interpolation [22] between the corresponding projections. The result is then transformed back by the classical Filtered Back-Projection (FBP) algorithm, which implements the inverse Radon transform. We will demonstrate that the modeling potential of this algorithm is much higher than that of the IR methods, as it is able to connect features that do not overlap. Moreover, our method is efficient to use even for interpolating between 3D density functions, as the alternative representation produced by the Radon transform remains 3D, while the classical shape-based interpolation of gray-scale volumes would require 4D IRs to calculate [12]. Distribution interpolation and the Radon transform are well-known tools that have been used separately in different application fields, but to the best of our knowledge, their combination and its application for shape-based interpolation has not been studied so far.

2   Related Work

In computer graphics, shape-based interpolation is usually applied for interpolating between the boundary contours of 2D shapes or morphing between the boundary surfaces of 3D objects. However, an automatic morphing between translucent objects [16], which are defined by volumetric density functions rather than explicit geometrical models is still a challenging task. It depends on subjective preferences whether a morphing algorithm should be fully automatic or user-controlled. We think that the advantages of these approaches are complementary and it depends on the given application which one to prefer. As we focus on automatic morphing, warping techniques that require user intervention [3,8,11,17] are out of the scope of this paper. Automatic morphing is favorable, for instance, if a shape-based interpolation is required between all pairs of consecutive slices in a huge volumetric data set, or a morphing needs to be performed between 3D objects that are completely different geometrically and corresponding features can hardly be specified.

2.1 Shape-Based Interpolation of Boundary Contours

Early shape-based interpolation techniques were proposed for medical imaging applications, where the goal was to reconstruct 3D shapes of different organs from 2D slices of CT or MRI scans [12,13,21]. For example, this is a typical application field, where an automatic processing is clearly an advantage, since a huge amount of voxel data is required to be efficiently processed preferably without any user interaction. As in a usual volumetric data set the inter-slice distance is higher than the distance between the pixels of the slices, additional intermediate slices need to be interpolated to produce an isotropic volume. The brute-force method is to directly interpolate between the original slices, and to apply the well-known Marching Cubes algorithm [19] to extract a boundary surface. However, this approach often results in severe staircase artifacts. To reproduce smooth boundary surfaces, shape-based interpolation techniques first detect boundary contours on the slices and then apply a more sophisticated contour-interpolation method. For contour interpolation, a variety of methods have been published that build a triangular mesh which connects the two consecutive contours [2,7,18,24]. Generally, this is a difficult task, since the problem of self-intersection and topologically different contours need to be carefully handled. Unlike the direct contour-interpolation techniques, the IR methods can easily avoid these problems. The basic idea is to interpolate between the functions that implicitly represent the consecutive contours, and extract intermediate contours from the interpolated function. As a shape-based interpolation method is required to handle the distance information somehow, it is a natural choice to use a Signed Distance Map (SDM) as an IR [5,14]. The pixels of a 2D SDM represent the distance to the nearest contour point, but inside the contour the sign is positive, while outside the contour it is negative. An intermediate contour is obtained by extracting the zero-crossing level set of the interpolated SDMs [12,13,21]. The SDM representation is efficient to calculate using the chamfering method [1].

2.2 Shape-Based Interpolation of Boundary Surfaces

Shape-based contour interpolation is straightforward to extend to surface interpolation [21]. A boundary surface of an object can be represented by a 3D SDM, where the voxels store the distance to the closest surface point. Similarly to the 2D SDMs, the sign is positive inside the object and negative outside the object. The interpolation of the 3D SDMs and the extraction of the intermediate boundary surface are done analogously to the 2D case. Although this method can produce transitions between topologically different objects, the transitions are often not smooth enough due to the discontinuous curvature of the SDMs. In order to achieve smoother transitions, variational interpolation was proposed [25], which constructs IRs of minimal aggregate curvature. The IRs are searched for as a linear combination of Radial Basis Functions (RBF) [6] and the coefficients are determined such that the curvature is minimized. This requires the solution of a large linear equation system, which is time-consuming for complex shapes.

Furthermore, in case of such constrained optimization problems, the coefficient matrix is prone to be ill-conditioned [26]; thus, its inversion by the proposed LU decomposition could easily become instable if the number of the unknown variables drastically increase. This is probably the reason why variational interpolation [25] has not been extended to gray-scale images or volumes.

2.3   Extension to Gray-Scale Images and Volumes

The shape-based interpolation of gray-scale images [12] is computationally much more expensive than the shape-based interpolation of contours as it requires 3D IRs to interpolate rather than only 2D IRs. Each image is considered to be a height field, which can alternatively be represented by a 3D SDM. The intermediate images are obtained as height fields extracted from the interpolated 3D SDMs. Thus, both the chamfering and the extraction of the zero-crossing level set require the processing of 3D volumes for each pair of consecutive images. A shape-based interpolation of gray-scale volumes [12] is even more expensive. Here, the height fields are defined over the 3D space; therefore, the corresponding SDMs are 4D data sets. Consequently, 4D data processing is required to obtain each single intermediate volume. The variational interpolation scheme [25] has not been adapted to gray-scale images or volumes yet, but using the same extension as for the SDMs, it would also require a one dimension higher object representation.

In contrast, our algorithm represents the gray-scale images and volumes by their Radon transforms, which are of the same dimensionality as the original data. Additionally, we perform efficient processing on the Radon transforms; thus, all the computation can be completed in a reasonable time. Moreover, as the smoothness of the transition is guaranteed by the distribution interpolation on the projections the Radon transform is evaluated for, a computationally expensive constrained optimization [9,20] is not necessary.

Our method is similar to displacement interpolation proposed by Bonneel et al. [4]. Using displacement interpolation, first, RBF decompositions of the source and target distributions are calculated. Then a mass transport optimization is performed, where each source particle is connected to one or more target particles, and the mass is transported along these links from the source distribution to the target distribution. The major advantage of our approach over displacement interpolation is that it completely avoids a mass transport optimization, but still provides a very similar behavior.

3   Multidimensional Distribution Interpolation

In this section, we describe how to extend 1D distribution interpolation [22] to higher dimensions by using the Radon transform.

3.1   Distribution Interpolation in 1D

Let us assume that we have two 1D density functions $f_0(x)$ and $f_1(x)$, and we want to interpolate between them. For example, Fig. 1 shows two Gaussian

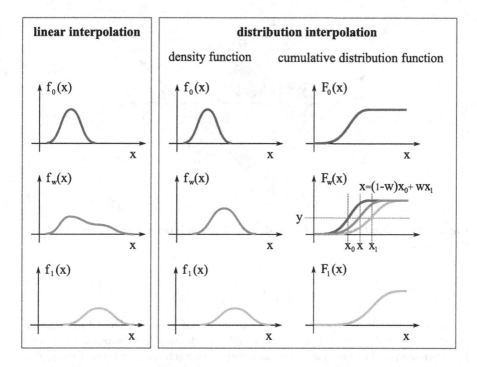

Fig. 1. Distribution interpolation in 1D.

density functions that are scaled and centered differently. In $f_0(x)$ the same amount of material is concentrated on the left side as in $f_1(x)$ on the right side. Therefore, it is a natural expectation that this mass is gradually moved from left to right in the intermediate interpolated density functions. Note that a simple linear interpolation along the y axis clearly does not fulfill this requirement. A distribution interpolation [22], however, does exactly what is expected from a shape-based interpolation technique. Instead of directly interpolating the density functions along the y axis, this method actually interpolates the Cumulative Distribution Functions (CDF) along the x axis. The CDFs for $f_0(x)$ and $f_1(x)$ are defined as follows:

$$F_0(x) = \int_{-\infty}^{x} f_0(x')dx', \qquad F_1(x) = \int_{-\infty}^{x} f_1(x')dx'. \tag{1}$$

The first step of the distribution interpolation is to find x_0 and x_1 such that $F_0(x_0) = F_1(x_1) = y$. The interpolated CDF $F_w(x)$ takes the same value y at a linearly interpolated position $x = (1 - w)x_0 + wx_1$. Positions x_0 and x_1 are simply obtained by inverting the CDFs:

$$F_w^{-1}(y) = x = (1 - w)x_0 + wx_1 = (1 - w)F_0^{-1}(y) + wF_1^{-1}(y). \tag{2}$$

The interpolated CDF $F_w(x)$ is completely defined by its inverse function $F_w^{-1}(y)$, and the corresponding interpolated density function is obtained by a simple derivation:

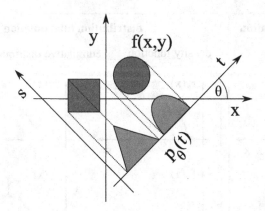

Fig. 2. The Radon transform of a 2D density function is defined as a set of 1D projections $p_\theta(t)$.

$$f_w(x) = \frac{dF_w(x)}{dx}. \tag{3}$$

Figure 1 shows the interpolated density function for $w = 1/2$. Distribution interpolation is usually applied between probability density functions, so their integrals are assumed to be equal to one. However, this scheme can be easily adapted to mass distributions if the distributions are normalized before the interpolation and the interpolated distributions are rescaled such that the a continuous transition of the total mass is guaranteed.

3.2 The Radon Transform

Note that the 1D distribution interpolation cannot be directly extended to higher dimensions. Since we propose an extension scheme that is based on the Radon transform (see Fig. 2), here, we briefly overview its evaluation and inversion. The Radon transform [10] of a 2D density function $f(x, y)$ is defined by a set of 1D projections $p_\theta(t)$:

$$p_\theta(t) = \int_{-\infty}^{\infty} \int_{-\infty}^{\infty} f(x, y)\delta(x\cos(\theta) + y\sin(\theta) - t)dxdy, \tag{4}$$

where δ is the Dirac delta and θ is the projection angle.

The Radon transform is invertible, and its inverse can be evaluated by the classical Filtered Back-Projection (FBP) algorithm [15], which consists of the following steps:

1. Fourier transform of the projections:

$$\hat{p}_\theta(\nu) = \int_{-\infty}^{\infty} p_\theta(t)e^{-i2\pi t\nu}dt. \tag{5}$$

2. Filtering in the frequency domain, where the frequency response of the filter is $|\nu|$:

$$\hat{q}_\theta(\nu) = \hat{p}_\theta(\nu) \cdot |\nu|. \tag{6}$$

3. Inverse Fourier transform of $\hat{q}_\theta(\nu)$:

$$q_\theta(t) = \int_{-\infty}^{\infty} \hat{q}_\theta(\nu)e^{i2\pi t\nu}d\nu. \tag{7}$$

4. Back-projection of the filtered projections $q_\theta(t)$:

$$f(x,y) = \int_0^{\pi} q_\theta(x\cos(\theta) + y\sin(\theta))d\theta. \tag{8}$$

3.3   2D Distribution Interpolation

Now assume that we want to interpolate between two 2D density functions $f_0(x,y)$ and $f_1(x,y)$. In order to guarantee that the projections of the interpolated density functions make a smooth transition between the projections of $f_0(x,y)$ and $f_1(x,y)$, we apply distribution interpolation on the 1D projections rather than a direct interpolation between the 2D density functions. The Radon transforms of $f_0(x,y)$ and $f_1(x,y)$ are denoted by $p_\theta^0(t)$ and $p_\theta^1(t)$, respectively. A distribution interpolation makes sense only if the projection functions are normalized before. Therefore, we need to calculate the integrals of $f_0(x,y)$ and $f_1(x,y)$:

$$s_0 = \int_{-\infty}^{\infty}\int_{-\infty}^{\infty} f_0(x,y)dxdy, \qquad s_1 = \int_{-\infty}^{\infty}\int_{-\infty}^{\infty} f_1(x,y)dxdy. \tag{9}$$

By using distribution interpolation between the normalized projections $p_\theta^0(t)/s_0$ and $p_\theta^1(t)/s_1$, we obtain normalized intermediate distributions $p_\theta^w(t)$. To ensure a smooth transition of the total mass between $f_0(x,y)$ and $f_1(x,y)$, the inverse Radon transform is performed on rescaled projections $p_\theta^w(t)((1-w)s_0 + ws_1)$. The result of the inverse Radon transform is the interpolated density function $f_w(x,y)$.

3.4   3D Distribution Interpolation

In order to interpolate between two 3D density functions $f_0(x,y,z)$ and $f_1(x,y,z)$, 2D projections need to be calculated:

$$p_\theta^0(t,z) = \int_{-\infty}^{\infty}\int_{-\infty}^{\infty} f_0(x,y,z)\delta(x\cos(\theta) + y\sin(\theta) - t)dxdy, \tag{10}$$

$$p_\theta^1(t,z) = \int_{-\infty}^{\infty}\int_{-\infty}^{\infty} f_1(x,y,z)\delta(x\cos(\theta) + y\sin(\theta) - t)dxdy.$$

For a fixed angle θ, $p_\theta^0(t,z)$ and $p_\theta^1(t,z)$ are, in fact, 2D density functions. Therefore, a 2D distribution interpolation can be applied between them as described in

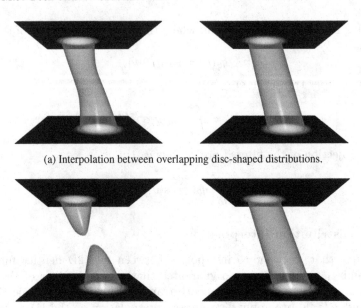

(a) Interpolation between overlapping disc-shaped distributions.

(b) Interpolation between non-overlapping disc-shaped distributions.

Fig. 3. Shape-based interpolation between overlapping and non-overlapping disc-shaped distributions. Left: Interpolation of the distance transforms. Right: Distribution interpolation of the Radon transforms.

Sect. 3.3. Afterwards, an intermediate 3D density function $f_w(x, y, z)$ is reconstructed from the interpolated 2D projections $p_\theta^w(t, z)$ by using the standard FBP algorithm, which is a de facto standard solution for this classical tomography reconstruction problem. To see that the 2D distribution interpolation of the projections yields a consistent and valid Radon transform of an intermediate density function, see the proof in the Appendix. In a practical implementation, a discrete approximation of the continuous integrals is applied. In order to avoid a loss of information, it is required that the total number of pixels in all discretized projections is not smaller than the number of the voxels in the discrete volumetric representations of the 3D density functions.

4   Applications

4.1   Shape-Based Interpolation of Gray-Scale Images

Since our method is based on a Radon Transform Interpolation, we refer to it as RTI. In contrast, the method by Udupa and Grevera [12] is based on a Distance Transform Interpolation; thus, we refer to it as DTI. We compare RTI to DTI for the following reasons:

1. DTI is a de facto standard for an automatic shape-based interpolation of gray-scale images.

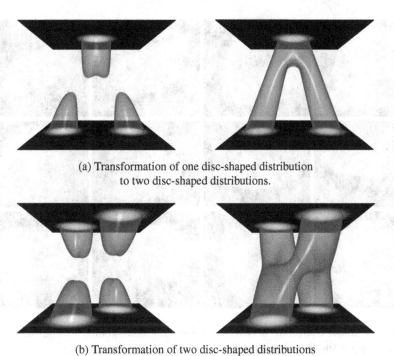

(a) Transformation of one disc-shaped distribution
to two disc-shaped distributions.

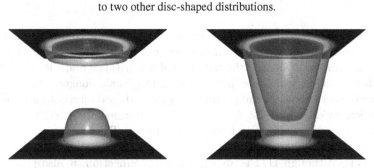

(b) Transformation of two disc-shaped distributions
to two other disc-shaped distributions.

(c) Interpolation between disc-shaped and ring-shaped distributions.

Fig. 4. Shape-based interpolation between significantly different 2D density distributions. Left: Interpolation of the distance transforms. Right:Distribution interpolation of the Radon transforms.

2. DTI is a general solution and not designed for a specific application field.
3. Although for contour interpolation the variational framework of Turk and O'Brien [25] provides smoother transitions than the interpolation of SDMs, its extension to gray-scale images is not trivial, and to the best of our knowledge, has not been published so far.

Note that, using DTI, the chamfering results in just an approximation of the Euclidean distance transform. Although the approximation can be improved

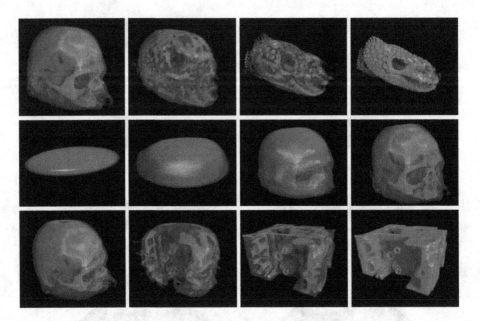

Fig. 5. Automatic morphing between different gray-scale volumes using distribution interpolation on the Radon transforms.

by larger chamfering windows, it significantly increases the computational costs. Therefore, to make our comparison independent from the precision of the distance transform, we evaluated the true Euclidean distance maps for analytically defined height fields, which are interpreted as gray-scale images.

Figures 3 and 4 show several examples for shape-based interpolations between different density distributions. We generated 64 intermediate slices, and rendered the resulting volume using direct volume rendering. The first example (Fig. 3a) is the easiest one, where we interpolate between two overlapping disc-shaped distributions. Although DTI is able to make a connection, it also produces an unexpected curvature. In contrast, RTI results in a perfect tubular connection. In the second example (Fig. 3b), only the distance between the disc-shaped distributions is increased such that they do not overlap anymore. Note that, in this case, DTI fails to make a connection, while RTI still provides a perfect transition. The third and fourth examples (Fig. 4a and b) well demonstrate that, unlike DTI, RTI can appropriately handle bifurcations. In the last example (Fig. 4c), again two density distributions are interpolated, which do not overlap. DTI cannot make a connection for this case either, while RTI is able to connect the ring-shaped and disc-shape density distributions forming a glass shape. These examples clearly show that RTI can be a reasonable alternative of DTI, for instance, in a modeling application.

4.2   Metamorphosis of Gray-Scale Volumes

In order to test our method also on gray-scale volumes, we implemented the entire algorithm on the GPU using CUDA. Note that, all the processing steps, such as the Radon transform, its inversion by FBP, and the distribution interpolation of the 1D projections are easy to map onto the parallel architecture of the GPU. For the frequency-domain ramp filtering, we used the CUDA FFT library. Figure 5 shows a couple of examples for 3D morphing. For each pair of volumes we generated 20 intermediate volumes of resolution 256^3, which took less than an hour on an nVidia Tesla M2070 graphics card. In contrast, we found it unfeasible to efficiently implement DTI for gray-scale volumes on the GPU. Although there exist fast GPU implementations for calculating the distance transform in 2D or 3D [23], applying DTI, the shape-based interpolation of gray-scale volumes would require 4D distance maps to calculate. For example, using floating-point arithmetics, for a volume of resolution 256^3, the corresponding 4D distance map would take $256^4 \times 4 = 16$ Gbytes of memory, which exceeds the capacity of recent graphics cards. Furthermore, the calculation of such a huge 4D distance map would involve a significant computational cost. Variational interpolation [25] is also very difficult to extend to gray-scale volumes. Using the extension scheme of Udupa and Grevera [12], the gray-scale volumes could be treated as height fields defined over the 3D space. Since these height fields can be represented by 4D volumes, the shape transformation between them would require a 4D RBF interpolation. To avoid the loss of fine details, at least to each voxel a 4D RBF needs to be assigned. Thus, for a pair of volumes with resolution 256^3, the number of unknown variables is $256^3 \times 2$. Consequently, the coefficient matrix of the corresponding linear equation system contains $256^6 \times 4$ elements, which makes the inversion practically impossible. Overall, we think that compared to the computational and storage costs of the previous methods, our solution is simple and efficient.

5   Conclusion

In this paper, we have introduced a novel algorithm for shape-based interpolation of gray-scale images and volumes. As far as we know, our technique is the first to combine distribution interpolation and the Radon transform. The major advantage of this combination is that the Radon transform does not require a one dimension higher representation than the original images and volumes. This is not the case for the previous IR methods, where the gray-scale images and volumes can alternatively be represented by 3D and 4D volumes, respectively. Moreover, we demonstrated that the distribution interpolation of the Radon transforms can make a smooth connection between non-overlapping features that the IR methods are typically not able to connect. Due to this advantageous properties, we think that our method represents a significant contribution to the field of shape-based interpolation.

Acknowledgements. This work was supported by OTKA K-101527. The Heloderma data set is from the Digital Morphology http://www.digimorph.org data archive. Special thanks to Dr. Jessica A. Maisano for making this data set available to us.

Appendix

Theorem 1. *Let us denote the Radon transforms of the 2D projections $p_\theta^0(t,z)$ and $p_\theta^1(t,z)$ of the density functions $f_0(x,y,z)$ and $f_1(x,y,z)$ by $p_{\theta,\varphi}^0(r)$ and $p_{\theta,\varphi}^1(r)$, respectively:*

$$p_{\theta,\varphi}^0(r) = \int_{-\infty}^{\infty} \int_{-\infty}^{\infty} p_\theta^0(t,z)\delta(t\cos(\varphi) + z\sin(\varphi) - r)dtdz, \tag{11}$$

$$p_{\theta,\varphi}^1(r) = \int_{-\infty}^{\infty} \int_{-\infty}^{\infty} p_\theta^1(t,z)\delta(t\cos(\varphi) + z\sin(\varphi) - r)dtdz.$$

Additionally, let us introduce operator D_w, which generates an interpolated 1D projection $p_{\theta,\varphi}(r)$ from the 1D projections $p_{\theta,\varphi}^0(r)$ and $p_{\theta,\varphi}^1(r)$ using distribution interpolation:

$$p_{\theta,\varphi}(r) = D_w[p_{\theta,\varphi}^0(r), p_{\theta,\varphi}^1(r)]. \tag{12}$$

The 2D inverse Radon transform $p_\theta(t,z)$ of $p_{\theta,\varphi}(r)$ yields consistent and valid projections of an intermediate density function $f(x,y,z)$ in a sense that

$$\int_{-\infty}^{\infty} p_{\theta_0}(t,z)dt = \int_{-\infty}^{\infty} p_{\theta_1}(t,z)dt, \tag{13}$$

for arbitrary pairs of θ_0 and θ_1 and for each value of z.

Proof. Note that $p_{\theta,\varphi}(r)$ is a consistent and valid 2D Radon transform of a projection $p_\theta(t,z)$ because

$$\int_{-\infty}^{\infty} p_{\theta,\varphi_0}(r)dr = \int_{-\infty}^{\infty} p_{\theta,\varphi_1}(r)dr = \tag{14}$$

$$(1-w) \int_{-\infty}^{\infty} \int_{-\infty}^{\infty} \int_{-\infty}^{\infty} f_0(x,y,z)dxdydz + w \int_{-\infty}^{\infty} \int_{-\infty}^{\infty} \int_{-\infty}^{\infty} f_1(x,y,z)dxdydz,$$

for arbitrary pairs of φ_0 and φ_1. Therefore, Eq. 13 can be expressed from $p_{\theta,\varphi}(r)$, where $\varphi = \pi/2$ and $r = z$ correspond exactly to projections onto the z axis:

$$\int_{-\infty}^{\infty} p_{\theta_0}(t,z)dt = p_{\theta_0,\frac{\pi}{2}}(z) = D_w[p_{\theta_0,\frac{\pi}{2}}^0(z), p_{\theta_0,\frac{\pi}{2}}^1(z)], \tag{15}$$

$$\int_{-\infty}^{\infty} p_{\theta_1}(t,z)dt = p_{\theta_1,\frac{\pi}{2}}(z) = D_w[p_{\theta_1,\frac{\pi}{2}}^0(z), p_{\theta_1,\frac{\pi}{2}}^1(z)].$$

Note that

$$p_{\theta_0,\frac{\pi}{2}}^0(z) = p_{\theta_1,\frac{\pi}{2}}^0(z) = \int_{-\infty}^{\infty} \int_{-\infty}^{\infty} f_0(x,y,z)dxdy, \tag{16}$$

and

$$p_{\theta_0, \frac{\pi}{2}}^1(z) = p_{\theta_1, \frac{\pi}{2}}^1(z) = \int_{-\infty}^{\infty} \int_{-\infty}^{\infty} f_1(x, y, z) dx dy. \tag{17}$$

Thus, $p_{\theta_0, \frac{\pi}{2}}(z)$ is equal to $p_{\theta_1, \frac{\pi}{2}}(z)$ because they are obtained by applying operator D_w on the same 1D projections. Consequently, Eq. 13 is satisfied for arbitrary pairs of θ_0 and θ_1 and for each value of z.

References

1. Butt, M.A., Maragos, P.: Optimum design of chamfer distance transforms. IEEE Trans. Image Process. **7**(10), 1477–1484 (1998)
2. Bajaj, C.L., Coyle, E.J., Lin, K.N.: Arbitrary topology shape reconstruction from planar cross sections. In: Proceedings of Graphical Models and Image Processing, pp. 524–543 (1996)
3. Beier, T., Neely, S.: Feature-based image metamorphosis. Comput. Graph. (Proc. SIGGRAPH) **26**(2), 35–42 (1992)
4. Bonneel, N., van de Panne, M., Paris, S., Heidrich, W.: Displacement interpolation using Lagrangian mass transport. In: Proceedings of SIGGRAPH Asia, pp. 158:1–158:12 (2011)
5. Borgefors, G.: Distance transformations in digital images. Comput. Vis. Graph. Image Process. **34**(3), 344–371 (1986)
6. Buhmann, M.: Radial Basis Functions: Theory and Implementations. Cambridge Monographs on Applied and Computational Mathematics. Cambridge University Press, Cambridge (2009)
7. Cheng, S.W., Dey, T.K.: Improved constructions of delaunay based contour surfaces. In: Proceedings of ACM Symposium on Solid Modeling and Applications, pp. 322–323 (1999)
8. Cohen-Or, D., Solomovic, A., Levin, D.: Three-dimensional distance field metamorphosis. ACM Trans. Graph. **17**(2), 116–141 (1998)
9. Csébfalvi, B., Neumann, L., Kanitsar, A., Gröller, E.: Smooth shape-based interpolation using the conjugate gradient method. In: Proceedings of Vision, Modeling, and Visualization, pp. 123–130 (2002)
10. Deans, S.R.: The Radon Transform and Some of Its Applications. Krieger Publishing, Malabar (1983)
11. Fang, S., Srinivasan, R., Raghavan, R., Richtsmeier, J.T.: Volume morphing and rendering - an integrated approach. Comput. Aided Geom. Des. **17**(1), 59–81 (2000)
12. Grevera, G.J., Udupa, J.K.: Shape-based interpolation of multidimensional greylevel images. IEEE Trans. Med. Imaging **15**(6), 881–892 (1996)
13. Herman, G.T., Zheng, J., Bucholtz, C.A.: Shape-based interpolation. IEEE Comput. Graph. Appl. **12**(3), 69–79 (1992)
14. Jones, M.W., Baerentzen, J.A., Sramek, M.: 3D distance fields: a survey of techniques and applications. IEEE Trans. Vis. Comput. Graph. **12**(4), 581–599 (2006)
15. Kak, A.C., Slaney, M.: Principles of Computerized Tomographic Imaging. IEEE Press, New York (1988)
16. Kniss, J., Premoze, S., Hansen, C., Ebert, D.: Interactive translucent volume rendering and procedural modeling. In: Proceedings of IEEE Visualization (VIS), pp. 109–116 (2002)

17. Lerios, A., Garfinkle, C.D., Levoy, M.: Feature-based volume metamorphosis. In: Proceedings of SIGGRAPH, pp. 449–456 (1995)
18. Liu, L., Bajaj, C., Deasy, J., Low, D.A., Ju, T.: Surface reconstruction from non-parallel curve networks. Comput. Graph. Forum **27**(2), 155–163 (2008)
19. Lorensen, W.E., Cline, H.E.: Marching cubes: a high resolution 3D surface construction algorithm. Comput. Graph. **21**(4), 163–169 (1987)
20. Neumann, L., Csébfalvi, B., Viola, I., Mlejnek, M., Gröller, E.: Feature-preserving volume filtering. In: VisSym 2002: Joint Eurographics - IEEE TCVG Symposium on Visualization, pp. 105–114 (2002)
21. Raya, S., Udupa, J.: Shape-based interpolation of multidimensional objects. IEEE Trans. Med. Imaging **9**(1), 32–42 (1990)
22. Read, A.L.: Linear interpolation of histograms. Nucl. Instr. Meth. **A425**, 357–360 (1999)
23. Schneider, J., Kraus, M., Westermann, R.: GPU-based real-time discrete Euclidean distance transforms with precise error bounds. In: Proceedings of International Conference on Computer Vision Theory and Applications (VISAPP), pp. 435–442 (2009)
24. Treece, G.M., Prager, R.W., Gee, A.H., Berman, L.H.: Surface interpolation from sparse cross-sections using region correspondence. IEEE Trans. Med. Imaging **19**(11), 1106–1114 (2000)
25. Turk, G., O'Brien, J.F.: Shape transformation using variational implicit functions. In: Proceedings of SIGGRAPH, pp. 335–342 (1999)
26. Turk, G., O'Brien, J.F.: Modelling with implicit surfaces that interpolate. ACM Trans. Graph. **21**(4), 855–873 (2002)

Multistory Floor Plan Generation and Room Labeling of Building Interiors from Laser Range Data

Eric Turner$^{(\boxtimes)}$ and Avideh Zakhor

Department of Electrical Engineering and Computer Sciences,
University of California, Berkeley, CA 94720, USA
{elturner,avz}@eecs.berkeley.edu
http://www-video.eecs.berkeley.edu/

Abstract. Automatic generation of building floor plans is useful in many emerging applications, including indoor navigation, augmented and virtual reality, as well as building energy simulation software. These applications require watertight models with limited complexity. In this paper, we present an approach that produces 2.5D extruded watertight models of building interiors from either 2D particle filter grid maps or full 3D point-clouds captured by mobile mapping systems. Our approach is to triangulate a 2D sampling of wall positions and separate these triangles into interior and exterior sets. We partition the interior volume of the building model by rooms, then simplify the model to reduce noise. Such labels are useful for building energy simulations involving thermal models, as well as for ensuring geometric accuracy of the resulting 3D model. We experimentally verify the performance of our proposed approach on a wide variety of buildings. Our approach is efficient enough to be used in real-time in conjunction with Simultaneous Localization and Mapping (SLAM) applications.

Keywords: Floor plan · Watertight modeling · Range data · LiDAR

1 Introduction

Indoor building modeling and floor plan generation are useful in many fields such as architecture and civil engineering. Green buildings and sustainable construction have increased the use of building energy simulation and analysis software, requiring building geometry as input. Even though existing energy simulation tools can accurately model the thermodynamic properties of building interiors, their performance is hindered by overly complex geometry models [6]. Indoor models can also be used for positioning in wide-area augmented reality applications, whereby low-complexity models enable low memory use for mobile client-side processing.

E. Turner—This research was conducted with Government support under and awarded by DoD, Air Force Office of Scientific Research, National Defense Science and Engineering Graduate (NDSEG) Fellowing, 32 CFR 168a.

© Springer International Publishing Switzerland 2015
S. Battiato et al. (Eds.): VISIGRAPP 2014, CCIS 550, pp. 29–44, 2015.
DOI: 10.1007/978-3-319-25117-2_3

In this paper, we present a technique for generating aesthetically pleasing, minimalist 2.5D models of indoor building environments. Such models are intended to capture the architectural elements of a building such as floors, walls, and ceilings while ignoring transient objects such as furniture. We generate our models by first computing a 2D floor-plan of the environment, then using estimated height information to extrude the floor-plan into a 3D building model.

Generating 3D models by extruding 2D floor-plans typically yield clean and aesthetically pleasing results. Even though such models may not capture the fine details of the environment, they still offer many advantages. As shown later, it is possible to generate sizable 2.5D extruded models at real-time speeds, enabling human operators to capture and navigate environments thoroughly and adaptively.

We also propose a technique to partition the interior environment rooms, yielding secondary features of buildings, such as locations of doorways. Room labeling is useful for many applications, such as fast rendering of models [8]. Furthermore, since energy simulation engines model heat and air flow within the building environment, they need accurate partitions of the interior spaces to represent distinct thermal zones [6].

In addition to exporting room labels, our proposed technique uses the computed labels to further improve the geometry of the model. Specifically, knowledge of room partitions can be exploited to reduce noise in the computed geometry while preserving fine details in doorways. Furthermore, since input height estimates are often noisy, using room labels to group these heights can provide substantial error reduction in the resulting extruded 3D meshes.

This paper is organized as follows. In Sect. 2, we describe related work to this research. Section 3 describes our proposed algorithm to generate floor plans from the specified input. In Sect. 4, we describe our approach to room labeling. In Sect. 5, we show how room labeling is used to reduce noise in the model. Section 6 describes how 2D floor plans are extruded into 2.5D models with height information. Section 7 demonstrates experimental results on a wide variety of building models. Lastly, in Sect. 8 we describe potential future work in this area.

2   Background

Modeling and navigation of indoor environments is a well-studied field. Due to cost of full 3D laser range finders, the majority of indoor modeling systems use 2D LiDAR scanners. Examples of such systems include autonomous unmanned vehicles [2,18] or systems worn by a human operator [4,7].

Most simultaneous localization and mapping (SLAM) systems use a horizontally-oriented 2D LiDAR scanner, which estimates the trajectory of the system, creating a 2D map of the environment [20]. The constructed 2D grid map is stored as a set of points in $\mathbb{R}^2$ that represent the primary features of the environment, such as walls and building architecture. Particle filtering approaches to localization typically result in real-time mapping [11,12] and can therefore benefit from a real-time floor plan generation algorithm that delivers a live map of the environment.

These mapping systems can also use additional scanners to create a dense 3D point-cloud representation of the environment geometry [14,19], which can be used to develop full 3D models [3,13]. Many applications are unable to use these 3D models due to their complexity and number of elements. For example, building energy simulations require watertight meshes that are also highly simplified in order to perform effectively [6].

To address this issue, a number of simplified building modeling algorithms have been developed, most of which assume vertical walls, rectified rooms, and axis-alignment [24]. Under these assumptions, fundamental features of the building can be identified, while ignoring minor details such as furniture or other clutter [1]. One of the major limitations of these techniques is that they are developed only for axis-aligned models. Often, such techniques correctly reconstruct major rooms while fundamentally changing the topology of minor areas, such as ignoring doorways, shapes of rooms, or small rooms entirely.

In this paper, we show that simple models can be generated with only 2.5D information, while preserving connectivity and geometry of building features, including doorways. Our approach generates a 2D floor plan of the building, then uses wall height information to generate a 3D extrusion of this floor plan. Such blueprint-to-model techniques have been well-studied [15,17], but rely on the original building blueprints as input. Our technique automatically generates the floor plan of the building and uses this information to create a 2.5D model of the environment.

Prior work on automatic floor plan generation use dense 3D point-clouds as input, and take advantage of the verticality of walls to perform histogram analysis to sample wall position estimates [16,21], which are in the same format as a grid map for particle filtering [10]. In situations where dense 3D point-clouds are available, we apply similar techniques to convert them to a 2D wall sampling.

A novel contribution of this paper is the use of room labeling to enhance building models, e.g. for thermal simulations of interior environments [6]. One motivation for existing work has been to capture line-of-sight information for fast rendering of building environments [8]. This technique requires axis-aligned rectilinear building geometry, which often is not a valid assumption. Others have partitioned building environments into submap segments with the goal of efficient localization and tracking [2]. This approach is meant to create easily recognizable subsections of the environment, whereas our proposed room labeling technique uses geometric features to capture semantic room definitions for both architectural and building energy simulation applications.

3 Floor Plan Generation

In this section, we present a technique to automatically generate accurate floor plan models at real-time speeds for indoor building environments. Section 3.1 describes the type of input for our approach, which can be generated from either 2D mapping systems or dense 3D point-clouds of environments. In Sect. 3.2, we discuss the way these input data are used to compute the interior space of the 2D floor-plan, which defines the resultant building geometry.

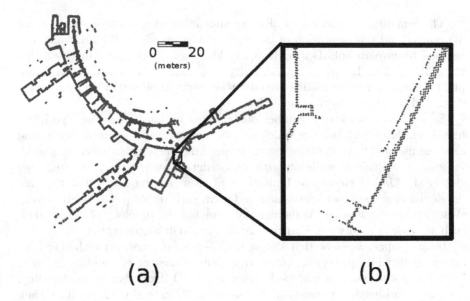

Fig. 1. Example input wall samples of hotel hallways and lobby generated from a particle filter system. (a) Wall samples of full model; (b) close up of wall in model.

3.1　Input Data

The input data used during floor plan generation consist of points in the (x,y) horizontal plane, which we call wall samples. These points depict locations of walls or vertical objects in the environment. We assume that interior environments satisfy "2.5-Dimensional" geometry: all walls are vertically aligned, while floors and ceilings are perfectly horizontal. In many application scenarios only 2D scanners operating in one plane are used, so this assumption is needed to extract 3D information about the environment. Many mapping systems use a horizontal LiDAR scanner to estimate a map of the area as a set of wall sample positions, while refining estimates for scanner poses. These mobile mapping systems often have additional sensors capable of estimating floor and ceiling heights at each pose [4,18]. The input to our algorithm is a set of 2D wall samples, where each sample is associated with the scanner pose that observed it, as well as estimates of the floor and ceiling heights at the wall sample location.

An alternate method of computing wall samples is to subsample a full 3D point-cloud to a set of representative 2D points [16,21]. This process cannot be done in a streaming fashion, but can provide more accurate estimates for wall positions than a real-time particle filter. Such an approach is useful when representing dense, highly complex point clouds with simple geometry. Under the 2.5D assumption of the environment, wall samples can be detected by projecting 3D points onto the horizontal plane. Horizontal areas with a high density of projected points are likely to correspond to vertical surfaces. Wall samples are classified by storing these projected points in a quadtree structure with resolution r. A resolution of 5 cm

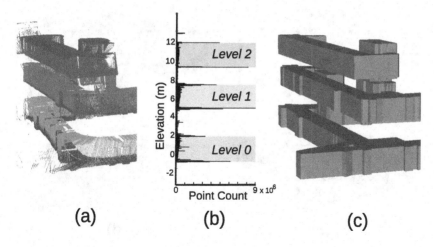

Fig. 2. An example point-cloud partitioning by height: (a) the input point-cloud, showing geometry for three levels; (b) the vertical histogram showing estimates of each building level height; (c) the produced mesh of this building scan.

typically results in sufficient detail in even the most cluttered environments. Each leaf node in this quadtree contains the 3D points that are projected onto its $r \times r$ area. A vertical histogram is computed using the original heights of these points. This histogram has bin-size r, and if a sufficient vertical coverage H is represented by at least `ceil`(H/r) bins, then the average (x, y) position of the leaf is considered a wall sample. The value of H may vary depending on application, but a length of 2 m works well to capture permanent wall features while ignoring furniture and other interior clutter.

If the wall samples are generated from point-clouds, then a histogram approach can be used to separate the point-cloud by levels [21]. Figure 2a shows an example point-cloud, colored by height, which contains multiple levels. By computing a histogram along the vertical-axis of the point-cloud, it is possible to find heights with high point density, which indicates the presence of a large horizontal surface. An example of this process is shown in Fig. 2b, where peaks in the histogram correspond to the floors and ceilings of each scanned level in the building. Points scanned from above, in a downward direction, are used to populate a histogram to estimate the position of each floor, and the histogram used to estimate the position of each ceiling is populated by points scanned from below. The local maxima of these two histograms show locations of likely candidates for floor and ceiling positions, which are used to estimate the number of scanned levels and the vertical extent of each level. Figure 2c shows the final extruded mesh with all three scanned levels.

The result is a set of wall samples $P \subseteq \mathbb{R}^2$, where each wall sample $p \in P$ is represented by its 2D position, the minimum and maximum height values of the points that sample represents, and the poses of the scanners that observed the sample location. As we discuss later, these scanner poses provide crucial

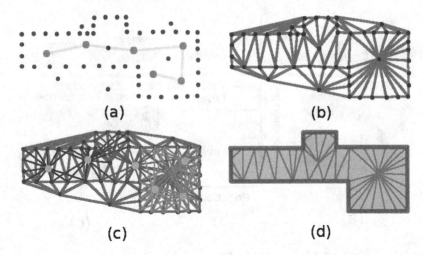

Fig. 3. Example of carving process to find interior triangles: (a) wall samples (in blue) with path of scanner (in green); (b) Delaunay Triangulation of wall samples; (c) laser scans from each pose (in red); (d) triangles that intersect with laser scans (in pink), used as interior triangles, with building model border (in blue) (Color figure online).

line-of-sight information that facilitate floor plan reconstruction. An example of such input for a hotel hallway is shown in Fig. 1. As shown, even though the walls are well sampled, noise in the localization estimate causes noisy wall samples with outliers.

3.2 Triangulation

We generate a floor plan by partitioning space into *interior* and *exterior* domains. The interior represents all open space in the environment, such as rooms and hallways, while the exterior represents all space outside of the building, space occupied by solid objects, or space that is unobservable. Once this partitioning is completed, as described below, the boundary lines between the interior and exterior are used to represent the exported walls of the floor plan.

The input samples are used to define a volumetric representation by generating a Delaunay Triangulation on the plane. Each triangle is labeled either interior or exterior by analyzing the line-of-sight information of each wall sample. Initially, all triangles are considered exterior. Each input wall sample, $p \in P$, is viewed by a set of scanner positions, $S_p \subseteq \mathbb{R}^2$. For every scanner position $s \in S_p$, the line segment (s, p) denotes the line-of-sight occurring from the scanner to the scanned point during data collection. No solid object can possibly intersect this line, since otherwise the scan would have been occluded. Thus, all triangles intersected by the line segment (s, p) are relabeled to be interior.

In order to prevent fine details from being removed, we check for occlusions when carving each line segment (s, p). If another wall sample p' is located in between the positions of s and p, then the line segment is truncated to (s, p').

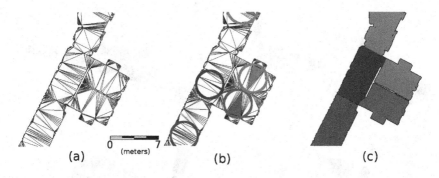

Fig. 4. Example room seed partitioning: (a) interior triangulation; (b) the room seed triangles, and their corresponding circumcircles; (c) room labels propagated to all other triangles.

Thus, no features captured by wall samples are ever fully carved away, preserving environment details. This process carves away the interior triangles with each captured scan. Since these scans are captured on a mobile scanner, the scanner poses are ordered in time. In order for the system to traverse the environment, the line segment between adjacent scanner poses must also intersect only interior space. In addition to carving via scanner-to-scan lines, the same carving process is performed with scanner-to-scanner line segments.

Figure 3 demonstrates an example of this process. Figure 3a shows the input wall samples, in blue, as well as the path of the mobile mapping system, in green. These points are triangulated, as shown in Fig. 3b. The line-of-sight information is analyzed from each pose of the system, demonstrated by the laser scans from each pose to its observed wall samples in Fig. 3c. The subset of triangles that are intersected by these laser scans are considered interior. The interior triangles are shown in pink in Fig. 3d, denoting the interior volume of the reconstructed building model. The border of this building model is shown in blue, denoting the estimated walls of the floor plan.

4 Room Labeling

Once the volume has been partitioned into interior and exterior domains, the boundary between these domains can be exported as a valid floor plan of the environment. Keeping volumetric information can also yield useful information, such as a partitioning of the interior into separate rooms.

We define a *room* to be a connected subset of the interior triangles in the building model. Ideally, a room is a large open space with small shared boundaries to the rest of the model. Detected rooms should match with real-world architecture, where separations between labeled rooms are located at doorways in the building. Since doors are often difficult to detect, or not even present, there is no strict mathematical definition for a room, so this labeling is heuristic in nature.

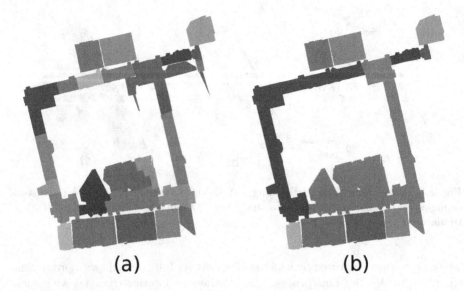

(a)   **(b)**

Fig. 5. Room labeling refinement example: (a) initial room labels; (b) converged room labels.

We model room labeling as a graph-cut problem. First, a rough estimate for the number of rooms and a seed triangle for each room is computed. A seed triangle is representative of a room, where every room to be modeled has one seed triangle. These seeds are used to partition the remainder of interior triangles into rooms. This process typically over-estimates the number of rooms, so prior knowledge of architectural compliance standards is used to evaluate each estimated room geometry. Using this analysis, the number of ill-formed rooms is reduced, providing an update on the original seed points. This process is repeated until the set of room seeds converges.

4.1 Forming Room Seeds

We use the Delaunay property of the triangulation to identify likely seed triangle locations for room labels. If we assume that the input wall samples represent a dense sampling of the building geometry, this property implies that the circum-circles of none of the interior triangles intersect the boundary walls of the carved floor plan, forcing these circles to represent only interior area. This make-up allows each triangle's circumradius to provide an estimate of the local feature size at its location on the floor plan boundary polygon. Given the example interior triangulation shown in Fig. 4a, the highlighted triangles in Fig. 4b show the chosen seed locations.

Triangles with larger circumradii are likely to be more representative of their rooms than those with smaller circumradii. We form the initial set of room seeds by finding all triangles whose circumcircles are local maxima. Specifically, given the set of interior triangles T, each triangle $t \in T$ has circumcircle c_t, which is tested against every other circumcircle in T that is intersected by c_t. If c_t has the largest radius of any intersecting circumcircle, then t is considered a seed for the room labeling. This process selects the largest triangles that encompass the space of rooms as the seeds for room labeling. Figure 4b shows example seed triangles and their corresponding circumcircles. The result is an estimate of the number of rooms and a rough location for each room.

4.2 Partitioning Room Labels

Let K be the number of room seeds found, with the seed triangles denoted as $t_1, t_2, \ldots, t_K$. We wish to partition all triangles in T into K rooms. This step can be performed as a graph-cut on the dual of the triangulation. Specifically, each triangle $t \in T$ is a node in the graph, and the edge weight between two abutting triangles is the length of their shared side. Performing a min-cut on this graph partitions rooms to minimize inter-room boundary length. In other words, rooms are defined to minimize the size of doors. This process propagates the room labels to every triangle, and the boundaries between rooms are composed of only the smallest edges in the triangulation T. The result of this process is shown in Fig. 4c.

4.3 Refining Rooms

Room labels partition T into a set of rooms $R = \{R_1, R_2, \ldots, R_K\}$, where each room R_i contains a disjoint subset of T and has seed triangle t_i. The initial room seeds over-estimate the number of rooms, since a room may have multiple local maxima. This case is especially true for long hallways, where the assumption that one triangle dominates the area of the room is invalid. An example is shown in Fig. 4c, where two lower rooms, shown in green and purple, are properly labeled, but their adjoining hallway is broken into three subsections. The solution is to selectively remove room seeds and redefine the partition.

A room is considered a candidate for merging if it shares a large perimeter with another room. Ideally, two rooms sharing a border too large to be a door should be considered the same room. By Americans with Disabilities Act Compliance Standards, a swinging door cannot exceed 48 inches in width [23]. Accounting for the possibility of double-doors, we use a threshold of 2.44 m, or 96 in., when considering boundaries between rooms. If two rooms share a border greater than this threshold, then the seed triangle with the smaller circumradius is discarded. This process reduces the value of K, the number of rooms, while keeping the interior triangulation T unchanged. With a reduced set of room seeds, existing room labels are discarded and the process of room partitioning is repeated. This iteration repeats until the room labeling converges.

Another way room labels are refined is by comparing the path of the mobile mapping system to the current room labeling for each iteration. The mobile scanning system does not necessarily traverse every room, and may only take superficial scans of room geometry passing by a room's open doorway. Since the room is not actually entered, the model is unlikely to capture sufficient geometry, and so only a small handful of wall samples are acquired for such a room. It is desirable to remove this poorly scanned area from the model rather than keeping it as part of the output. After each round of room partitioning, if none of the triangles in a room R_i are intersected by the scanner's path, then we infer that room has not been entered. The elements of R_i are removed from the interior triangulation T. Since the topology of the building model is changed, the set of room seeds is recomputed in this event and room labeling is restarted. This process will also remove areas that are falsely identified as rooms, such as ghost geometry generated by windows and reflective surfaces, which cause rooms to be replicated outside the actual model.

Figure 5 shows an example of the room refinement process for the hallways and classrooms in an academic building. Figure 5a shows the initial room seeds that were found based on circumcircle analysis of Sect. 4.1. The hallways of this building are represented by several room labels, but after room label refinement as shown in Fig. 5b, the hallways are appropriately classified. Additionally, rooms that are insufficiently scanned and represented with triangulation artifacts are removed from the model in the manner described above.

5   Simplification

The interior building model is represented as a triangulation of wall samples, which densely represent the building geometry. In many applications, it is useful to reduce the complexity of this representation, so that each wall is represented by a single line segment. This step is often desirable in order to attenuate noise in the input wall samples or to classify the walls of a room for application-specific purposes. The goal is to simplify the wall geometry while preserving the general shape and features of the building model.

We opt to simplify walls using a variant of QEM [9]. Since this mesh is in the plane, only vertices incident to the model boundary are considered for simplification. The error matrix Q_v of each boundary vertex v is used to compute the sum of squared displacement error from each adjoining line along the boundary polygon. Since error is measured via distance away from a line in 2D, each Q_v has size 3×3, and is defined as:

$$Q_v = \sum_{l \in lines(v)} E_l \tag{1}$$

where E_l is defined from the line equation $ax + by + c = 0$, with $a^2 + b^2 = 1$:

$$E_l = \begin{bmatrix} a^2 & ab & ac \\ ab & b^2 & bc \\ ac & bc & c^2 \end{bmatrix} \tag{2}$$

The simplification of the boundary proceeds in a similar manner to QEM, but if a wall vertex v is contained in multiple rooms or if it is connected by an edge to a vertex that is contained in multiple rooms, then it is not simplified. This constraint is used to preserve the fine details of doorways between rooms, while freely simplifying walls that are fully contained within one room. Wall edges are iteratively simplified until no simplification produces error of less than the original wall sampling resolution, r. Thus, walls are simplified while preserving any geometry features of the building interior.

Since we are interested in preserving the 2D triangulation T of the building model, in addition to the boundary polygon, every edge simplification is performed by collapsing an interior triangle. This computation simplifies the boundary polygon of the model while still preserving the room labeling of the model's volume. These triangle collapses do not preserve the Delaunay property of the triangulation, but do preserve the boundaries between room volumes, which is more desirable in the output.

6   Height Extrusion

As mentioned in Sect. 3.1, each input wall sample also references the vertical extent for the observed scans at that location. This information can be used to convert the labeled 2D interior building model to a 2.5D extruded model, by using the minimum and maximum height values for each scan as an estimate of the floor and ceiling heights, respectively.

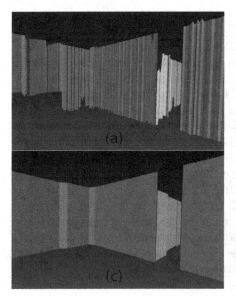
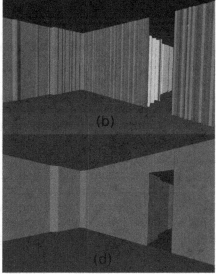

Fig. 6. Example of creating a 3D extruded mesh from 2D wall samples: (a) walls of generated floor plan with estimated height ranges; (b) floor and ceiling heights are grouped by room; (c) simplification performed on walls; (d) floor and ceiling triangles added to create a watertight mesh.

(a)   (b)

Fig. 7. Comparison of models from (a) our approach with (b) existing approach [22].

(a)   (b)

Fig. 8. Interior view of 3D extruded reconstructed model: (a) without and (b) with texture-mapping [5].

Since these wall samples are collected using 2D planar scanners in an environment containing clutter, the minimum and maximum heights associated with each point are noisy. Figure 6a shows an example room with these initial heights. To produce aesthetically-pleasing models, each room uses a single floor height and a single ceiling height. This assumption is reasonable since the goal of this processing is to produce a simplified building mesh. This step demonstrates the utility of room labeling to modeling. The height range for each room is computed from the median floor and ceiling height values of that room's vertices. An example is shown in Fig. 6b and the corresponding result from the simplification process from Sect. 5 is demonstrated in Fig. 6c.

The 2D triangulation of a room is then used to create the floor and ceiling mesh for that room, with the boundary edges of the triangulation extruded to create rectangular vertical wall segments. The result is a watertight 3D mesh of the building, capturing the permanent geometry in an efficient number of triangles. Figure 6d shows an example of this watertight extruded geometry, including the effects of wall boundary simplification on the resulting extruded mesh.

7 Results

Our approach works well on a variety of test cases, spanning several model types including offices, hotels, and university buildings. For the largest models, total

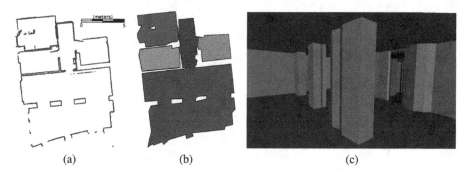

(a)   (b)   (c)

Fig. 9. Apartment complex office: (a) Input represented by 3,462 wall samples; (b) generates floor plan with 5 rooms; (c) extruded 3D mesh represented with 512 triangles. Total processing time required is 1.2 s.

processing time to compute an extruded 3D model from 2D wall samples is under 10 s. Most of this time is spent on carving interior triangles, which can be performed real-time in a streaming manner during data acquisition, which typically lasts several minutes.

Our 2.5D approach produces simplified models when compared to surface reconstruction techniques that preserve fine detail with more complex output. Specifically, our method omits interior clutter such as furniture since it uses wall samples as input. Figure 7 compares the models resulting from our 2.5D method with that of an existing 3D building modeling technique [22] for the hotel hallways shown in Fig. 1. The two methods result in 2,944 triangles and 4.1 millions triangles, respectively.

Since these models were generated with a system that captures imagery in addition to laser range points, these models can also be texture-mapped with the scenery of the environment [5]. Figure 8 depicts the hallways of an academic building with and without texturing.

Next, we show sample models resulting from our proposed method in multiple environments. For all the models shown in this paper, the scale is in units of meters, and the resolution is 5 cm. Figure 9 shows a small test model of an apartment office complex and Fig. 10 denotes a hotel lobby, hallways, and side rooms. The vast majority of this model is labeled as one room, consisting of the hallways of the building. Since no part of these hallways are separated by doors, this result is desirable. This model is also the largest example output, covering over 260 m of hallways. An interior of the 3D extruded model for this dataset is shown in Fig. 7a. Figure 11 represents an academic research lab, including conference rooms and student cubicles. The upper portion of the center room, shown in blue, is a kitchenette area, with a counter-top. Since the counter was not sufficiently captured by the wall samples, it is not represented in the 2.5D extrusion of the model. The room in the upper-left of this figure contains student cubicles. While the cubicle walls were scanned and some appear in the wall sampling, they also do not meet our height threshold and are not fully represented in the final floorplan.

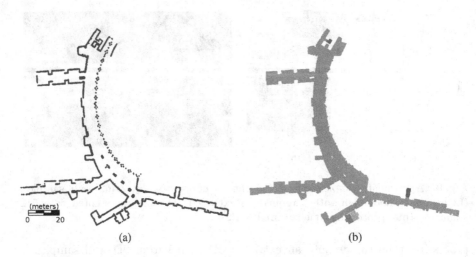

(a)   (b)

Fig. 10. Hotel lobby and hallways: (a) Input represented by 33,582 wall samples; (b) generates floor plan with 5 rooms. Extruded 3D mesh represented with 5,012 triangles. Total processing time required is 8.5 s.

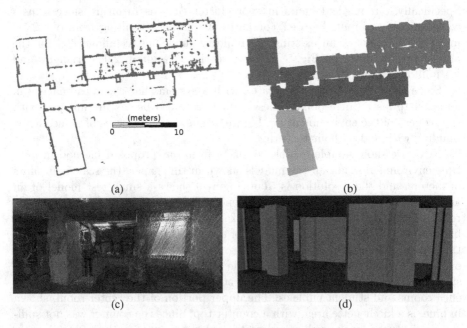

(a)   (b)

(c)   (d)

Fig. 11. University office area: (a) Input represented by 12,183 wall samples; (b) generates floor plan with 4 rooms; (c) original point-cloud of scanned area; (d) extruded 3D mesh represented with 4,912 triangles. Processing time required to convert wall samples to mesh is 7 s.

8   Conclusion

We demonstrate an efficient approach to automatically generate floor plans of building interiors at real-time speeds. Classifying and labeling the rooms within each generated floor plan allows for simplification schemes that can preserve fine details at doorways. These room labels allow for accurate 2.5D extrusion from noisy floor and ceiling height estimates of the input points. The resulting model is suitable for visualization, simulation, and navigation applications. Current limitations of this algorithm include the verticality assumption made about observed building features. If the horizontal cross-section of an environment changes dramatically between different heights, the modeling techniques presented in this paper does not accurately portray the actual geometry. Such limitations could be overcome by observing more information about each wall sample than just (x, y) position and height ranges. If 3D normal estimates could be made about surfaces, this information may allow better filtering of outlier wall samples, or to infer building geometry that was poorly scanned.

References

1. Adan, A., Huber, D.: 3D reconstruction of interior wall surfaces under occlusion and clutter. In: International Conference on 3D Imaging, Modeling, Processing, Visualization and Transmission, pp. 275–281 (2011)
2. Brunskill, E., Kollar, T., Roy, N.: Topological mapping using spectral clustering and classification. In: International Conference on Intelligent Robots and Systems, pp. 2491–3496, October 2007
3. Chauve, A.L., Labatut, P., Pons, J.P.: Robust piecewise-planar 3D reconstruction and completion from large-scale unstructured point data. In: CVPR (2010)
4. Chen, G., Kua, J., Shum, S., Naikal, N., Carlberg, M., Zakhor, A.: Indoor localization algorithms for a human-operated backpack system. In: 3D Data Processing, Visualization, and Transmission (2010)
5. Cheng, P., Anderson, M., He, S., Zakhor, A.: Texture mapping 3D planar models of indoor environments with noisy camera poses. In: SPIE Electronic Imaging Conference (2013)
6. Crawley, B.B., Kawrie, L.K., Pedersen, C.O., Winkelmann, F.C.: Energyplus: energy simulation program. ASHRAE **42**(4), 49–56 (2000)
7. Fallon, M.F., Johannsson, H., Brookshire, J., Teller, S., Leonard, J.J.: Sensor fusion for flexible human-portable building-scale mapping. In: Intelligence Robots and Systems, pp. 4405–4412 (2012)
8. Funkhouser, T.A., Sequin, C.H., Teller, S.J.: Management of large amounts of data in interactive building walkthroughs. In: Proceedings of the 1992 Symposium on Interactive 3D Graphics, pp. 11–21 (1992)
9. Garland, M., Heckbert, P.S.: Surface simplification using quadric error metrics. In: SIGGRAPH pp. 209–216 (1997)
10. Grisetti, G., Stachniss, C., Burgard, W.: Improving grid-based slam with rao-blackwellized particle filters by adaptive proposals and selective resampling. In: Proceedings of IEEE International Conference of Robotics and Automation, pp. 2443–2448 (2005)

11. Grisetti, G., Stachniss, C., Burgard, W.: Improved techniques for grid mapping with rao-blackwellized particle filters. IEEE Trans. Robot. **23**(1), 34–46 (2007)
12. Hahnel, D., Gurgard, W., Fox, D., Thrun, S.: An efficient fastslam algorithm for generating maps of large-scale cyclic environments from raw laser range measurements. In: International Conference on Intelligent Robots and Systems, vol. 1, pp. 206–211 (2003)
13. Holenstein, C., Zlot, R., Bosse, M.: Watertight surface reconstruction of caves from 3D laser data. In: IEEE/RSJ International Conference on Intelligent Robots and Systems (2011)
14. Kua, J., Corso, N., Zakhor, A.: Automatic loop closure detection using multiple cameras for 3D indoor localization. In: IS&T/SPIE Electronic Imaging (2012)
15. Lewis, R., Sequin, C.: Generation of 3D building models from 2D architectural plans. Comput. Aided Des. **30**(10), 765–779 (1998)
16. Okorn, B., Xiong, X., Akinci, B., Huber, D.: Toward automated modeling of floor plans. In: 3DPVT (2009)
17. Or, S.H., Wong, K.H., kin Yu, Y., yuan Chang, M.M.: Highly automatic approach to architectural floorplan image understanding and model generation. Pattern Recogn. (2005)
18. Shen, S., Michael, N., Kumar, V.: Autonomous multi-floor indoor navigation with a computationally constrained mav. In: IEEE International Conference on Robotics and Automation, pp. 20–25 (2011)
19. Smith, M., Posner, I., Newman, P.: Adaptive compression for 3D laser data. Int. J. Robot. Res. **30**(7), 914–935 (2011)
20. Thrun, S., Burgard, W., Fox, D.: Probabilistic Robotics. MIT Press, Cambridge (2005)
21. Turner, E., Zakhor, A.: Watertight as-built architectural floor plans generated from laser range data. In: 3DimPVT, October 2012
22. Turner, E., Zakhor, A.: Watertight planar surface meshing of indoor point-clouds with voxel carving. In: 3DV 2013 (2013)
23. U.S. Architectural and Transportation Barriers Compliance Board, 1331 F Street N.W. Suite 1000 Washington D.C. 20004-1111: Americans with Disabilities Act, July 1990. aNSI A117.1-1980
24. Xiao, J., Furukawa, Y.: Reconstructing the world's museums. In: Fitzgibbon, A., Lazebnik, S., Perona, P., Sato, Y., Schmid, C. (eds.) ECCV 2012, Part I. LNCS, vol. 7572, pp. 668–681. Springer, Heidelberg (2012)

Semantic Composition of Language-Integrated Shaders

Georg Haaser[(✉)], Harald Steinlechner, Michael May, Michael Schwärzler,
Stefan Maierhofer, and Robert Tobler

VRVis Research Center, Donau-City-Strasse 1, Vienna, Austria
haaser@vrvis.at
http://www.vrvis.at

Abstract. In order to simplify shader programming we propose a system to specify composable shaders in a functional way directly in typical implementation languages of modern rendering frameworks. In contrast to existing pipeline shader frameworks, our system exposes a radically simplified pipeline, which we purposefully aligned with our basic intuition of shaders as compositions of per-primitive and per-pixel operations. By programming the shaders in the host language, we additionally remove the complexity of handling different programming languages for shaders and the rest of the framework.

The resulting simplicity lends itself to structure modules purely based on their semantic, instead of dealing with structure enforced by specific versions of graphics APIs. Thus our system offers great flexibility when it comes to reusing and combining shaders with completely different semantics, or when targeting different graphics APIs: our high level shaders can be automatically translated into the shading language of the backend (e.g. HLSL, GLSL, CG).

Keywords: Shader · Composition · Rendering · Language · Embedded

1 Introduction

Implementing shaders for hardware-accelerated rasterization frameworks like DirectX or OpenGL has become an important part of developing rendering systems. Even though the flexibility and possibilities in graphics development have drastically improved with the introduction of these shaders over the last few years, recent advances in "CPU-based" programming languages and software engineering are often not reflected in shader-programming. Especially the limitations in terms of shader management in larger software projects cause the tasks of *combining shader effects, targeting different hardware, supporting older API versions* or *optimizing these shader permutations* to become extremely time-consuming, tiresome, and error-prone.

The C-style definition of a single shader stage program is only simple during the primary creation process: As soon as such an effect has to be *combined*

© Springer International Publishing Switzerland 2015
S. Battiato et al. (Eds.): VISIGRAPP 2014, CCIS 550, pp. 45–61, 2015.
DOI: 10.1007/978-3-319-25117-2_4

with other shaders to generate the desired final surface illumination, or has to be used on another API version or target platform, programmers either tend to build large, complex Über-Shader constructs with computationally expensive dynamic branching techniques, or manage hundreds of shader combinations and permutations manually. Object-oriented approaches [4,9,12] and novel Shader Model 5 functionality (e.g. `interfaces` [16]) extend procedural languages with abstractions like interfaces and limit code duplication via inheritance. Inheritance as mechanism for composition however has shortcomings in terms of ad hoc compositions and re-usability, as each composition has to be stated explicitly (see Sect. 2) and in terms of extensibility, as extension points have to be anticipated by providing abstract or virtual methods (Fig. 1).

Fig. 1. Example built with composed shader modules (from left to right): transformation and per-pixel lighting, transformation and texturing, transformation/normal mapping/texturing and lighting, transformation/normal mapping/texturing/point sprite generation and lighting, transformation/normal mapping/texturing/point sprites/thick line generation and lighting.

We propose a novel shader programming model that emphasizes the *semantic* and simple programmability of a shader, based on the following ideas:

- *abstract shader stages*: by freeing the shader modules from concrete pipeline stages, we let the programmer specify what he wants his shader modules to do in an abstract, backend-independent manner
- *composition via semantic input/output types*: with the introduction of semantically annotated input and output types, these types encode the semantic of what is computed by each shader module, and thus composition operators can be built, that combine the modules according to the semantic
- *fine-tuning of semantics*: by providing more detailed information for the semantic types such as computation rates, the programmer can exactly specify the semantic of his shader
- *programming in the host language*: by using the host language as a shader language and providing automatic translation into the shader language of the backend, we significantly simplify shader programming

Using these ideas to provide a system that automatically combines modules based on their semantic, we overcome the combinatorial explosion of typical shader systems where each and every combination has to be specified explicitly: Modules are typically expressed only in their most general form, and can be composed either statically as hard-coded expressions, or programmatically, which is useful to generate shaders based on runtime information or whole families of related shaders. Our high-level shader-code requires a specific functionality to be defined only once—no matter how often it is combined with other shaders and on how many target platforms it is deployed—while unneeded calculations are automatically eliminated. We leave the error-prone task of finding the optimal shader stage for each computation to the machine, which automatically maps shaders onto specific pipeline architectures (e.g. DirectX), performs global and local optimizations and code generation for distinct shader permutations, and finally emits a backend shader program (e.g. HLSL) comparable to hand-crafted code.

2 Background

The ancestors of today's shading languages are Cook's shade trees [2] and Perlin's image synthesizer [18]. Cook's shade trees classify independent aspects like lighting, surface and volume into separate modules called shading processes. As a mechanism for composition each process is represented as an expression tree which supports grafting of commonly used expressions into other processes. However, the underlying model of computation which is purely declarative allows for no conditional control flow like loops as well as no mutable state. Perlin's image synthesizer is based on imperative procedures and therefore dissolves these limitations, but abandons the idea of logically independent shading processes. Procedures work on streams of fragments, and describe shading computations after hidden surface removal.

The most prevalent shader languages for real-time rendering (Cg [10], HLSL [17], and GLSL [7]) follow the *shader-per-stage* approach. Similarly to Perlin's image synthesizer each stage works on streams of objects like *vertices*, *primitives* or *fragments*. As a consequence they directly reflect the various pipeline stages of the hardware in the language itself. Although there are little restrictions in terms of algorithms that can be formulated, a corresponding shader function must be provided for each of the stages.

Progressive Sampling. Shader languages like HLSL provide procedures as their main structuring mechanism. Über-Shaders usually implement the sum of all desired features and use ad-hoc mechanisms like macros and plain text processing for specialization and feature selection.

Metaprogramming frameworks [9,12–14] overcome the lack of language level abstractions by utilizing meta-programming and macros. LibSh [12] provides an embedded language in C++, utilizing its features like objects and templates for combining shaders. McCool et al. [11] extends LibSh with algebraic combinators *connection* and *combination* which provides an expressive basis for combining shader functions.

Elliot [3] proposes Vertigo, an embedded domain specific language written in Haskell that provides combinators in a very natural way. Based on these combinators, an implementation of a sophisticated shading infrastructure comparable to RenderMan Shading Language (RSL) [6] is demonstrated, including a subsequent compilation process which creates vertex- and fragment-shader programs.

Abstract shade trees [15] are based on a visual programming approach for shaders, and also provide automatic linkage of shader parameters as well as semantic operations like vector basis conversion. Although different shader components compose well, geometry shaders and tessellation are not treated at all. Trapp et al. [22] structures GLSL shader code into code fragments, each typed with predefined semantics. Code fragments may be composed at run-time and compiled to Über-Shaders. Of course Über-Shaders suffer from bad performance. Like other metaprogramming approaches the system cannot provide proper semantic analysis and cross-fragment optimization.

Towards Pipeline Shaders. The RenderMan Shading Language by Hanrahan and Lawson [6] combines the expressiveness of Perlin's image synthesizer with independent shader processes introduced by Cook. The concept resembles object-oriented classes, whereby each virtual method corresponds to an entry point called by the render system. Subclasses like *surface*, *light* and *volume* may be attached to surfaces. Furthermore RSL extends the concept with computation rates, i.e. the notion of inputs varying two different rates: `uniform` and `varying`. Specialized control-flow constructs provide mechanisms for communication between shaders.

A further refinement for computation rates was introduced by Proudfoot et al. [19] in their Stanford Real-Time Shading Language (RTSL): `constant`, `primitive group`, `vertex` and `fragment`, where the last two rates directly corresponded to the stages of early programmable GPUs. Like Cook's shade trees [2], RTSL programs are purely declarative and can therefore be represented as DAGs, which affects expressiveness (e.g. limited data dependent control flow).

Renaissance [1] takes a more general approach and represents different shader pipeline stages as single functional shader programs. Parameters implicitly correspond to different computations rates. Compilation automatically lifts expressions into the earliest possible pipeline stage while maintaining semantics. However, Renaissance lacks support for structuring monolithic shader programs into well defined reusable modules, and no semantics for lifting expressions to groupwise shader stages (e.g. geometry shaders) are presented.

Foley and Hanrahan introduce Spark [4], a *pipeline shader* approach based on RTSL [19]. Its two-layer approach uses declarative shader graphs on top of procedural subroutines and therefore combines the approaches of Cook [2] and Perlin [18]. Spark expands RSL's idea of treating a shader in an object-oriented way by using extending, virtual-, and abstract identifiers for compositing and customizing shaders. Rate-qualifiers and conversions between different rates are extensible and thus defined individually by each supported pipeline. Different modules may be composed by using mixin inheritance. Like other Über-Shader approaches before, Spark does not solve the combinatorial explosion problem because each composition must be stated explicitly.

3   Design

A Shader as a Pixel-Valued Function. Shader programming targets a highly parallel execution environment, where shading can be performed independently for each surface point, therefore functional programming is a natural match for specifying shaders [2]. Although parts of a shader can be programmed in procedural style using local variables and loops, a complete shader program only has a single output value—the target pixel—and can thus be viewed as a single function. By using tail recursion instead of loops, and higher-order functions for control-flow it is even possible to map any procedural shader program to a purely functional representation. Rennaissance [1] is an example of such a functional approach to shader programming.

Since the output value of a shader expression for a single pixel can be an aggregate of multiple simple values (e.g. it can contain a colour, a depth value, etc.), we use the term *shader module* to denote a shader function with multiple input and output values. Multiple output values are programmatically handled by returning a single structure containing the individual output values.

Although our approach is based on the composition of such shader modules, and thus retains the expressiveness and extensibility of a functional design (which goes beyond what is possible with the specialised control flow elements introduced in RSL [6]), we have included control flow functions that are modelled on imperative languages in order to cater to shader programmers that are used to imperative shader languages. Details on these control flow functions are given in Sect. 4.

Semantic Composition of Shader Modules. Combining shader modules that are formulated as expressions can be done in a pipeline approach, by routing the output of one component into the input of another component. In order to

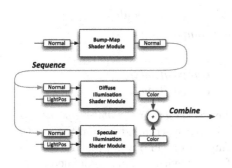

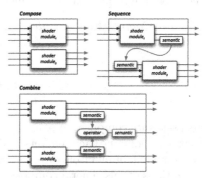

Fig. 2. The output values of three simple shaders can be combined with two composition operators: *Sequence* and *Combine(operator +).*

Fig. 3. The three semantic-aware basic composition operators for shader composition. For illustration purposes, the *Sequence* function is depicted with only one semantic routed between its shader module arguments.

derive the necessary composition functions for combining shader modules we will look at a simple example that combines three shader modules with two composition operators, namely *Sequence* and *Combine(operator)* that are applied to the individual output fields (see Fig. 2). Note, that the different composition operators need to be applied to different types of input and output (*Normal, Color, LightPos*). We use the term *semantic* for these types, as they go beyond the typical notion of data types in a language: both *Normal* and *LightPos* are represented as float vectors, but this does not capture their semantic.

Generalizing from this example, we define the following three basic composition operators for semantic shader composition, that operate on arbitrary shader modules each with one or more semantic input types and one or more semantic output types (see Fig. 3):

Compose($module_1, module_2$) composes the output of the two shader modules. All output semantics of the two input shader modules must be different.

Sequence($module_1, module_2, \{semantic_1, ..., semantic_M\}$) combines the specified semantics of the supplied shader modules in sequence. The remaining output semantics of the two input shaders must be different.

Combine($module_1, module_2, semantic, operator$) applies the supplied binary operator to the output of the two input shaders with the given semantic to return a value of the same semantic. The remaining output semantics of the two input shader modules must be different.

For all composition operators, the input semantics of the two input shader modules are allowed to be either partially or completely equal. In this case, the same value is supplied to both shaders. These basic composition operators add the concept of semantic-specific operations to the usual composition functions used in functional languages. On top of these basic composition operators we can now define a more general composition function that sequences, combines, and composes multiple shader modules based on their semantic:

$$Composition(module_1, \quad module_2, \quad ... \quad module_N,$$
$$semantic_1 \; : composition_1,$$
$$\vdots \qquad\qquad \vdots$$
$$semantic_M : composition_M,$$
$$default \quad : composition_{default})$$

where a separate composition operator (either *Sequence* or *Combine(operator)*) is specified to combine each semantic and *Compose* is wrapped around the result.

We provide a number of convenience compositions in our approach, that are specializations of this general composition function with various predefined function and operator arguments. For convenience we also predefine simple shaders for changing semantics (e.g. *Pos* → *Color*). An example composition can be found in Sect. 5.

Our approach of automatically combining shaders based on their semantically tagged inputs and outputs is inherently more flexible than a static object-oriented approach as implemented by Spark [4]:

- The object-oriented way of extending functionality by overriding virtual functions requires, that each possible extension point needs to be foreseen by the implementer of the base shaders. Since only a limited number of possible ways of extending functionality can be provided in a typical design of such base shaders, the extensibility of such an object-oriented approach is necessarily limited.
- Due to the static way of combining and extending shaders, each and every new combination of simple shader functions must be explicitly and manually implemented. Since the number of combinations of simple shaders is exponential in the number of shaders, this leads to a combinatorial explosion that cannot be handled by a static approach. The use of a composition function as shown above, makes it possible to automatically combine simple shaders based on the geometry that needs to be rendered: the rendering framework can analyse the properties of the geometry, and combine only the simple shaders that are actually needed for rendering the combination of properties encountered.

All composition possibilities offered by a static object-oriented approach can be easily built using a sub-set of the available functionality in our meta-function approach:

- Each virtual method corresponds to a semantic tag: different simple shaders can perform different operations on the input with the same semantic tag. Changing the implementation of one virtual method thus corresponds to replacing one of the simple shaders in a composition of multiple shaders.
- The effect inheritance in the object-oriented approach can be realised using the *combine* composition operator on two simple shaders that correspond to the base-class implementation and the overriding implementation. By using a function that ignores the result of the simple shader corresponding to the base-class the result of the combination corresponds to the result of the overriding simple shader.

Thus our approach provides a superset of the functionality provided by the object-oriented approach, and the additional functionality eliminates the large number of shader combinations that have to be manually specified.

Abstract Stages. The various stages in the shader pipeline can be viewed as optimizations on the single pixel-value function, in order to reduce the number of evaluations of various expressions (for an example see Fig. 4).

Although in principle, every shader could be formulated as a single function that returns a pixel value, this would require the implicit interpolation that is performed between the vertex and fragment stages of the shader pipeline (see Fig. 4) to be explicitly specified in this function. In order to overcome this inconvenience, we propose to retain the notion of shader stages, but as opposed to the multiple hardware stages we only specify two abstract stages, that turn out to be sufficient in practice :

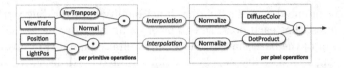

Fig. 4. Optimizing the evaluation of a shader expression by evaluating the parts at different stages. The interpolation functionality is provided by the hardware stages.

Per-pixel Operations: Also called per-fragment operations, these are all the operations that need to be performed for each pixel or fragment. Typically this includes all traditional shading operations that affect the material of an object. In functional notation, these operations perform the mapping: *per pixel parameters → per pixel output*.

Per-primitive Operations: All operations performed for each primitive (e.g. triangle or line). This typically encompasses geometric transformations. In functional notation, these operations perform the following mapping: *per primitive parameters → per primitive output*.

Thus all the simple shaders that can be composed in our framework consist of explicitly specified per-pixel and/or per-primitive operations, and thus each of our simple shaders can be viewed as either a partial or a fully specified, but still abstract pipeline of operations (see Fig. 5). Since we do not explicitly specify operations for a specific hardware pipeline, all our shaders are still specified in an abstract manner, and need to be explicitly mapped onto the hardware stages of a concrete pipeline.

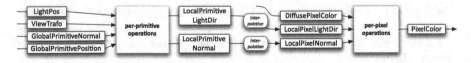

Fig. 5. A simple shader module consists of explicitly specified per-primitive and per-pixel operations each with their semantic inputs and outputs. This shows the example from Fig. 4 tagged with semantic inputs and outputs.

Note that in our current implementation the tessellation stages are represented in a very hardware-like manner and can therefore not be composed with each other. They can however still be composed with other per-primitive operations. In order to overcome this limitation one would have to define tessellation functionality in a more abstract way, where a programmer would need to explicitly subdivide each input primitive. Mapping such a program to the hardware stages would require very sophisticated analysis and may result in inefficient code. Furthermore the need for composition of tessellation functions turned out to be very rare in practice. We therefore decided not to introduce such an abstraction in our current implementation.

This has some consequences for the composition operators defined in the previous section: specifically, a per-pixel output of a shader module cannot serve as an input for a per-primitive shader module. If each shader module that consisted of a per-geometry stage and a per-pixel stage was viewed as a monolithic block with no allowed change in the routing of data between the per-geometry and the per-pixel stage, this would lead to significant limitations on which shader-modules could be combined.

In order to avoid that, we combine the stages of our shader modules individually, i.e. the composition works independently on the per-primitive and on the per-pixel stage. This makes it possible to combine shader modules as if they were single stage modules, and retain the intended functionality programmed in the different stages. The semantically tagged inputs and output of the individual stages are available for automatic composition with other shader modules.

Thus we extend our concept of shader modules to encompass whole shader pipelines, on which our composition operators work, retaining the semantic input and output types between the abstract stages, which is beyond the functionality of the algebraic combinators of McCool et al. [11].

Optimization. Naïvely mapping such high-level shaders onto the hardware stages of a concrete pipeline such as the DirectX 11 pipeline results in a lot of overhead due to a number of possible inefficiencies. In order to provide comparable performance in our system, we perform several optimisations highlighted in the following section.

Dead Code Elimination. Since our modules are programmed to be maximally reusable, they are implemented to cover the most general case, and thus provide a large number of semantic outputs which can be used by other modules that are placed later in a composition. Thus it is of vital importance for the overall performance to identify all unused outputs, and eliminate these from the composition result: this is done by starting with all used pixel outputs and tracking all necessary inputs back through the pipeline. All unused inputs and outputs are removed, essentially performing a dead code elimination step.

Backend Stage Mapping. A typical hardware shader pipeline has a number of stages that can be used to perform the operations in our pipelined shaders. Based on the abstract stage, the following optimisation steps are performed:

Per-primitive Operations: Since the current (DirectX 11) backend basically offers three stages for geometry processing (vertex shader, tesselation shader, and geometry shader) with different capabilities and associated computation rates our composed modules need to be mapped onto these efficiently. Since the computation rates for these hardware-stages are generally unknown (we don't know what our inputs will look like) we decided to move all operations to the earliest stage possible, with the underlying assumption, that each operation is thus performed at the lowest rate. It is, for example, possible to move geometry shader operations, which are equally performed for all vertices, to the vertex shader. Similar rules can be derived for the other hardware-stages, an overview of these rules can be found in Sect. 4. A number of additional limitations (due to hardware capabilities) are introduced, e.g., the tessellation-shader needs to be done prior to a geometry-shader, etc.

Per-pixel Operations: Pixel or fragment-shaders are divided into two parts: the first part represents calculations invariant respective to rasterizer-interpolation, which can therefore be performed per primitive (in general the faster solution). The second part consists of all operations that can only be performed in a pixel shader. This splitting may cause additional traffic for the rasterizer-interpolation since this may need e.g. the interpolation of a vector instead of a scalar. Since these costs are hard to estimate we move only operations where the interpolated type does not exceed the result type in size.

Shader Module Specialization. Since a single generated shader module might still cover a number of different input-setups (textured vs. non-textured, etc.) using shader-control-flow we provide methods for simply specialising a shader module using contextual information (e.g. there are no textures available, etc.) The shader modules are then partially evaluated using this information and recompiled for the backend. If, for example, a geometry does not contain normals the corresponding shader modules are optimised to eliminate any code that accesses normals of the geometry, thereby improving rendering performance.

Shader Module Unification. Since shaders are compositions of abstract modules it's relatively easy to find common operations for them using the high level information provided by the composition operators. If the rendering performance can be improved by reducing switches between shaders, two shader modules can be unified using simple control flow, adding a parameter to select the shader module as an additional input to the combined shader module.

Common Subexpression Elimination. Although common subexpression elimination results in optimal code respective to the number of operations, additional temporary variables stressing the *HLSL*-compiler need to be introduced. In optimizing compilers, sophisticated analysis carefully choose subexpressions to be considered for code motion (e.g. [8]) in order to limit temporaries. Our system in contrast heuristically eliminates expressions exceeding a syntactic complexity threshold. These complexities are based on estimated complexities for all intrinsic functions which are simply summed for each expression. With this simple scheme the *HLSL*-compiler does a good job in optimizing shaders while maintaining good compile-time performance.

Constant/Uniform Calculations. All computations resulting in a constant value (for each draw call) can be pre-calculated by the rendering system. Since a brute force approach would result in a large number of uniform-parameters only calculations exceeding a certain complexity (as mentioned above) are considered.

Arithmetic Optimizations. Since there are only very few restrictions on how to compose shaders (i.e. outputs and inputs must match), it is possible to introduce unnecessary calculations through these compositions (e.g. `normalize(` `normalize(vector))`, $(a-a)$, etc.). Similarly to tree parsers [5], used for instruction selection in code generators our optimizer maintains a set of expression patterns with associated rewrite rules and some estimated cost. Notably, our system also considers domain-specific knowledge as a variables vector-basis for further optimization. As an example, `ViewMatrix*ModelMatrix` is transformed to use the uniform `ModelViewMatrix` in order to eliminate expensive matrix multiplications.

User-Guided Simplification. Additional contextual information can be specified for shader inputs values. As an example, the user may annotate the vertex colors to be constant or the normals to be constant per face.

Using these annotations the backend stage mapping can perform further optimisations by moving operations to earlier stages. Together with redundancy removal, dead code elimination, and constant/uniform calculations this can lead to significantly simplified shaders. As an example, it is unnecessary to interpolate face normals in a shader, when the normals are known to be constant per primitive.

Further Optimization Possibilities. Our abstract pipeline representation is general enough to support completely different approaches like perceptual simplification methods [21], or automatic approaches exploiting temporal coherence [20], which we will pursue in the future.

4 Implementation

Our shading language is implemented as an extension of an existing rendering framework written in C#. Of course it is possible to implement our expression tree based approach with any language that provides abstract data types, the use of anonymous functions significantly reduces the syntactic overhead. A C++ 11 implementation would be equivalent to our approach, while a Java implementation would use anonymous classes instead of anonymous functions.

Fig. 6. Shader effects composed with our approach. From left to right: raytraced reflections with simple texturing, a composition of an illumination and shadow mapping shader, and a subsivision shader (all from Sect. 5), as well as a screenshot from our lighting design application (see Sect. 6) demonstrating texturing and reflective materials with environment mapping.

The expression trees in our approach are created and combined using so called *shader types*, which represent predefined data types available to shaders (e.g. vectors, matrices, textures, aso.). Each shader type provides methods (e.g. the operators + and − or the dot product) that do not actually perform operations, but build an expression tree for the corresponding operations. Thus each expression that specifies a shader module, returns the complete expression tree for that shader module upon execution.

Due to the flexibility of our functional-style implementation we are not limited to predefined control flow statements such as the RenderMan Shading Language (RSL) [6] : we provide higher-order functions that encapsulate conditional evaluation and loops. This makes it possible to integrate conditions and loops into expression trees in typical implementation languages of rendering frameworks, even if it is not possible to overload intrinsic language constructs such as the conditional evaluation operator *condition* ? *value$_1$* : *value$_2$* and the *for loop* for shader types:

```
var floatVal = Fun.IfThenElse<Float>(c < 1.0f, c, 1.0f);
var initial = new { Index = Int.Zero, Val = Float.Zero };

var diffuse = Fun.Loop(initial, i => i.Index < lightCount,
        i => { var dir = light[i.Index] - worldPos.XYZ;
              return new { Index = i.Index + 1, Val = i.Val + normal.Dot(dir.Normalized) }; });
```

As mentioned in Sect. 3, the evaluation of each expression is moved to the earliest possible stage in any given hardware pipeline. In the following list we give the conditions for performing the indicated optimizations:

PixelShader → GeometryShader: An expression can be moved, if it comprises a linear function. Note that functions can be linear under specific circumstances, e.g. if one function argument is a constant.

GeometryShader → DomainShader/VertexShader: If the same function is applied to all vertex-dependent inputs (i.e. it appears for each of the vertices), the function-expression with its arguments can be moved.

DomainShader → HullShader: If an expression does not contain the tessellation coordinate (i.e. the *domain location*) it can be moved.

DomainShader → HullShaderConstantFunction: Similar to *Domain-Shader → HullShader*.

HullShader → VertexShader: Identical pre-conditions to GeometryShader → VertexShader stage.

Although we focussed on the Rules for DirectX 11 and OpenGL 4 we also implemented an experimental backend for our OpenCL based raytracer, which only supports one shader-stage computing the color for a primitive at a certain coordinate. Due to our abstract stage interpretation the modules could easily be mapped onto this stage when possible (features like tessellation are currently not supported by the raytracer)

The first step of the compilation process is the creation of a single pipeline for the completely combined shader modules. This abstract pipeline is then processed using the optimisation stages shown in Sect. 3. The output of the optimisation process is a complete hardware shader in the shading language of the backend: in our case a complete HLSL shader for the DirectX 11 backend.

5   Examples

In order to demonstrate the applicability of our approach to common techniques, we provide code excerpts for per-pixel lighting, shadow mapping and subdivision (see Fig. 6). Furthermore, we demonstrate the interactive capabilities of our system in the accompanying video.

The following example shows the full implementation of a basic transformation pipeline with per-pixel lighting. Note that user-defined parameters are communicated by name (e.g. "LightPosition"), and vertices of primitives supplied to the primitive shader can be accessed using iterators (e.g. .DoByVertex). We also provide swizzle operators (e.g. .XYZ) including constants (letter O is zero, letter I is one) for all vector types.

```
public class PerPixelLighting : Module {

  public class Vertex {
    public Float4 Pos = Varying.Position;
    public Float3 Normal = Varying.Normal;
    public Float4 WorldPos = Varying.WorldPosition;
    public Float4 Color = Varying.Color;
  }

  public class Pixel {
    public Float4 Color = Varying.Color;
    public Pixel(Float3 color, Float alpha = 1) {
      Color = new Float4(color, alpha);
    }
  }
}
```

```
public Fragment<Pixel> Shader(AnyPrimitive<Vertex>input){
  var transformed = input.DoByVertex(v => {
    v.Pos = Uniform.ModelViewProjTrafo * v.Pos;
    v.Normal = Uniform.NormalTrafo * v.Normal;
    v.WorldPos = Uniform.ModelTrafo * v.Pos;
    return v;
  });

  return transformed.Rasterize(f => {
    var dir = (Uniform.LightPositions[0]
             - f.WorldPos).XYZ.Normalized;
    return new Pixel(dir.Dot(f.Normal.Normalized)*
                     f.Color.XYZ,f.Color.W);
  });
}
```

Geometry vertex colors are automatically bound to the *Varying.Color* input. Here we demonstrate the composability of modules by combining per-pixel lighting from the previous listing with a simple shadow mapping module.

```
public class ShadowMapping : Module {
  public Float4x4 ShadowMapTrafo; // transformation
  public Texture2D ShadowMap;

  private Float3 GetShadowTexCoord(Float4 worldPos) {
    var p = ShadowMapTrafo*worldPos; var pp=p.XYZ/p.W;
    var tc = new Float3((Float2.II + pp.XY)*0.5f,pp.Z);
    tc.Y = 1 - tc.Y; return tc;
  }
```

```
public Fragment<Pixel>Shader(AnyPrimitive<Vertex>input){
  return input.Rasterize(f => {
    var mytc = GetShadowTexCoord(f.WorldPos);
    var smValue = ShadowMap.SampleCmp(mytc.XY, mytc.Z);
    return new Pixel(f.Color.XYZ * smValue, f.Color.W);
  });
}}
```

Both modules can be simply composed in the following way:

```
...
var sg = ...                        // some scene graph node
var shadowMapping = new ShadowMapping();
var surface = Composition.Sequence.Compose(new PerPixelLighting(), shadowMapping);
shadowSurface.ShadowMap = renderTarget.DepthTexture;
shadowSurface.ShadowMapTrafo = ... // the transformation value
sg = sg.Surface(surface);
...
```

The code shown in this listing assigns *values* to *module inputs* by implicitly creating uniform inputs in the backend code and setting their values using the renderer infrastructure. These values can thus be changed at runtime.

```
public Float GetFactor(Float4 p0, Float4 p1) {
  var len = (p1.XYZ - p0.XYZ).Length; return Float.Clamp(len / MaxLineLength, 1, 64);
}

public Triangle<Vertex> Shader(Triangle<Vertex> input) {
  return input.DoByPrimitive( tri => {    // tessellation is defined as in OpenGL 4/DirectX 11
    var f0 = GetFactor(tri.P1.Pos, tri.P2.Pos); var f1 = GetFactor(tri.P2.Pos, tri.P0.Pos);
    var f2 = GetFactor(tri.P0.Pos, tri.P1.Pos); var factors = new TessellationFactors();
    factors.EdgeFactors = new[] { f0, f1, f2 }; factors.InnerFactors[0] = (f0 + f1 + f2) / 3.0f;
    return factors;
  },
  (tri, constant, crd) => return new Vertex() { Pos = Float4.Lerp(tri.P0.Pos, tri.P1.Pos, tri.P2.Pos, crd) });
}
```

The previous example shows a simple subdivision operation mapped to the DirectX 11 tessellation stages. The first anonymous function which calculates the tessellation factors can return any custom type inheriting from **Tessellation Factors**. The **constant** argument in the interpolation function then refers to that type. Although it would theoretically be possible to create arbitrary output triangles for a given input patch using the DirectX/OpenGl tessellation stages we decided to expose the functionality as provided by our main backend. The first lambda function basically corresponds to the Hull-/TessellationControl-Shader and the second to the Domain-/TessellationEvaluation-Shader.

6   Analysis

A Real-World Comparison. We evaluated our concept by re-implementing all shaders used in a production-quality real-world application for lighting design: it uses shaders for computing global illumination, for drawing lines and points in debugging and editing views, and for rendering a number of different materials with diffuse, and specular components and environment maps for realistic looking reflections. The complexity of shaders ranges from simple flat-shading all the way to a global-illumination shader that needs to perform polygon clipping for each rendered pixel (see Fig. 6 as well as the demonstration in the accompanying video). The original HLSL implementation had 37 modules with a total of 2324 lines and compiled in 7.5 s. Our new semantically composed shaders with the same functionality consists of 20 modules with a total of 1148 lines and compiled in 11.8 s. Thus we were able to roughly halve the number of code-lines as well as the number of modules, while maintaining comparable shader-compile-times and shader performance. Due to the high level of reuse enabled by our shader modules, the shader modules implemented for this application are a lot less specific, and can therefore be reused in a number of future applications.

Performance. We evaluated our performance using three sample shaders:

Skinning: Renders two animated simple meshes replicated 50 times without hardware instancing with standard skinning and diffuse lighting, where skinning is performed with a maximum of four bone influences per vertex.

Tesselation: Subdivides all 871414 triangles of the Stanford dragon to 1-pixel-sized triangles using simple interpolation of positions, normals and light direction, and applies diffuse lighting.

Shadow Mapping: Renders a shadow mapped Standard dragon using a pre-calculated shadow map. A 9×9 Gaussian filter implemented in the pixel shader is used to blur the shadow.

All our tests were performed on an Intel(R) Core(TM) i7 CPU 930 @ 2.80 GHz system with 12 GB of RAM and a GeForce GTX 480 graphics card.

In our first evaluation (see Table 1) we compared the final rendering speed of manually coded shaders with our new shaders. We found that in most cases our

Table 1. Performance results in frames per second (fps) comparing our new semantic shaders with manually coded shaders.

Shader	Manual	New	Factor
Skinning	35 fps	35 fps	1.00
Tesselation	178 fps	189 fps	1.06
Shadow-map	209 fps	209 fps	1.00

Table 2. Compile times in milliseconds (ms) of manually written HLSL code and our new semantic shader code. The times for the new shaders are the sum of generation of HLSL from our source and the HLSL compile time.

Shader	Manual	New	Factor
Skinning	140 ms	217 ms (78+139)	1.54
Tessellation	46 ms	81 ms (33+48)	1.77
Shadow-map	85 ms	110 ms (24+86)	1.30

system produced equally fast or faster shader code. We rely on the aggressive optimizations in the compiler backend (HLSL or GLSL) to overcome some of the deficiencies in our code generator. The tessellation example was slightly faster since some calculations performed in the DomainShader were automatically moved to the vertex shader.

Our second evaluation (see Table 2) shows the compile times for manually coded shaders compared to the sum of code generation and compile times for our new shaders. The additional generation of HLSL from our source code results in compile-times that are less than 2 times longer for the given examples, an acceptable overhead given the increased flexibility and reusability.

7 Conclusion and Future Work

We have demonstrated a powerful framework for combining simple shader modules into complex shaders, by providing semantic composition of modules, an abstract model of two shader stages and a powerful optimizing backend for mapping this programming model onto existing hardware pipelines. With only moderate increase in shader compilation time, this results in a significant reduction of code while retaining the execution performance of hand-coded shaders. Shader composition is performed fast enough for interactively combining shaders, allowing rapid shader development in an explorative manner. Due to the high reusability of the shader modules of our new approach, we have already started to build a comprehensive library of modules that can be freely combined and provide a framework for rapid development of rendering applications that minimises the necessity of writing shader modules.

In a number of use cases such as volume rendering and global illumination, we found that using a pure expression based language—although possible—can be somewhat inconvenient, since it requires a complete reformulation of some algorithms that can be easily expressed in procedural languages. We therefore plan to add support for imperative shader fragments while maintaining composability.

Acknowledgements. We would like to thank Manuel Wieser for providing 3D models, especially *Eigi, The Dinosaur*. The competence center VRVis is funded by BMVIT, BMWFJ, and City of Vienna (ZIT) within the scope of COMET Competence Centers for Excellent Technologies. The program COMET is managed by FFG.

References

1. Austin, C.A.: Renaissance: a functional shading language. Master's thesis, Iowa State University, Ames, Iowa, USA (2005). http://www.aegisknight.org/hci_portfolio/thesis.pdf
2. Cook, R.L.: Shade trees. SIGGRAPH Comput. Graph. **18**(3), 223–231 (1984). http://doi.acm.org/10.1145/964965.808602
3. Elliott, C.: Programming graphics processors functionally. Proceedings of the 2004 ACM SIGPLAN Workshop on Haskell. Haskell 2004, pp. 45–56. ACM, New York (2004)
4. Foley, T., Hanrahan, P.: Spark: modular, composable shaders for graphics hardware. ACM Trans. Graph. **30**(4), 107:1–107:12 (2011)
5. Fraser, C.W., Henry, R.R., Proebsting, T.A.: BURG: fast optimal instruction selection and tree parsing. SIGPLAN Not. **27**(4), 68–76 (1992). http://doi.acm.org/10.1145/131080.131089
6. Hanrahan, P., Lawson, J.: A language for shading and lighting calculations. SIGGRAPH Comput. Graph. **24**(4), 289–298 (1990)
7. Kessenich, J., Baldwin, D., Rost, R.: OpenGL Shading Language, v 4.3 (2012). http://www.opengl.org/documentation/glsl/. Accessed 23 October 2012
8. Knoop, J., Rüthing, O., Steffen, B.: Lazy code motion. SIGPLAN Not. **27**(7), 224–234 (1992). http://doi.acm.org/10.1145/143103.143136
9. Kuck, R., Wesche, G.: A framework for object-oriented shader design. In: Bebis, G., Boyle, R., Parvin, B., Koracin, D., Kuno, Y., Wang, J., Wang, J.-X., Wang, J., Pajarola, R., Lindstrom, P., Hinkenjann, A., Encarnação, M.L., Silva, C.T., Coming, D. (eds.) ISVC 2009, Part I. LNCS, vol. 5875, pp. 1019–1030. Springer, Heidelberg (2009)
10. Mark, W.R., Glanville, R.S., Akeley, K., Kilgard, M.J.: Cg: a system for programming graphics hardware in a C-like language. ACM Trans. Graph. **22**(3), 896–907 (2003)
11. McCool, M., Du Toit, S., Popa, T., Chan, B., Moule, K.: Shader algebra. ACM SIGGRAPH 2004 Papers. SIGGRAPH 2004, pp. 787–795. ACM, New York (2004)
12. McCool, M.D., Qin, Z., Popa, T.S.: Shader metaprogramming. Proceedings of the ACM SIGGRAPH/EUROGRAPHICS Conference on Graphics Hardware, HWWS 2002, pp. 57–68. Eurograph. Assoc, Aire-la-Ville (2002)
13. McCool, M.D., Toit, S.D.: Metaprogramming GPUs with Sh. A K Peters, Stanford (2004)
14. McGuire, M.: The SuperShader. In: Shader X4: Advanced Rendering Techniques, chap. 8.1, pp. 485–498. Cengage Learning Emea (2005). http://www.cs.brown.edu/research/graphics/games/SuperShader/index.html
15. McGuire, M., Stathis, G., Pfister, H., Krishnamurthi, S.: Abstract shade trees. Proceedings of the 2006 Symposium on Interactive 3D Graphics and Games. I3D 2006, pp. 79–86. ACM, New York (2006)
16. Microsoft: Shader model 5 DirectX HLSL (2010). http://msdn.microsoft.com/en-us/library/windows/desktop/ff471356%28v=vs.85%29.aspx. Accessed 23 October 2012
17. Microsoft: Programming Guide for HLSL (2012). http://msdn.microsoft.com/en-us/library/bb509635(v=VS.85).aspx. Accessed 23 October 2012
18. Perlin, K.: An image synthesizer. Proceedings of the 12th Annual Conference on Computer Graphics and Interactive Techniques. SIGGRAPH 1985, pp. 287–296. ACM, N.Y. (1985)

19. Proudfoot, K., Mark, W.R., Tzvetkov, S., Hanrahan, P.: A real-time procedural shading system for programmable graphics hardware. Proceedings of the 28th Annual Conference on Computer Graphics and Interactive Techniques. SIGGRAPH 2001, pp. 159–170. ACM, New York (2001)

20. Sitthi-Amorn, P., Lawrence, J., Yang, L., Sander, P.V., Nehab, D., Xi, J.: Automated reprojection-based pixel shader optimization. ACM Trans. Graph. **27**(5), 127:1–127:11 (2008). http://doi.acm.org/10.1145/1409060.1409080

21. Sitthi-Amorn, P., Modly, N., Weimer, W., Lawrence, J.: Genetic programming for shader simplification. ACM Trans. Graph. **30**(6), 152:1–152:12 (2011). http://doi.acm.org/10.1145/2070781.2024186

22. Trapp, M., Döllner, J.: Automated combination of real-time shader programs. In: Cignoni, P., Sochor, J. (eds.) Eurographics 2007 Shortpaper, pp. 53–56. Eurograph. Assoc. (2007)

Simulated Virtual Crowds Coupled
with Camera-Tracked Humans

Ivan Rivalcoba[1]([⊠]), Oriam De Gyves[1], Isaac Rudomin[2], and Nuria Pelechano[3]

[1] Department of Computer Science, Tecnológico de Monterrey, Mexico City, Mexico
{ivan.rivalcoba,odegyves}@gmail.com
[2] Computer Sciences, Barcelona Supercomputing Center, Barcelona, Spain
rudomin.isaac@gmail.com
[3] Llenguatges I Sistemes Informàtics, Universitat Politècnica de Catalunya,
Barcelona, Spain
npelechano@lsi.upc.edu

Abstract. Our objective with this paper is to show how we can couple a group of real people and a simulated crowd of virtual humans. We attach group behaviors to the simulated humans to get a plausible reaction to real people. We use a two stage system: in the first stage, a group of people are segmented from a live video, then a human detector algorithm extracts the positions of the people in the video, which are finally used to feed the second stage, the simulation system. The positions obtained by this process allow the second module to render the real humans as avatars in the scene, while the behavior of additional virtual humans is determined by using a simulation based on a social forces model. Developing the method required three specific contributions: a GPU implementation of the codebook algorithm that includes an auxiliary codebook to improve the background subtraction against illumination changes; the use of semantic local binary patterns as a human descriptor; the parallelization of a social forces model, in which we solve a case of agents merging with each other. The experimental results show how a large virtual crowd reacts to over a dozen humans in a real environment.

1 Introduction

Real-time crowd simulation is a research area that is growing rapidly and has become one of the main research directions in computer games, movies and virtual reality [27]. Applications include urban planning, security and entertainment, among others [8]. It has been studied extensively but only in a few cases the computer generated crowds react to real humans. Most research is focused on the behavior of virtual humans, that may be data driven, rule based or socially inspired. The end result is almost every time a crowd interacting only with its virtual environment. In the human computer interaction literature there is some research on the interaction between virtual humans and a specific kind of real humans: end-users, but most are limited to one or two users and a small number of virtual humans.

© Springer International Publishing Switzerland 2015
S. Battiato et al. (Eds.): VISIGRAPP 2014, CCIS 550, pp. 62–77, 2015.
DOI: 10.1007/978-3-319-25117-2_5

A simulation system which includes some real humans (e.g. pedestrians) in crowds of agents, that react to the real humans would have several potential applications in special-effects for films and video games. One goal of this research is to visualize an authoring tool that would use a camera looking down on a group of actors to reduce the time of animation production in the context of crowd generation. One could also conceive games where several real players lead teams of virtual team-mates. Other applications would allow researchers to evaluate their behavior algorithms by directly comparing them with real human behaviors.

The aim of the present work is to perform the coupling between a group of real people and a simulated crowd of virtual humans. To achieve this, we propose a two-stage interactive system. In the first stage, named the vision system, pedestrians are segmented and tracked from a live video. The positions of the captured pedestrians are used in the second module, the simulation system. The simulation system is then responsible for the reaction of the virtual crowd to the tracked real humans.

The rest of this paper is organized as follows. A brief overview of the related work is presented in Sect. 2. Section 3 describes the data and formulation we used to couple real humans with a simulated virtual crowd. The experiments and results are presented in Sect. 4 and finally, we conclude and present the limitations and future work in Sect. 5.

2   Related Work

A crucial part of making the system suitable for real-time interaction is the time consumed by the recognition, simulation and rendering tasks. To get a deeper understanding of the system, this section describes the challenges and the state of the art related to computer vision and crowd simulation.

There has been a large amount of research using vision for human detection. Viola et al. [25] described a machine learning approach for visual object recognition and also introduced the *"integral images"*, a popular technique to accelerate the calculus of many descriptors. The same authors adapted the technique to detect pedestrians from a video sequence by taking advantage of the motion appearance of a walking person and using *AdaBoost* for the training process and construction of the classifier [26]. Dalal et al. [7] performed a complete study of *Histogram of Oriented Gradients* (HOG) applied to the representation of humans. This method offers good results for pedestrian detection by evaluating local histograms of image gradient orientations over a dense normalized overlapping grid, avoiding the common problem of illumination variance. Tuzel et al. [24] detected pedestrians by representing an image region as covariance matrices of spatial locations, intensities and derivatives, to name a few.

In most cases, the detection systems include a pre-processing phase, which consists of the scene's background removal. This is the case of the work presented by Banerjee et al. [2], which combined an adaptive background modeling with HOG features. The authors distinguish between the stationary parts and the moving regions, which are considered as *Regions of Interest* (ROI). HOG are applied to classify whether such region is a human or not. Bhuvaneswari et al. [5] presented another approach called *edgelet features*, which are short segments belonging to a line or a curve. The aforementioned works dealt with side perspective scenes, however they are not recommended when dealing with crowd analysis, due to the occlusion presented when the crowd size increases, causing the accuracy of the recognition systems to decrease.

Lengvenis et al. [13] presented the passenger counter system created for and installed in the Kaunas public city transport. Unlike other works based on edge features, this paper uses the vertical and horizontal projection of the scene using a top-view perspective, also known as bird's eye view. The results reported an accuracy of 86 % for a single passenger getting on or off the bus, however it could not detect people passing each other or getting on a bus together and presented problems with lighting changes. Ozturk et al. [20] proposed a system to determine the head and body orientation of humans in top-view scenes , but it has the same problems that the work by Lengvenis. Their system used a shape context based approach to detect basic body orientation and proposed an optical flow based on *Scaled Invariant Feature Transform* (SIFT) features, however the occlusions were left for future work making this algorithm unsuitable for crowded scenes.

On the other hand, there is also extensive research that focuses on modeling accurate behaviors for virtual agents. Researchers have studied crowd simulations in both microscopic and macroscopic levels; macroscopic simulations are concerned about the realistic movement of the crowd as a mass while microscopic simulations focus on the realistic movement of the individuals. One of the first research works focused on autonomous virtual agents and their behaviors are the steering behaviors by [22]. According to Reynolds' research, a bird is aware of three aspects during flocking, itself, two or three nearest neighbors, and the rest of the flock.

Helbing et al. presented the social forces model, where each agent tries to move at a desired velocity while applying friction and repulsion forces against obstacles and other agents [10]. According to the authors, social forces are well suited for pedestrians in normal, or known, situations. The social forces model has been extended to support groups of related pedestrians by Moussaid et al. [17]. In our work we have modeled the behavior of virtual pedestrians using this algorithm. This model incorporates a group force that makes it possible to simulate group formation. In his paper the size of the groups is determined statistically according to observations from video sequences. Millan et al. used attraction and repulsion forces codified in textures to steer the agents through the environment [16]. In this approach every individual agent has its own set of goals depending on which agents were nearby, but in some cases the forces may cancel each other, leaving the agents *stuck* in the environment.

The GAMMA group [4] extended the *Velocity Obstacles* (VO) into *Reciprocal Velocity Obstacles* (RVO). RVO assumes that every agent in the simulation is steered using the same rules, which eliminates the oscillations in the paths of the VOs. Bleiweiss [6] parallelized the RVO algorithm and implemented it on CUDA, obtaining a 4.8x speedup over the original implementation. Van den Berg et al. [3]

further extended RVO into *Reciprocal n-body Collision Avoidance*. The algorithm reduced the problem to solving a low-dimensional linear program. By using this algorithm, the authors were able to simulate thousands of pedestrians in real-time.

Tsai-Yen et al. [15] presented a computer simulation crowded by virtual users and virtual humans. Each group of virtual humans is led by an intelligent group leader and a world manager can interactively assign a goal configuration to a group leader through a graphical user interface at the virtual environment server. Thalmann et al. [27] presented an interactive navigation system in which a user is able to control an avatar in the crowd through a natural interface, but the interaction remains between a single user and the virtual characters. Pelechano [21] proposed an experiment to closely study the behavior of people interacting with a virtual crowd. A virtual scenario is created to simulate a cocktail party. The main goal of the experiment was to examine whether participants interacting with a virtual crowd would react to the virtual crowd as they would do in a similar real situation.

3   System Architecture

The system we describe in this article is designed to work in real-time even when using consumer grade hardware, while also being capable of detecting and tracking humans from a video sequence and simulating thousands of virtual agents which react under the influence of the real pedestrian detected. A diagram showing the application flow can be seen in Fig. 1.

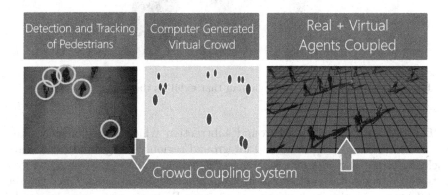

Fig. 1. This figure shows an overview of the system; the data captured by the detection phase feeds the simulation phase. The simulated crowd is able to react to the real humans.

3.1   Vision System

The system that performs the human detection consists of 5 stages (Fig. 2).

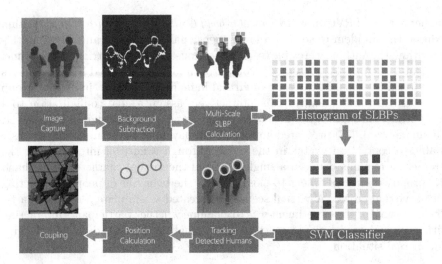

Fig. 2. The vision stages that produce the set of positions that will be sent to the simulation system.

Fig. 3. Head-Shoulder part of a human that exhibits their omega-like shape.

The first stage is the background subtraction which is responsible splitting the scene into background and foreground regions. Foreground regions are labeled as *Regions of interest* (ROI), and this labeling allows the system to concentrate the computational efforts in the human detection tasks. Once the set of ROI have been obtained, in the second stage ROIs are passed one by one to the feature extraction module which generates a vector of features that models the object detected. Each one of these vectors is then sent to a *support vector machine* (SVM) that performs the classification into human or not human. We decided to detect the head and shoulders section of the body due to the fact that the human head remains constant over all the scene [18], unlike other parts of the body. The head and shoulders section exhibit an Ω like shape (see Fig. 3). This property makes the head and shoulders the most stable parts of the body to be detected and tracked.

In the rest of this section we explain each stage in more detail.

To perform background subtraction we propose a modified version of the codebook algorithm presented by [12]. The codebook algorithm adopts a clustering and quantization technique to construct a background model from a long observation of sequences of pixel intensities. Each pixel owns its codebook set, that is made of codewords produced on the training stage; these codewords are constructed based on two parameters: *brightness* and *color distortion*.

If the brightness of a given input pixel $x\,(R,G,B)$ at a certain time t, expressed as x_t, is defined as the norm of the vector in the **RGB** color space described by (1), and the brightness $v_i\,(R,G,B)$ for the codeword c_i is also defined as the norm of the vector in the **RGB** color space, as shown in (2), then the color distortion between two pixels, x_t and v_i, is the length Δ of the opposite side of the angle formed by x_t and v_i, as seen in Fig. 4. This function is described in (3) and (4):

$$\|x_t\|^2 = R^2 + G^2 + B^2 \tag{1}$$

$$\|v_i\|^2 = \bar{R}_i^2 + \bar{G}_i^2 + \bar{B}_i^2 \tag{2}$$

$$colordist(x_t, v_i) = \Delta = \|x_t\|^2 - \|x_t\|^2 \cos^2\alpha \tag{3}$$

$$\|x_t\|^2 \cos^2\alpha = \frac{\langle x_t, v_i \rangle^2}{\|v_i\|^2} \tag{4}$$

$$\langle x_t, v_i \rangle^2 = \left(\bar{R}_i R + \bar{G}_i G + \bar{B}_i B\right) \tag{5}$$

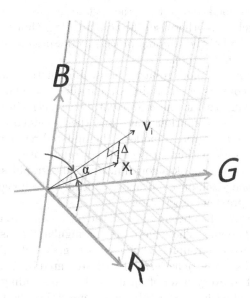

Fig. 4. This figure shows the model used to calculate the color distortion Δ between two pixels x_t and v_i.

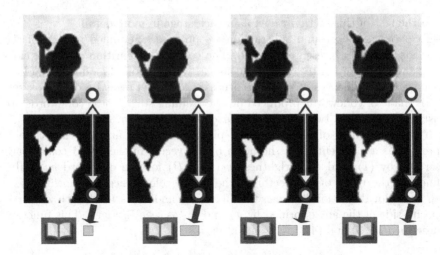

Fig. 5. To maintain the background model and make it robust against intensity changes, a second code book is used and maintained by those regions with no movement. This image shows, in black, all the regions that feed the auxiliary codebook, represented by the icon below the images.

A codeword could be seen as a box that stores information about the pixel intensities; every codeword is generated through the observation of each pixel over a period of time, called training stage. The training may take one or two minutes. If there is no codeword, already in the codebook, that meets the similarity measures described above, then the system creates a new one and then is stored in the codebook. Otherwise, the training algorithm counts every codeword match. At the end of the training process, codewords with few occurrences are discarded from the codebook.

To make the algorithm more robust against illumination changes, the modification we propose consists of maintaining an auxiliary codebook that generates codewords only for regions with no movement. Once a certain time has elapsed, this codebook is cleaned of spurious data, and then is added to the main codebook. This allows the background model to be updated regularly, see Fig. 5.

To perform this task within an acceptable frame rate, we implemented a parallel version of the codebook algorithm, using the graphics hardware. Each thread works over one pixel, once a frame is captured, the image is processed on the GPU using the previous description.

Once the background has been removed, the next step is to extract the blobs from the foreground regions. These blobs are rectangular regions that contain the humans, and are converted from RGB color space to gray scale. To model a human we use a variant of *local binary patterns* (LBP) presented by [19]. The LBP operator assigns a label to every pixel of an image by thresholding the 3×3 neighborhood of each pixel with its own value and considering the result as a binary number. E.g., given a pixel (x_c, y_c), the LBP of that pixel is calculated using (6),

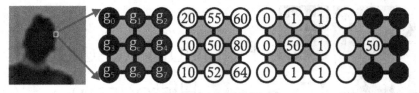

Binary: 01101011
Decimal: 107

Fig. 6. An illustration of the basic LBP operator.

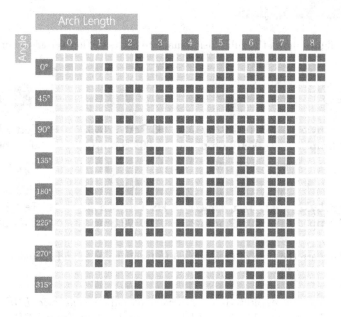

Fig. 7. This figure shows the only 58 possible S-LBPs of a pixel.

where P corresponds to the number of sampling points, R is the radius of the neighborhood to be sampled, g_p represents each sampling point and g_c represents the pixel (x_c, y_c); the histogram of the numbers obtained along all the image is used as the descriptor. Figure 6 shows an illustration of the basic LBP.

$$LBP_{P,R}(x_c, y_c) = \sum_{p=0}^{P-1} s(g_p - g_c)2^p \qquad (6)$$

$$s(x) = \begin{cases} 1, & x \geq 0 \\ 0, & otherwise \end{cases} \qquad (7)$$

Although LBPs have many advantages over other gradient-based features, the basic LBP is not robust enough to be used as human descriptor: however

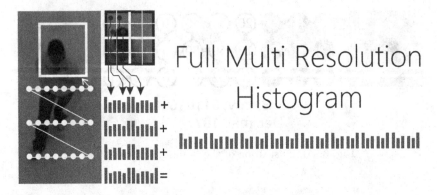

Fig. 8. This figure shows the subdivision made to the detection kernel to compute the histogram.

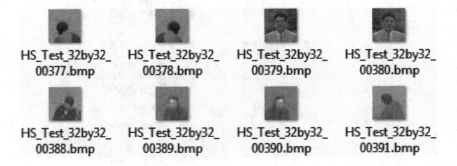

Fig. 9. These are the type of images used to train the head and shoulder detector.

its strengths can be exploited using a variant called *semantic LBP* (S-LBP) proposed by Huang et al. [11]. S-LBPs are made of uniform LBPs. An LBP is called uniform if only at most two bitwise transitions between 0 and 1 are presented, in other words several continuous "1" bits form an arc on the sampling circle; these arcs can be represented with its principle direction and arc length (see Fig. 7); by using this strategy, the dimension of the feature vector is reduced. In our work we use a detection kernel of 32×32 pixels size which produces a 1160D vector; all the S-LBPs are used to compute an histogram, each histogram represents the feature vector of each region. To speed up the computation of S-LBPs, we propose the use of a look up table to classify the type of S-LBP.

As it has been demonstrated by Tuzel et al. [24] the ensemble of variable-size sub-windows can greatly promote detection efficiency. For this reason, we adopted a multi resolution approach similar to [1]. We divided each detection kernel into local regions, then the S-LBPs descriptors are extracted for each region independently. The vector "v" obtained for each sub-window is normalized using the following equation:

$$L1 - sqrt : f = \sqrt{\frac{v}{\|v\|_1 + e}} \tag{8}$$

The descriptors are then concatenated to form a global descriptor vector, see Fig. 8. As previously mentioned, a 1160D vector is generated using a detection kernel of 32×32 pixels, these vectors feed an SVM classifier that decide if the image of the kernel belongs to a person or not.

We use the OpenCV SVM library to train and classify the head and shoulders part of the human body, using the same dataset employed by [14]. This dataset uses top view shots of humans' heads and shoulders, as seen in Fig. 9.

One important observation is that the pedestrians exhibit a deformation due to the camera perspective, making difficult to tune up the classifier [23]. To overcome this issue, and improve the detection results, we divide the camera view on different segments in function by the distance between the camera and the human. Figure 10 shows the depth map and the deformation suffered by a group of humans as they move along the scene.

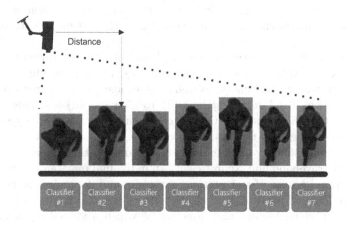

Fig. 10. Segmentation of the scene in function of the distance between the camera and the human.

Once a human head is detected, a point is associated with it. Each detected point is tracked over every frame using the pyramidal version of the Lucas Kanade tracker implemented on the OpenCV library. All the points detected are then stored into a texture to be used in the simulation stage; this step is detailed in the next section.

3.2 Simulation System

OpenGL 4.3 introduced *Compute Shaders* to its API, which enable its shading language to perform general computing in the graphics hardware. Compute Shaders provide an alternative approach to CUDA from Nvidia and OpenCL from the Khronos Group for General-Purpose Computing on Graphics Processing Units (GPGPU). Unlike CUDA and OpenCL, Compute Shaders are similar to other shader stages in the graphics pipeline and are able to read from textures, images and buffers and write to images and buffers without mapping to other data structures or switching between graphics and compute contexts. For our simulation system, we use Compute Shaders in order to simulate the agents and use a method based on social forces to steer the agents around the environment. All the logic related to agent control is performed in the graphics hardware, which allows us to perform large scale simulations with thousands of agents in real-time. For the interoperability between rendering and simulation, we store all the data required for the simulation in texture memory. Data captured by the vision system (Sect. 3.1) is stored in the same textures that the simulated agents use.

Most of the research works only consider the basic attributes of a pedestrian. A crowd is a heterogeneous group of people, each person with a different set of physical and psychological features that make them unique. The crowd dynamic would probably change depending on the number of older pedestrians in it, the concentration of aggressive pedestrians, or a combination of both. We assign physical and psychological characteristics to the virtual pedestrians in our simulations.

The physical characteristics we use are based on studies on real pedestrians and their average walking velocities regarding age and gender. Our simulations comprise both females and males, as well as three age groups, young pedestrians (15 to 30 years old), adult pedestrians (31 to 65 years old) and elderly pedestrians (more than 65 years old). On the other hand, we base our psychological characteristics using trait theory, specifically the Eysenck 3-factor model, which categorizes the personality of an individual according to three main factors, Psychoticism, Extraversion and Neuroticism. An individual may tend to exhibit one of these factors because of testosterone, serotonin and dopamine, respectively.

The physical and psychological characteristics are used to compute the maximum speed, preferred speed and direction at which a virtual agent is able to walk at any given time. This maximum velocity is computed per agent during the initialization of the system. Planning horizon represents how far ahead an agent plans its movement. According to the data we gathered, aggressive agents should have the minimal planning horizon of the three personalities, since they try to maintain its route without many changes and their lack of respect towards other pedestrian's personal spaces. On the other hand, shy agents have the highest value of the three personalities, since they are easily diverted from their routes and try to avoid contact with other agents as much as possible. The agent radius, the number of neighbors and their distance to the agent are other values used to compute the preferred velocity and direction of the agents.

To steer the simulated agents we developed a parallel implementation of the work by Moussaid et al. [17]. Three data textures are required for the simulation: position, velocity and target textures. Each texture element (*texel*) i holds the data of the agent or pedestrian i. To update the positions of the simulated agents, we use (9), (10) and (11).

$$\boldsymbol{f}_i = \boldsymbol{f}_i^d + \boldsymbol{f}_i^n + \boldsymbol{f}_i^o + \boldsymbol{f}_i^g \tag{9}$$

$$\boldsymbol{v}_{inew} = \boldsymbol{v}_i + \boldsymbol{f}_i \cdot \Delta t \tag{10}$$

$$p_{inew} = p_i + v_{inew} \cdot \Delta t + \frac{f_i \cdot \Delta t^2}{2} \tag{11}$$

Where f_i represents the acceleration of the agent i in the current frame, and it is affected by its internal motivations f_i^d, its neighbors f_i^n, the nearest obstacles f_i^o and its group dynamic f_i^g. The new velocity v_{inew} is obtained by adding the acceleration to the previous velocity v_i of the agent. Finally, the position of the agent is updated with both its acceleration and its velocity.

In his original research, Helbing describes a *body force* and a *sliding force* between agents for counteracting body compression and impeding relative tangential motion, respectively [9]. These forces are part of the force between agents, f_i^n. Through experimentation, we found that the tangential velocity difference, $\Delta v_{ji}^t = (v_j - v_i) \cdot t_{ij}$, used in the sliding force is $\Delta v_{ji}^t = 0$ if the agents are facing each other in opposite directions with the same velocity. This leads to the agents getting stuck in tests like the bidirectional movement. In case $\Delta v_{ji} = v_j - v_i \leq \epsilon$ then we assign $\Delta v_{ji} = v$, where v represents a vector constant that satisfies $\|v\| \neq 0$. The magnitude of v is proportional to the sliding force, and a greater sliding force makes the agents evade each other faster. A bidirectional test performed using this modification to the tangential velocity difference can be seen in Fig. 11.

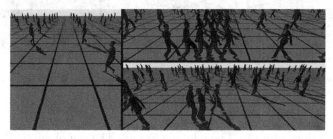

Fig. 11. Bidirectional test (left). The original tangential velocity difference may be zero when agents face each other with the same velocity, leaving the agents stuck in a line (top right). Bidirectional test performed with the modified tangential velocity difference (bottom right).

4   Results

Results show that this system can perform a real-time coupling between a group of tracked humans and a computer simulated group of virtual characters. Simulated agents can react to the captured pedestrians in different ways. Every reaction that is possible between virtual agents, is possible between virtual agents and real pedestrians. Figure 12 shows a scene from a video feed in which pedestrians are being captured and coupled with a simulated virtual crowd; captured pedestrians are highlighted in white. In this figure, two scenarios are presented. First, a scenario in which the virtual agents only react to the real pedestrians by evading collisions with them. In the other scenario, the virtual agents flee from the real pedestrians. In both cases, we have tuned the group social forces to

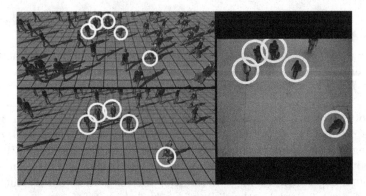

Fig. 12. Captured pedestrians (right). Coupling of a simulated virtual crowd with the captured pedestrians (top left). The same scene, but the virtual agents flee from the captured pedestrians (bottom left).

Fig. 13. This image shows the formations produced using group social forces, the group formations are up to 3 people.

produce formations of up to 3 people, see Fig. 13. The use of an auxiliary codebook for background modeling made the system capable to correctly segment the scene in the presence of illumination changes. The combination of semantic LBPs and the subdivision of the scene into segments allow handling the deformations suffered by the humans as they move away from the camera.

The usage of special cameras was not necessary, any standard web camera fully meets the minimum technical specifications to make the system reliable. The dimension of the processed images were 640 by 480 pixels. The top view perspective -also called bird's eye view- was used to capture as many pedestrians as possible, minimizing the probability of occurrence of occlusions. In our experiments, we were able to capture over a dozen pedestrians while keeping interactive frame rates.

We achieved a frame rate above 60 frames per second, enough to be considered a real-time system. Using the system, we were able to simulate thousands of autonomous agents. The use of the social forces model on the simulation side, produces a plausible effect of integration between the two worlds.

5 Conclusions and Future Work

The computer interaction powered by camera technology is becoming more popular over the years. In our case, there are several challenges to overcome, such as background subtraction against illumination changes, human detection, tracking groups in crowded scenes, and fitting behaviors on virtual characters, to name a few.

Taking into account the above, we have presented a system that allows a large virtual crowd to react to real humans by tracking them on video using only consumer grade hardware. This system is capable of detecting and tracking over a dozen humans in a scene, and simulating thousands of virtual agents. It allows coupling between the tracked humans and the simulated agents while maintaining real-time frame rates.

The system presented in this paper could lead to a system that could test new behavior algorithms for virtual agents, whether they are data driven or rule based, because it allows a direct visual comparison with real humans. It may also allow more accurate urban planning and security simulations since some part of the population in the crowd is constituted of real people. In the entertainment industry, this system provides a contribution to the workspace of applications that seek the coupling of the real world and a virtual world.

We are working in making both systems, vision and simulation, totally asynchronous from each other. When the vision data is not ready to be processed by the simulation system, the captured pedestrians can be simulated using the same behaviors of the agents.

Allowing the systems to work independently has two main advantages, the first one is that each system can be executed in a different platform -this would allow the vision system to be implemented in an embedded device such as a security camera while the simulation system is running in a remote computer. The second advantage of this setup is the ability to detect errors in the behaviors used in the simulation mode when the vision system updates the data of the tracked humans. Having a direct visual feedback between errors in simulated and real data means a better benchmark for a new simulation behavior.

References

1. Ahonen, T.: Face description with local binary patterns: application to face recognition. Pattern Anal. Mach. Intell. **28**(12), 41–2037 (2006). http://www.ncbi.nlm.nih.gov/pubmed/17108377ieeexplore.ieee.org/xpls/abs_all.jsp?arnumber=1717463
2. Banerjee, P., Sengupta, S.: Human motion detection and tracking for video surveillance. In: Proceedings of the National Conference of Tracking and Video Surveillance Activity Analysis, pp. 88–92 (2008)
3. Van den Berg, J., Guy, S.J., Lin, M., Manocha, D.: Reciprocal n-body collision avoidance. Robot. Res. **70**, 3–19 (2011). http://www.springerlink.com/index/15814853H6002Q67.pdf
4. Van den Berg, J., Manocha, D.: Reciprocal velocity obstacles for real-time multi-agent navigation. In: 2008 IEEE International Conference on Robotics and Automation, pp. 1928–1935. IEEE, May 2008. http://ieeexplore.ieee.org/lpdocs/epic03/wrapper.htm?arnumber=4543489

5. Bhuvaneswari, K., Rauf, H.A.: Edgelet based human detection and tracking by combined segmentation and soft decision. In: Control, Automation, Communication and Energy Conservation, 4–9 June 2009. http://ieeexplore.ieee.org/xpls/abs_all.jsp?arnumber=5204487

6. Bleiweiss, A.: Multi agent navigation on GPU. White paper, GDC (2009). http://www.cs.uu.nl/docs/vakken/mcrs/papers/28.pdf

7. Dalal, N., Triggs, B.: Histograms of oriented gradients for human detection. In: 2005 IEEE Computer Society Conference on Computer Vision and Pattern Recognition (CVPR 2005), vol. 1, pp. 886–893 (2005). http://ieeexplore.ieee.org/lpdocs/epic03/wrapper.htm?arnumber=1467360

8. De Gyves, O., Toledo, L., Rudomín, I.: Comportamientos en simulación de multitudes : revisión del estado del arte. Res. Comput. Sci. **62**, 319–334 (2013). Special Issue: Avances en Inteligencia Artificial

9. Helbing, D., Farkas, I., Vicsek, T.: Simulating dynamical features of escape panic. Nature **407**(6803), 90–487 (2000). http://www.ncbi.nlm.nih.gov/pubmed/11028994

10. Helbing, D., Molnár, P.: Social force model for pedestrian dynamics. Phys. Rev. E **51**(5), 4282–4286 (1995). http://link.aps.org/doi/10.1103/PhysRevE.51.4282

11. Huang, T.: Discriminative local binary patterns for human detection in personal album. In: 2008 IEEE Conference on Computer Vision and Pattern Recognition. pp. 1–8. IEEE, June 2008. http://ieeexplore.ieee.org/lpdocs/epic03/wrapper.htm?arnumber=4587800

12. Kim, K., Chalidabhongse, T.H., Harwood, D., Davis, L.: Real-time foregroundbackground segmentation using codebook model. Real-Time Imaging **11**(3), 172–185 (2005)

13. Lengvenis, P., Simutis, R., Vaitkus, V., Maskeliunas, R.: Application of computer vision systems for passenger counting in public transport. Electron. Electr. Eng. **19**(3), 69–72 (2013). http://www.eejournal.ktu.lt/index.php/elt/article/view/1232

14. Li, M., Zhang, Z., Huang, K., Tan, T.: Rapid and robust human detection and tracking based on omega-shape features. In: 2009 16th IEEE International Conference on Image Processing (ICIP), pp. 2545–2548. IEEE, November 2009. http://ieeexplore.ieee.org/lpdocs/epic03/wrapper.htm?arnumber=5414008

15. Li, T.Y., Wen Lin, J., Liu, Y.L., Ming Hsu, C.: Interactively directing virtual crowds in a virtual environment. Conf. Artif. Real Telexistence vol. 10 (2002). http://dspace2.lib.nccu.edu.tw/bitstream/140.119/15022/1/59.pdf

16. Millan, E., Hernandez, B., Rudomin, I.: Large crowds of autonomous animated characters using fragment shaders and level of detail. In: Engel, W. (ed.) ShaderX5: Advanced Rendering Techniques, chap. Beyond Pix, pp. 501–510. Charles River Media (2006). http://www.shaderx5.com/TOC.html

17. Moussaïd, M., Perozo, N., Garnier, S., Helbing, D., Theraulaz, G.: The walking behaviour of pedestrian social groups and its impact on crowd dynamics. PloS one **5**(4), e10047 (2010)

18. Mukherjee, S., Das, K.: Omega model for human detection and counting for application in smart surveillance system. Int. J. Adv. Comput. Sci. Appl. **4**(2), 167–172 (2013). arXiv preprint arXiv:1303.0633

19. Ojala, T.: Multiresolution gray-scale and rotation invariant texture classification with local binary patterns. Pattern Anal. Mach. Intell. **24**(7), 971–987 (2002). http://ieeexplore.ieee.org/xpls/abs_all.jsp?arnumber=1017623

20. Ozturk, O., Yamasaki, T., Aizawa, K.: Tracking of humans and estimation of body/head orientation from top-view single camera for visual focus of attention analysis. In: Computer Vision Workshops (ICCV Workshops), pp. 1020–1027 September 2009. http://ieeexplore.ieee.org/lpdocs/epic03/wrapper. htm?arnumber=5457590ieeexplore.ieee.org/xpls/abs_all.jsp?arnumber=5457590

21. Pelechano, N., Stocker, C.: Being a part of the crowd: towards validating VR crowds using presence. In: Proceedings of the 7th International Joint Conference on Autonomous Agents and Multiagent Systems (AAMAS), pp. 12–16 (2008). http://dl.acm.org/citation.cfm?id=1402407

22. Reynolds, C.W.: Flocks, herds and schools: a distributed behavioral model. ACM SIGGRAPH Comput. Graph. **21**(4), 25–34 (1987). http://portal.acm.org/ citation.cfm?doid=37402.37406

23. Rivalcoba, I.J., Rudomin, I.: Segmentación de peatones a partir de vistas aéreas. Res. Comput. Sci. **62**, 129–230 (2013)

24. Tuzel, O., Porikli, F., Meer, P.: Human detection via classification on Riemannian manifolds. In: 2007 IEEE Conference on Computer Vision and Pattern Recognition, pp. 1–8 June 2007. http://ieeexplore.ieee.org/lpdocs/epic03/wrapper.htm? arnumber=4270222

25. Viola, P., Jones, M.: Rapid object detection using a boosted cascade of simple features. In: Proceedings of the 2001 IEEE Computer Society Conference on Computer Vision and Pattern Recognition. CVPR 2001, vol. 1, pp. I-511–I-518 (2001). http://ieeexplore.ieee.org/lpdocs/epic03/wrapper.htm?arnumber=990517

26. Viola, P., Jones, M., Snow, D.: Detecting pedestrians using patterns of motion and appearance. Int. Conf. Comput. Vision **63**(2), 153–161 (2003). http://ieeexplore.ieee.org/xpls/abs_all.jsp?arnumber=1238422

27. Wang, Y., Dubey, R., Magnenat-Thalmann, N., Thalmann, D.: An immersive multi-agent system for interactive applications. Vis. Comput. **29**(5), 323–332 (2012). http://link.springer.com/10.1007/s00371-012-0735-7

Multi-kernel Ray Traversal for Graphics Processing Units

Thomas Schiffer[1]([✉]) and Dieter W. Fellner[1,2]

[1] Institut für ComputerGraphik und Wissensvisualisierung, TU Graz,
Inffeldgasse 16c, Graz, Austria
{t.schiffer,d.fellner}@cgv.tugraz.at
http://www.cgv.tugraz.at
[2] TU Darmstadt und Fraunhofer IGD, Darmstadt, Germany

Abstract. Ray tracing is a very popular family of algorithms that are used to compute images with high visual quality. One of its core challenges is designing an efficient mapping of ray traversal computations to massively parallel hardware architectures like modern algorithms graphics processing units (GPUs).

In this paper we investigate the performance of state-of-the-art ray traversal algorithms on GPUs and discuss their potentials and limitations. Based on this analysis, a novel ray traversal scheme called batch tracing is proposed. It subdivides the task into multiple kernels, each of which is designed for efficient parallel execution. Our algorithm achieves comparable performance to current approaches and represents a promising direction for future research.

Keywords: Ray tracing · SIMT · Parallelism · Graphics processing units

1 Introduction

Ray tracing is a widely used algorithm to compute highly realistic renderings of complex scenes. Due to its huge computational requirements massively parallel hardware architectures like modern graphics processing units (GPUs) have become attractive target platforms for implementations. We investigate high performance ray tracing on NVidia GPUs, but many of our contributions and analysis may also apply to other wide-SIMD hardware systems with similar characteristics. In this paper, we use bounding volume hierarchies (BVHs) as acceleration structure for ray traversal and we solely focus on the task of intersecting a ray with a scene containing geometric primitives only and do not include shading and other operations into our discussion.

In general, we evaluate our algorithms on ray loads generated by a path tracer with a fixed maximum path length of 3. Our test scenes (Fig. 1) contain only purely diffuse materials, from which rays bounce off to a completely random direction of the hemisphere. So, the generated ray data covers a broad spectrum

© Springer International Publishing Switzerland 2015
S. Battiato et al. (Eds.): VISIGRAPP 2014, CCIS 550, pp. 78–93, 2015.
DOI: 10.1007/978-3-319-25117-2_6

of coherency ranging from coherent primary rays to highly incoherent ones after two diffuse bounces.

For a diligent performance assessment, we use a diverse set of NVidia GeForce GPUs consisting of a low-end Fermi chip (GT 540M), a high-end Fermi chip (GTX 590) and a high-end Kepler-based product (GTX 680).

Our paper has the following structure: First, we provide an analysis of the state-of-the-art ray tracing algorithms and their characteristics that seem to leave significant room for improvement. We subsequently discuss our novel algorithmic approach called batch tracing that addresses the current problems. Finally, we analyze the practical implementation as well as some optimizations and point out possible directions for future research.

2   Previous Work

This paper targets massively parallel hardware architectures like NVidia's Compute Unified Device Architecture (CUDA) that was initially presented by Lindholm et al. [9]. CUDA hardware is based on a single-instruction, multiple-thread (SIMT) model, which extends the commonly known single-instruction, multiple-data (SIMD) paradigm. On SIMT hardware, threads are divided into small, equally-sized groups of elements called warps (current NVidia GPUs batch 32 threads in one warp). Contrary to SIMD, execution divergence of threads in the same warp is handled in hardware and thus transparent to the programmer. To avoid starvation of the numerous powerful computing cores of GPUs, L1 and L2 caches were introduced with the advent of the more recent generation of hardware called Fermi [10]. The latest architecture of GPU hardware termed Kepler [11] contains several performance improvements concerning atomic operations, kernel execution and scheduling.

In 2009, Aila et al. [2] presented efficient depth-first BVH traversal methods, which still constitute the state-of-the-art approach for GPU ray tracing. Their implementation uses a single kernel containing traversal and intersection, that is run for each input ray in parallel. More recently they presented some improvements, algorithmic variants to increase the SIMT efficiency and updated results for NVidia's Kepler architecture in [3]. They also reported that ray tracing potentially generates irregular workloads, especially for incoherent rays. To handle these uneven distributions of work, compaction and reordering have been employed in the context of shading computations [8], ray-primitive intersection tasks [12] and GPU path tracing [14]. Garanzha et al. propose a breadth-first BVH traversal approach in [6], which is implemented using a pipeline of multiple GPU kernels. While coherent rays can be handled very efficiently using frustra-based traversal optimizations, the performance for incoherent ray loads significantly falls below the depth-first ray tracing algorithms as reported in [5].

3   The Quest for Efficiency

While the SIMT model is comfortable for the programmer (e.g. no tedious masking operations have to be implemented), the hardware still has to schedule the

Fig. 1. Ray tracing-based renderings of our test scenes approximating diffuse illumination. From left to right: Sibenik (80 K triangles), Conference (282 K triangles) and Museumhall (1470 K triangles).

current instruction for all threads of the warp. Diverging code paths within a warp have to be serialized and lead to redundant instruction issues for non-participating threads. If code executed on SIMT hardware exhibits much intra-warp divergence, a considerable amount of computational bandwidth will be wasted. In accordance with Sect. 1 of [2], we use the term SIMT efficiency, which denotes the percentage of actually useful operations performed by the active threads related to the total amount of issued operations per warp to quantify this wastage in our experiments. As computational bandwidth continues to increase faster than memory bandwidth, SIMT efficiency is one of the key factors for high performance on current and most probably also future GPU hardware platforms.

Beside computational power, memory accesses and their efficiency are a crucial issue too. Threads of a warp should access memory in a coherent fashion, in order to get optimal performance by coalescing their requests. However, traditional ray tracing algorithms (as well as many others) are known to generate incoherent access patterns that potentially waste a substantial amount of memory bandwidth as discussed in [1]. The caches of GPUs can help to improve the situation considerably as discussed by Aila et al. in [2,3], where they describe how recent improvements of the GPU cache hierarchy affect the overall ray tracing performance.

In general, GPUs are optimized for workloads that distribute the efforts evenly among the active threads. As reported in [2], ray tracing algorithms generate potentially unbalanced workloads, especially for incoherent rays. This fact poses substantial challenges to the hardware schedulers and can still be a major source for inefficiency as noted by Tzeng et al. [13]. To support the scheduling hardware and achieve a more favorable distribution of work, compaction and task reordering steps are explicitly included in algorithms. This paradigm has been successfully applied to shading computations of scenes with strongly differing shader complexity [8], where shaders of the same type are grouped together to allow more coherent warp execution. In the context of GPU path tracing, Wald applied a compaction step to the rays after the construction of each path segment in order to remove inactive rays from the subsequent computations [14].

3.1 Depth-First Traversals

The most commonly used approach for GPU ray tracing is based on a depth-first traversal of a BVH as discussed in [2]. Their approach is based on a monolithic design, which combines BVH traversal and intersection of geometric primitives into a single kernel. This kernel is executed for each ray of an input array in massively parallel fashion. Given two different rays contained in the same warp, potential inefficiencies now stem from the fact that either they require different operations (e.g. one ray needs to execute a traversal step, while the other ray needs to perform primitive intersection) or the sequences of required operations have a different length (e.g. one ray misses the root node of the BVH, while the other ray does not). These two fundamental problems have a negative impact on SIMT efficiency and lead to an uneven distribution of work, especially for incoherent ray loads.

Table 1. SIMT efficiency percentages for traversal (T) and intersection (I) as well as kernel's warp execution efficiency (K) of the two best performing ray tracing kernels *fermi_speculative_while_while* (from [2]) and *kepler_dynamic_fetch* (from [3]) using different thresholds of active rays for dynamic fetching of work. The numbers are averages of multiple views taken of the Sibenik and the Museum scene for different generations of rays.

Scene	Rays	while-while			dyn-fetch 25 %			dyn-fetch 50 %			dyn-fetch 75 %		
		T	I	K	T	I	K	T	I	K	T	I	K
Sibenik	1. gen	87.0	61.3	80.1	86.4	59.9	79.2	86.4	60.0	79.4	86.3	60.0	79.1
	2. gen	46.7	27.7	40.5	49.2	30.1	43.0	49.7	30.5	43.3	50.4	31.4	43.9
	3. gen	38.4	23.6	33.8	41.0	26.2	36.2	42.1	27.6	37.2	42.5	28.2	37.6
Museum	1. gen	65.4	45.8	59.0	66.2	46.2	59.4	66.8	47.1	60.3	65.9	47.7	59.8
	2. gen	32.6	19.8	28.8	37.9	28.7	34.8	39.7	31.7	36.6	40.3	33.2	37.2
	3. gen	27.8	16.6	24.5	33.6	25.8	31.2	35.3	28.7	32.9	36.3	30.3	33.9

To mitigate the effects of irregular work distribution, Aila et al. investigated the concept of dynamic work fetching. Given a large number of input tasks, the idea is to process them using an (usually significantly smaller) amount of worker threads that fetch new input items until all input elements have been processed. In this paradigm, the fetching of a new task can happen immediately after the current one has been finished or depending on a more general condition (e.g. all threads of a warp have completed their tasks). This dynamic software scheduling was introduced in Sect. 3.3 of [2] called persistent threads, where they used dynamic fetching on a per-warp basis to balance the shortcomings of the hardware scheduler on Tesla generation hardware. In [3] they suggest fetching new rays on Kepler hardware, if more than 40 % of the threads of a warp have finished their work. It is important to note that dynamic fetching of work attempts to increase warp utilization and thus potentially SIMT efficiency at the expense of memory bandwidth, because it tends to generate incoherent access patterns when fetching single or only a few new work items.

To have a baseline for our further experiments, we evaluated the SIMT efficiency of the two best performing ray tracing kernels $fermi\_speculative\_while\_while$ (from [2]) and $kepler\_dynamic\_fetch$ (from [3]) using different thresholds for the dynamic fetching paradigm. We directly used large parts of Aila et al.'s kernels with minor adaptations for our evaluation framework. Table 1 shows SIMT efficiency percentages of the traversal and the intersection code parts and the kernel's overall warp execution efficiency as reported by the CUDA profiler obtained by averaging over multiple views of the test scenes. The SIMT efficiency for traversal and intersection have been computed by inserting additional code into the kernel, which counts the number of operations that are executed by the kernel and that are actually issued on the hardware using atomic instructions. However, the percentages in Table 1 vary from what Aila et al. reported in Table 1 of [2], since we use different view points and a different BVH builder. In [3], Aila et al. note that fine-grained dynamic work fetching on Kepler improves the performance by around ten percent, especially for incoherent rays. This fact can be well explained by looking at the corresponding efficiency percentages in Table 1. For coherent primary rays the percentages are already quite high and hardly change, so there is no notable overall performance gain. Incoherent ray loads, however, definitely benefit from this optimization. A newly fetched ray can either require the same sequence of operations as the others and thus increases, or require a different sequence of operations and thus decreases the SIMT efficiency of a warp. Assuming equal probabilities for traversal and intersection, it is therefore probable that low percentages of incoherent rays are increased as indicated by the measurements.

We also experimented with speculative traversal as proposed in Sect. 4 of [2] using a postpone buffer of small size of up to three elements. Comparing ordinary and speculative traversals, we observed an increase in traversal efficiency of up to 10 % for coherent and up to 50 % for incoherent rays, which becomes smaller for larger buffer sizes. Please note that the kernels shown in Table 1 already use a fixed size postpone buffer of one element. So, enlarging the speculative buffer yields just a slight increase of up to 10 % in traversal efficiency for incoherent rays, but no overall performance improvement in our experiments. Also, we noted a slight decrease in intersection efficiency for buffer sizes larger than one for ray tracing kernels based on the while-while paradigm, which is in accordance with the results in Table 1 of [2]. These results hint that in a monolithic design, SIMT efficiency of traversal and intersection are correlated and cannot be increased arbitrarily without adversely impacting the other. Pantaleoni et al. (in Sect. 4.4 of [12]) address this problem by redistributing intersection tasks of the active rays among all rays of the warp. After the intersection they perform a reduction step to obtain the closest intersection for each ray. This structural modifications of the monolithic approach result in a significantly increased SIMT efficiency of intersection (50 %–60 % instead of about 25 %), but is used only for specialized spherical sampling rays in an out-of-core ray tracing system.

3.2 Breadth-First Traversals

A radically different approach to GPU-based ray tracing was presented by Garanzha et al. in [6]. They implemented a ray tracing algorithm that traverses their specialized BVH structure in a breadth-first manner including some key modifications. In a first step before the actual traversal, the input rays are sorted and partitioned into coherent groups bounded by frusta. Then, traversal starts at the root node. At the currently processed tree level, all active frustra (active means intersecting a part of the BVH) are intersected with the inner nodes and these results are propagated to the next lower level. Traversal stops at the leaf nodes and yields lists of intersected leaves for each frustum. Subsequently, all rays of each active frustum are tested for intersection with the primitives contained in each leaf to obtain the final results. As described in [5], each of these steps is implemented using multiple kernels resulting in a rather long pipeline. Although we did not provide an actual implementation of their algorithm, we still want to note several important characteristics of their approach.

First of all, we argue that large parts of their implementation can be assumed to exhibit high SIMT efficiency. Each kernel is carefully designed to perform a single task (e.g. frustum intersection) and complex operations are broken down into smaller components (e.g. ray sorting exploiting existing coherency), which can again be efficiently implemented. Secondly, this approach also tends to generate workloads that are significantly more coherent and regular than the ones of depth-first traversal. There appear to be no sources for such major inefficiencies in the traversal phase and persistent threads are used to balance the workloads in the intersection stage, since the length of the corresponding list of leaf nodes may vary significantly. We believe that the performance wins over monolithic traversals reported on rather coherent rays are not only due to potentially efficient multi-kernel implementation, but also largely due to their clever frusta-traversal based optimization.

Despite these favorable properties, the approach also possesses a major algorithmic inefficiency. In the traversal stage, a complete traversal of the acceleration structure is performed for each frustum regardless of the number of the actually necessary operations. Thus, a lot of likely redundant work per ray is carried out, since the ray may intersect a primitive that is referenced by the first leaf that is encountered in traversal. This node over-fetching potentially leads to a lot of redundant instructions and memory accesses. For coherent rays, this drawback apparently does not outweigh the benefits from the efficient traversal and intersection stages. For incoherent rays, however, the set of the traversed nodes grows considerably and the node over-fetching dominates the all aforementioned benefits resulting in a significantly reduced overall performance.

4 Multi-kernel Batch Traversal

Based on the room for improvement that is left by current state-of-the-art ray tracing algorithms, we propose a novel approach called batch traversal. It is designed to achieve the following objectives:

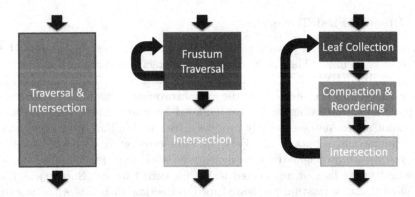

Fig. 2. High-level system designs for monolithic depth-first (left), breadth-first (middle) and our batch tracing approach (right). Batch tracing consists of multiple phases, which are repeatedly executed until all rays have terminated. Blocks denote logical parts of the algorithms, black arrows denote the control flow.

- Given the low percentages for incoherent rays of depth-first traversals shown in Table 1, our algorithm should notably increase SIMT efficiency.
- Unlike breadth-first traversal, the algorithm should not exhibit substantial inefficiencies in handling ray loads of varying coherency.

The components of our algorithm and their interplay are shown in Fig. 2 (right) together with depth-first (left) and breadth-first (middle) traversals. Like in breadth-first traversal, our approach splits the ray tracing task into several algorithmic stages implemented in multiple kernels to increase SIMT efficiency of the single code parts. During leaf collection stage, the ray traverses the BVH similar to depth-first traversal and collects a number of intersected leaf nodes called intersection candidates. In the subsequent phase, the collected intersection candidates are reorganized to achieve efficient execution of the intersection stage and the active rays are compacted for the next execution of the traversal stage. In the intersection phase the primitives contained in the intersection candidates are tested for intersection with the ray and the results are used to update the rays. These stages are executed in a loop until all input rays have terminated. Contrary to breadth-first traversal, our approach performs partial BVH traversals and intersection testing alternately in order to avoid collecting a large number of redundant intersection candidates.

4.1 Leaf Collection

As mentioned before, the leaf collection stage performs a partial depth-first traversal of the BVH similar to the monolithic ray tracing algorithms and is implemented in a single kernel. If a leaf is encountered during traversal, it is stored into the buffer of the corresponding ray and traversal continues immediately as shown in Algorithm 1. A ray terminates this stage if the whole BVH has been traversed or a certain maximum number of collected leaves has been found. The efficiency of this stage depends on the number of intersection candidates that have to be expected for a given ray. The second crucial question is how the collection of these leaves should be distributed optimally among the various iterations.

Algorithm 1. Leaf Collection Stage.

---

while ray not terminated **do**
    while node is inner node **do**
        traverse to the next node
    end while
    store leaf into buffer
    if maximum number of leaves collected **then**
        return
    end if
end while

As the overall number of collected leaves for a ray is influenced by many factors (e.g. input ray distribution and scene geometry), we provide probabilistic bounds backed up by experiments. To estimate the size of the intersection candidate set, we assume a uniformly distributed ray load and that our input scene consists of N objects O_i. Each O_i is enclosed by a bounding box B_i and all objects are enclosed by a scene bounding box W. Furthermore, let $B = \cup B_i$ be the union of all bounding boxes and $p(n)$ be the probability for a random line to intersect n objects, given it intersects W. Using Proposition 5 from [4], a tight upper bound for $p(n)$ is then given by Eq. 1 (SA denotes surface area).

$$p(n) \leq \frac{SA(B)}{nSA(W)} \tag{1}$$

This probabilistic bound shows that the size of a set of intersection candidates for a random ray can be expected to be reasonably small. In practice, we can hope for an even smaller candidate set, because we are only interested in intersections occurring before the first hit of the ray.

To assess the quality of these theoretical considerations in practice, we have evaluated the size of the intersection candidate set for multiple views of our test scenes. The obtained results prove our probabilistic assumption that for a large number of rays the size of the candidate is small (<4 for 70 % and <8 for 90 % of the rays) regardless of their distribution. Figure 3 shows a histogram of the size of the intersection candidate set for different ray generations for a view of the Sibenik and a view of the Museumhall scenes. Firstly, the histogram varies significantly for small set sizes (e.g. up to 8 for Fig. 3) depending on the distribution of the input rays (e.g. viewpoint). However, the distribution of the larger candidate set sizes exhibits no dependence on the input ray distribution approaching zero for all scenes and viewpoints. Additionally, the overall shape of the histogram curves is also similar for all tested ray loads and input scenes. Although specific

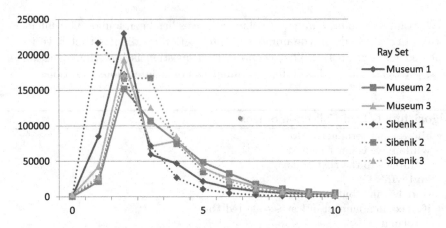

Fig. 3. Histogram of size of the intersection candidate set for rays of different generations (1–3) for a view of the Sibenik and a view of the Museumhall scene. The intersection candidate set of a ray consists of all leaves that are pierced by this ray from its origin to its first hit point. The X axis shows the size of the intersection candidate set, the Y axis displays the corresponding number of rays. The graphs for larger candidate set sizes all are very close to the X axis, thus only set sizes up to 10 are shown.

ray-geometry configurations might cause exceptionally large intersection candidate lists, these cases can still be handled by our implementation in a reasonably efficient manner.

The second important open issue that remains is, how the collection of the leaves should be distributed among the various passes. It is challenging to provide a short and optimal solution, since choosing the number of intersection candidates per iteration is a trade-off. Collecting a large number of leaves per iteration reduces the total number of required passes and thus potentially increases overall performance. However, a lot of potentially redundant traversal and intersection operations might be introduced, since most of the rays have only got a small candidate set. To counter this effect, a small number of intersection candidates per pass can be chosen to minimize the overhead, potentially increasing the total number of iterations required. We finally came up with the following heuristic, which is backed up by numerous experiments: We start with a small number (e.g. 2 for GT 540M) in the first pass and increase the size of the intersection candidate set gradually over the next few passes. This avoids much redundant work in the first passes and also keeps the number of total passes in an acceptable range.

4.2 Compaction and Reordering

The compaction and reordering stage performs various housekeeping tasks in a single kernel. Firstly, we perform a compaction of the rays that remain active after the execution of the leaf collection stage. We simply map the active rays to consecutive slots using parallel reduction based on atomic instructions for shared and global memory. This removes already terminated rays and helps to maintain a high efficiency during the leaf collection stage. Additionally, this stage is responsible for managing the leaf collection buffer. At the start of our batch tracing algorithm, this buffer is allocated once with a size that is a fixed multiple of the number of input rays (e.g. number of rays times eight elements).

In each leaf collection pass we partition this buffer in equal parts among the remaining active rays. This layout can flexibly accommodate a wide range of thresholds for the number of collected leaves without the need to perform costly memory reallocations. Secondly, a parallel grid-wide reduction is made to compute the total number of intersection tasks and to assign these tasks to threads for the subsequent intersection stage. Again, rays, which have collected leaves during the previous stage, are mapped to consecutive slots for efficiency reasons. For storing the different mappings generated by work compaction, index arrays are allocated at the start of the algorithm. They keep track of the original indices for all active rays and intersection tasks and help to save costly movements of larger data structures (e.g. input ray data or traversal stacks).

4.3 Intersection

Our implementation of the intersection stage maps the intersection tasks of one ray to one thread, which iteratively processes its array of collected leaves. For each leaf, all contained primitives are intersected with the ray and the results are stored for use in the subsequent phases of the algorithm.

5 Results and Discussion

In this section, we evaluate the previously discussed batch tracing algorithm and discuss some optimizations. To see how the total running time is distributed we profiled our implementation for different scenes and ray loads. Around 70 % of the time is spent in the leaf collection, which makes this stage the primary target for optimizations. Intersection computations constitute 25 % of total time and compaction and reordering 5 %, while negligible overhead is caused by memory allocations and CPU-GPU communication.

5.1 Optimized Leaf Collection

In order to assess the efficiency of the leaf collection stage, we analyzed the resulting distribution of workloads. As the number of operations for different rays may vary considerably, we implemented an optimized version that uses dynamic fetching of work to counter the effects of this imbalance. If the SIMT efficiency of a warp drops below a certain threshold, already terminated rays get replaced with new input rays. We experimented with different threshold percentages and empirically determined an optimal threshold of 25 % for the GTX 590. For the GT 540M dynamic fetching is only beneficial if performed on a per-warp basis, i.e. when the last active ray of a warp terminates, a whole group of new 32 work items is fetched. For a detailed comparison with monolithic traversal approaches we implemented an extended kernel based on $fermi\_speculative\_while\_while$ that uses dynamic fetching of rays in the sense of [3].

Table 2. Ray tracing performance for our monolithic dynamic fetch kernel (DF), the batch tracing baseline implementation (Batch) and its improved variant using dynamic fetching again (Bt + DF). Numbers are relative to the performance of our implementation of $fermi\_speculative\_while\_while$ and were averaged over multiple views of the Museum scene. The largest speed-ups are shown in bold numbers.

GPU	Ray type	DF	Batch	Bt + DF
GT 540M	1. gen	**1.13**	0.97	1.04
	2. gen	1.46	1.42	**1.54**
	3. gen	**1.55**	1.42	**1.55**
GTX 590	1. gen	**1.09**	0.78	0.79
	2. gen	**1.42**	1.10	1.17
	3. gen	**1.51**	1.12	1.21

Table 2 shows the ray tracing performance for the Museum scene of our monolithic dynamic fetch kernel (DF), the batch tracing baseline implementation (Batch) and its improved variant using dynamic fetching (Batch + DF) relative to the performance of the $fermi\_speculative\_while\_while$ kernel. Our optimized batch tracing algorithm (Batch + DF) achieves comparable performance on the 540 GT, but is consistently outperformed on the 590 GTX by our monolithic dynamic fetch kernel. The low-end card with few compute cores seems to benefit from our batch tracing approach especially in combination with per-warp dynamic fetching, because irregular workloads can be hardly balanced out by the hardware. On Kepler hardware batch traversal is around 70 % slower than our fastest monolithic kernel regardless of the coherency of the input rays.

While dynamic fetching increases the performance of the monolithic ray traversal notably, only small speedups (up to 10 %) are achieved for the leaf collection stage. Contrary to the obtained results, we expected larger benefits (at least in a range similar to the difference between $fermi\_speculative\_while\_while$ and its dynamic work fetching variant) from this optimization: Since the leaf collection kernel contains no intersection code, dynamic fetching should deliver large increases in SIMT efficiency and also improve overall performance. In order to find a plausible cause for this discrepancy, we measured the warp execution efficiency of the leaf collection stage. Table 3 shows the percentages obtained by batch tracing using different variants of dynamic fetching (per-warp and using different thresholds) in comparison to the highest SIMT efficiency values for traversal achieved by state-of-the-art monolithic kernels in the Museum scene. As we expected, the efficiency of the BVH traversal is slightly increased for primary

Table 3. SIMT efficiency percentages for traversal of monolithic kernels (Mono) and for the leaf collection stage of batch tracing using dynamic fetching per-warp (Warp DF) and with different thresholds (DF 25 % and DF 50 %) in the Museum scene. For the first column, we use the highest percentages achieved by any monolithic kernels, including all variants of dynamic fetching (see also Table 1).

Ray type	Mono	Warp DF	DF 25	DF 50
1. gen	66.8	66.5	72.1	69.3
2. gen	40.3	36.7	52.3	56.8
3. gen	36.3	32.6	50.3	54.6

rays (up to 10 %) and substantially enhanced for high order rays (up to almost 60 %).

Since these improvements do not result in an overall performance win, we decided to profile our implementation comprehensively. The analysis revealed that using dynamic fetching for our leaf collection kernels results in an increased number of instruction issues due to memory replays. This fact strongly hints that the performance of this stage is limited by the caches of the GPU. As the available DRAM bandwidth is not entirely used up, we believe that the unexpected results are caused by the unfavorable memory access pattern resulting from traversal of incoherent rays. This leads to an exhausted cache subsystem that cannot service all the memory requests on time resulting in severe latency that nullifies most of the benefits of the increased SIMT efficiency.

To verify the aforementioned assumptions experimentally and to identify primary performance limiters we use selective over-clocking of the processing cores and the memory subsystem (e.g. the performance of memory-bounded kernels will benefit from an increased memory clock rate). In fact, our leaf collection kernels hit the memory wall much earlier than we expected, given the fact that they perform no intersection computations and thus have a smaller memory working set compared to monolithic traversal. While we obtained an optimal threshold of 25 % for dynamic fetching on the GTX 590, it only decreases performance on the Kepler hardware. In our opinion, this shows a drawback of separating traversal and intersection: Leaf collection kernels cannot hide the occurring memory latencies as well as a monolithic kernel, which can potentially schedule additional operations while waiting for memory fetches (e.g. execute primitive intersection while waiting for a BVH node memory access). These observations raise the question for alternative memory layouts and acceleration structures, which better suit this kind of traversal algorithm.

5.2 Traversal Performance Limiters

However, the analysis of our monolithic ray tracing kernels revealed that memory performance becomes a major challenge for this kind of algorithms, too. As discussed by Aila et al. [2], traditional depth-first traversal are commonly believed to be compute-bound. This is confirmed by our results and holds for Fermi hardware regardless of whether dynamic fetching is used or not. The Kepler-based GTX 680, however, provides a significantly increased amount of computational power, while memory bandwidth has grown only moderately in comparison. Our speculative depth-first traversal kernel is apparently still compute-bound as shown in Fig. 4.

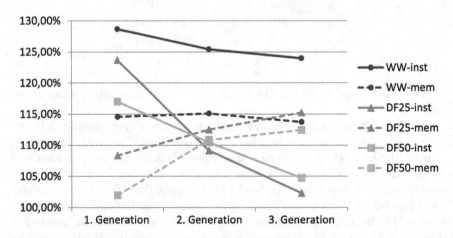

Fig. 4. Kernel performance limiter analysis for monolithic ray traversals on Kepler-based 680 GTX. We evaluate our fastest monolithic kernel (WW) using different thresholds for dynamic fetching (DF25 and DF50). Series suffixed with "-instr" denote that processor clock has been increased by 20 %, "-mem" means that the memory clock has been over-clocked by 20 %. The graphs show obtained performance relative to the default settings of the device.

The *kepler_dynamic_fetch* kernel similarly exhibits compute-bound behavior for coherent ray loads. With diminishing coherence of the input rays, the process of dynamic work fetching puts an increased load of incoherent memory requests on the memory subsystem, which in turn becomes the dominating factor (compared to computational bandwidth). Contrary to the speculations in [3], this shows that monolithic state-of-the-art traversal already tends to be limited by memory system performance, at least for incoherent rays. Since a significant decrease in the instruction-to-byte ratio in future GPUs cannot be expected, memory bandwidth and cache throughput will become a more important performance factor for traditional algorithms. Additional investigations using the CUDA profiler revealed that the traversal kernels (monolithic and leaf collection for batch traversal) do not achieve close to peak compute or memory bandwidth, which indicates substantial latency issues in accordance with the results of [7].

5.3   Intersection Stage

Table 4 shows the SIMT efficiency of our implementation of the intersection stage in comparison to the highest numbers that have been achieved by monolithic traversals.

Table 4. SIMT efficiency percentages of primitive intersection for monolithic kernels (Mono) and our batch tracing implementation (Batch) for the Museum scene.

Ray type	Mono	Batch
1. gen	47.7	58.3
2. gen	33.2	47.1
3. gen	30.3	45.4

Our approach improves the efficiency of primitive intersection notably which is now in the range of what has been achieved by intersection task reordering [12]. As the leaf collection pass usually submits an amount of redundantly collected leaves to the traversal stage, we tried to avoid all unnecessary primitive tests by storing the entry distance of a ray along with every intersection candidate. The ray always examines this value before any contained primitives are tested. However, this optimization yielded no noticeable performance improvements and might pay off only for more complex geometric primitives. Our basic mapping which lets one ray work on one intersection candidate set generates potentially uneven workloads, since the number of primitives contained in a single leaf may vary as well as the size of the intersection candidate set. Measurements hint that our intersection implementation is memory-bounded and we suppose that it would benefit significantly from a more elaborate reordering of the intersection tasks.

6   Conclusion and Future Work

Given our comprehensive analysis of current ray traversal algorithms, we believe that state-of-the-art approaches leave a lot of room for future algorithmic improvements. A prominent example is the SIMT efficiency of depth-first traversal algorithms, which waste a considerable amount of computational bandwidth when dealing with incoherent rays. However, also the limits of traditional paradigms become obvious, which puts the research focus on alternative traversal methods like the batch tracing algorithm proposed in this paper.

Although the provided implementation cannot quite compete with mature and heavily optimized monolithic traversal, our multi-kernel method possesses appealing characteristics, which makes it an attractive direction for further research. A central point is to find an acceleration structure that permits increased traversal/leaf collection performance. To alleviate the latency issues it might be beneficial to decompose the parts of the algorithm even further yielding more efficient building blocks. Furthermore, we would like to optimize the intersection stage of our algorithm and investigate efficient handling of multiple types of geometric primitives.

Acknowledgements. We thank Marko Dabrovic for providing the Sibenik cathedral model and the University of Utah for the Fairy Scene. Many thanks also go to Timo Aila for making his testing and benchmarking framework publicly available. We also want to thank the anonymous reviewers for their valuable comments.

References

1. Aila, T., Karras, T.: Architecture considerations for tracing incoherent rays. In: Proceedings of the Conference on High Performance Graphics, HPG 2010, pp. 113–122. Eurographics Association, Saarbrücken (2010). http://portal.acm.org/citation.cfm?id=1921479.1921497
2. Aila, T., Laine, S.: Understanding the efficiency of ray traversal on GPUs. In: Proceedings of the Conference on High Performance Graphics 2009, HPG 2009, pp. 145–149. ACM, New York (2009). http://doi.acm.org/10.1145/1572769.1572792
3. Aila, T., Laine, S., Karras, T.: Understanding the efficiency of ray traversal on GPUs - kepler and fermi addendum. NVIDIA Technical report NVR-2012-02, NVIDIA Corporation, June 2012
4. Cazals, F., Sbert, M.: Some integral geometry tools to estimate the complexity of 3D scenes. Technical report, iMAGIS/GRAVIR-IMAG, Grenoble, France, Departament dInformtica i Matemtica Aplicada, Universitat de Girona, Spain (1997)
5. Garanzha, K.: Fast ray sorting and breadth-first packet traversal for GPU ray tracing. Oral presentation at EG2010 (2010)
6. Garanzha, K., Loop, C.: Fast ray sorting and breadth-first packet traversal for GPU ray tracing. In: Proceedings of the Eurographics, EG 2010, pp. 289–298. Eurographics Association (2010). http://research.microsoft.com/en-us/um/people/cloop/GaranzhaLoop2010.pdf
7. Guthe, M.: Latency considerations of depth-first GPU ray tracing. In: Galin, E., Wand, M. (eds.) EG 2014 - Short Papers, pp. 53–56. Eurographics Association, Strasbourg (2014). http://diglib.eg.org/EG/DL/conf/EG2014/short/053-056.pdf
8. Hoberock, J., Lu, V., Jia, Y., Hart, J.C.: Stream compaction for deferred shading. In: Proceedings of the Conference on High Performance Graphics 2009, HPG 2009, pp. 173–180. ACM, New York (2009). http://doi.acm.org/10.1145/1572769.1572797
9. Lindholm, E., Nickolls, J., Oberman, S., Montrym, J.: NVIDIA tesla: a unified graphics and computing architecture. IEEE Micro **28**(2), 39–55 (2008)
10. NVIDIA: Nvidia's Next Generation Cuda compute architecture: Fermi (2009)
11. NVidia: Nvidia gk110 architecture whitepaper (2012). http://www.nvidia.com/content/PDF/kepler/NVIDIA-Kepler-GK110-Architecture-Whitepaper.pdf
12. Pantaleoni, J., Fascione, L., Hill, M., Aila, T.: Pantaray: fast ray-traced occlusion caching of massive scenes. In: ACM SIGGRAPH 2010 papers, SIGGRAPH 2010, pp. 37:1–37:10. ACM, New York (2010). http://doi.acm.org/10.1145/1833349.1778774

13. Tzeng, S., Patney, A., Owens, J.D.: Task management for irregular-parallel workloads on the GPU. In: Doggett, M., Laine, S., Hunt, W. (eds.) High Performance Graphics, pp. 29–37. Eurographics Association (2010). http://dl.acm.org/citation.cfm?id=1921485
14. Wald, I.: Active thread compaction for GPU path tracing. In: Proceedings of the ACM SIGGRAPH Symposium on High Performance Graphics, HPG 2011, pp. 51–58. ACM, New York (2011). http://doi.acm.org/10.1145/2018323.2018331

Information Visualization Theory and Applications

User's Interpretations of Features
in Visualization

Aqeel Al-Naser[1], Masroor Rasheed[2], Duncan Irving[2], and John Brooke[1(✉)]

[1] School of Computer Science, The University of Manchester, Manchester, UK
john.brooke@manchester.ac.uk
[2] Teradata Corporation, London, UK

Abstract. Visualization is often used to identify features of interest in a dataset. The identification of features cannot be fully automated and the subjective interpretation of the user is involved in the identification of the feature. There can be many such interpretations, both from a single user as s/he explores the data, and also in collaborations. Managing all these interpretations is problematic. We propose a novel visualization architecture that addresses this problem. We illustrate our method by examining how geoseismic data is interpreted, since this application presents all of the issues above.

Keywords: Geospatial visualization · Data acquisition and management · Provenance · Data exploration · Query-driven visualization

1 Introduction

One of the most powerful benefits that visualization brings to data analysis is the ability to harness the intuition of the user in the process of understanding the data. In many cases, human intuition is supported by algorithms that help to identify and highlight features of the data. However, it can often be the case that the algorithms cannot completely identify the features of interest. Human intuition must complete the process, and given the nature of intuition this can be a source of differing interpretations depending on the human expert. This may particularly occur in data that is noisy or visually complex. Examples of such data are found in medical imaging and in the field that is the topic of this paper, interpretation of geophysical seismic imaging data [1].

At some stage, collaborative visualization may be required for experts to discuss and reconcile these different interpretations. However the visualization pipeline as currently perceived is unidirectional, from the data to the rendered graphical view. In this paper we propose a reverse direction to the pipeline, so that amendments to the data made using the visualization tools can be stored back with the original data as first class data objects. This makes it much easier to view and share multiple interpretations over multiple visualization sessions, as is required for analysis of complex features. We also need to track the provenance of such interpretations; sometimes earlier interpretations may need to be

© Springer International Publishing Switzerland 2015
S. Battiato et al. (Eds.): VISIGRAPP 2014, CCIS 550, pp. 97–114, 2015.
DOI: 10.1007/978-3-319-25117-2_7

revisited. Conventionally, users' interpretations are stored as geometric objects separately from the data and possibly on different local machines; therefore keeping track of these interpretations becomes very complex and error prone.

We propose a novel method of creating the objects that underlie such visual interpretations, in such a way that the information contained in the interpretation can be stored as metadata alongside the original data. In addition to the ability of users to experiment with local views of data, this provides support for the considered results of such experiments to be stored, shared and re-used. In our proposed architecture, users' interpretations can flow in the reverse direction from the usual pipeline, back from the user's interaction with the visual presentation of the objects to the original data source of the pipeline. Our work is motivated by surveys of users who work in geographically distributed teams on complex datasets and over an extended period. This brings a major challenge in the management of different interpretations that eventually have to be reconciled to form a consensus opinion about features of the data. This consensus is then used to plan complex procedures, for example in medical surgery.

We demonstrate the applicability of our methods to the interpretation of geoseismic surveys. This is a field that has received a great deal of attention in terms of research (e.g. [2–8]) as well as software development such as Avizo Earth [9], GeoProbe [10], and Petrel [11]. The data in this field is noisy and the features to be extracted have a very complex spatial structure owing to processes of buckling, folding and fracturing [1]. This makes a purely automated approach to feature extraction very difficult to achieve. The expert interpretation is very central to the definition of the features and the considerations outlined above are of critical importance [12]. The interpretations are of critical importance in terms of financial implications.

This paper presents a continuation of prototype that we have previously published [13,14]; but in this paper we present a more fully featured system. The structure of the paper is as follows. In Sect. 2 we review related work. In Sect. 3 we present a survey of requirements from practitioners in geoscience. In Sect. 4 we show the abstract principles of our visualization architecture. In Sect. 5 we describe how we implement these principles in the field of geoscience. In Sect. 6 we present the results of our case studies for evaluation. In Sect. 7 we describe extended functionality developed as a result of the evaluation. Section 8 presents conclusions and plans for future work.

2   Background and Related Work

2.1   The Visualization Pipeline

The visualization pipeline builds the visual objects presented to the user in the form of a data processing workflow that starts from the original data right to the rendering on the display device. This basic formulation has proved very durable and has undergone extensive elaboration since its formulation for over twenty years [15]. A basic visualization pipeline features the following modules in the order of execution: reader → geometry generator → renderer. Improvements and elaborations have been proposed to address variety of issues such as visualizing multiple datasets, visualizing large datasets, and enhancing performance and efficiency.

A fuller description of data by metadata enhanced the power of visualization by allowing a more full-featured view of the data, taking into account its special properties and allowing users flexibility in creating visual objects. Using metadata, users can select a region or multiple regions to process, for example this allowed Ahrens et al. to visualize large-scale datasets using parallel data streaming [16]. In addition to regional information, a time dimension can be added to metadata, adding time control to the visualization pipeline [17]. The usefulness of metadata was further developed with the introduction of query-driven visualization [18,19]. Query-driven pipelines require the following technologies: file indexing, a query language and a metadata processing mechanism to pass queries. For fast retrieval, indexing technologies are used, such as FastBit [20,21] (based on compressed bitmap indexing) and hashing functions. To handle massive datasets efficiently, the visualization pipeline can be executed in parallel over multiple nodes. This data parallelism can be implemented with a variety of parallel methods, for example by MapReduce [22]. Database management systems (DBMS) are also capable of parallel execution of analysis but were not widely used for the purpose of visualization. A comparison between MapReduce and DBMS was presented by Pavlo et al. [23]; the authors suggest that DBMS has a performance advantage over MapReduce while the latter is easier to setup and to use.

2.2   Data Provenance and Management

Provenance, or *lineage*, is metadata that allows retrieval of the history of the data and how it was derived. Simmhan et al. [24] and Ikeda et al. [25] in their surveys categorize provenance techniques in different characteristics, two of which are (1) subject, or type, of provenance and (2) granularity. Data provenance can be of two types: (1) *data-oriented*, or *where-lineage*, which is a provenance applied on the data explicitly answering the question of which dataset(s) contributed to produce the output, and (2) *process-oriented*, or *how-lineage*, which is a provenance applied on the process answering the question of how the resulting output was produced from the input data. Each of the two types can be applied on one of two granularity levels: (1) coarse-grained (schema level) and (2) fine-grained (instance-level). The latter deals with individual data items separately.

The first software to bring the concept of provenance into visualization was *VisTrails* [26,27]. In VisTrails, the changes to the pipeline and parameter values are captured, allowing to review and compare previous versions. Thus, following the taxonomy of provenance by Simmhan et al. [24], VisTrails can be classified as a *process-oriented* model. We share a similar aim in respect to maintaining data provenance, but our method differs in the type of metadata that is being captured from the user's visual exploration. We adopt a *data-oriented* provenance model by explicitly storing metadata of the data points of the interpreted features. This is due to the nature of the data we deal with in this paper; a *process-oriented* model is not sufficient in our case as interpreted features are mainly a result of users' visual understanding. This was highlighted in a recent study on the value of data management in the oil and gas industry [28]. The study showed that *human expertise* contributed by 32.7 % towards understanding the subsurface data for the purpose of interpretation; the other elements were *data* which contributed by 38 %, *tools* which contributed by 15.1 %, and *process* which contributed by 13.7 %. This means that *human expertise* is essential to interpret the *data*, with a minor help from some *tools* (algorithms). Despite such impressive development, the evolution of the visualization pipeline has not (to our knowledge) developed a reverse direction to directly link the users' understanding of the data back into the dataset.

Data management is of a concern for systems with an integrated visualization. MIDAS [29], a digital archiving system for medical images, adopts a "data-centric" solution for massive dataset storage and computing to solve issues like moving data across networks for sharing and visualization. The visualization of MIDAS was originally provided through its web interface. Later, Jomier et al. [30] integrated *ParaViewWeb*, a remote rendering service based on *ParaView*, with MIDAS to directly visualize the centrally located large datasets. In the oil and gas industry, Schlumberger's *Studio E&P* [31], an extension to its commercial software Petrel [11], allows a multi-user collaboration via a central data repository. Users import a dataset to visualize and interpret from *Studio E&P* to their local *Petrel*, then commit their changes back to the central *Studio E&P*. These two recent examples, from the fields of medical imaging and oil and gas, show us a trend toward centralizing users' data to allow an efficient collaboration and ease data management. These solutions offer a provenance of a *coarse-grained* granularity, which deals with individual files as the smallest provenance unit. In this paper, we continue with this direction of a data-centric architecture via further integration into the visualization architecture, based on a *fine-grained* granularity, applying provenance to tuples in the database.

2.3   Seismic Imaging Data—The Oil and Gas Industry

In this paper, we apply our method to seismic imaging data from the oil and gas industry. To acquire seismic data, acoustic waves are artificially generated on the earth's surface and reflect from subsurface geological layers and structures. Due to the variation of material properties, these waves are reflected back to the surface and their amplitudes and travel time are recorded via receivers [32]. This data is then processed to generate a 2D or 3D image illustrating the subsurface layers. A 2D seismic profile consists of multiple vertical *traces*. Each *trace* holds the recorded amplitudes sampled at, typically, every four milliseconds. Seismic imaging is then interpreted by a geoscientist to extract geological features such as horizons and faults. This interpretation potentially identifies hydrocarbon traps: oil or gas. In 2011, the consumption of fossil fuels (oil and gas) accounted for 56 % of the world's primary energy as calculated by BP [33]. By 2030, it is projected that the demand of fossil fuels will continue to grow [34]. Thus, this industry requires to manage and visualize its massive datasets more efficiently.

The SEG-Y format has been used by the industry to store seismic data since mid 1970 s. SEG-Y structure consists of a *textual file header*, a *binary file header*, and multiple trace records. Each trace record consists of a binary *trace header* and trace data containing multiple sample values; more details can be found in the SEG-Y Data Exchange Format (revision 1) [35].

Data in a SEG-Y file is stored sequentially and therefore retrieval of seismic data for a 3D visualization could negatively affect interactivity. For this reason, seismic visualization and interpretation applications, such as Petrel [11], offer an internal format which stores seismic data in multi-resolution bricks for fast access; this is based on the *Octreemizer* technique by Plate et al. [2]. This has been a successful approach in visualizing very large seismic data. However, data management is still a challenge, mainly in managing multi-user interpretations and moving data between users and applications; this was confirmed to us through discussions with geoscientists from the oil and gas industry and through a survey which we present in Sect. 3. Current seismic applications often use proprietary internal formats and also represent and store interpreted surfaces such as horizons and faults in objects stored separately from the seismic data.

2.4   Other Related Work

In order to maintain users' data provenance coherently with the data, we need to create data structures that contain both; thus the original data becomes progressively enhanced as the users visualize it. An architecture needs to be developed that can incorporate this enhancement in a scalable manner. Al-Naser et al. [36] were first to introduce the concept of feature-aware parallel queries to a database in order to create a volume in real time ready for direct volume rendering. In this approach, features—which are classically represented by meshes—are stored as points tagged into the database; thus queries are "feature-aware." Their work was inspired by Brooke et al. [37] who discussed the importance of data locality in visualizing large datasets and exploited the (then) recently available programmable GPUs for direct rendering without the creation of geometric objects based on meshing. This definition of "feature-aware" differs from that used by Zhang and Zhao [38] in which an approximation is applied for time-varying mesh-based surfaces to generate multi-resolution animation models while preserving objects' features.

With the rapid advances in the capabilities of GPUs, direct volume rendering techniques such as *3D texture slicing* [39] and *GPU-based ray casting* [40] have become more efficient for interactive visualization on a large uniform grid. The latter was inspired by the introduction of shading languages such as OpenGL Shading Language (GLSL). We exploit such standard techniques in our architecture to support the primary claim in this paper; we plan to utilize other advanced techniques in future to deal with different data structures, e.g. unstructured spatial data.

3 Survey

We recently made a non-public survey of 18 senior staff from oil and gas companies (5 geoscientists and 13 IT staff who explicitly support and maintain subsurface data, hardware, and software infrastructures) and 6 geoscientists from a university (5 postgraduate students and 1 senior staff). The purpose of the survey was to understand the following:

1. the importance of a collaborative environment and data provenance for seismic interpretation,
2. the differences between a seismic interpretation environment in universities and in the industry,
3. the challenges IT staff encountered to fulfil the need of seismic visualization and interpretation in an efficient data management manner.

The *collaboration* in this context refers to the ability of users to share their results of interpretations and work together to produce such results. Ten out of eleven of the participated geoscientists perceived that a collaborative visualization and interpretation environment is "very important". The participants from the industry added that collaboration "raises productivity." We observed that the need of a collaboration was more obvious to geoscientists in industry than in universities where the work is almost performed individually. The same ratio also perceived that it is overall challenging to collaborate on seismic interpretation using existing software. The geoscientists from the industry highlighted this challenge on sharing their interpretations with other teams; for example between the interpreters and engineers. In such an industry, it is often that each team uses different software, and each software uses its proprietary internal format. All participated geoscientists perceived that data provenance, including history tracking, is important in seismic interpretation; most of the participants from the industry added that history tracking "raises efficiency." Looking into the challenge of history tracking, geoscientists from the universities did not find this challenging while the industry saw that it is "manageable for recent interpretations but a bit challenging for very old datasets;" this perhaps clarifies the former finding.

The response from the IT staff highlighted the technical challenges and thus area of improvements, where geoscientists might not perceive this in the same way, since it is the job of the IT staff to do this for them. Ten out of thirteen IT staff perceived that users cannot immediately access all subsurface data for visualization or interpretation in a high availability fashion; but data access needs some initial preparation. They share with the geoscientists the perception of challenging collaboration; 46 % found that sharing users' results with any other staff members is a time consuming task as it requires to export then import data to be shared. The participants suggested that to enhance the infrastructure, for a collaboration that maximizes productivity, the industry need to introduce a "more integrated data environment", "use cloud-based technology", "standardize the workflow between all users", and "have shared databases among all disciplines (teams)." In addition, they evaluated the move of data between application as it negatively affect productivity; 62 % of them emphasised this as a "high" effect and it "needs a great attention." Eleven out of thirteen believed that centralizing subsurface datasets would improve data management, but this "has been challenging up to now." Some of their suggested reasons for not centralizing datasets were; the proprietary internal format of multiple applications and the lack of a well integrated environment in the industry. The need of a "standard data repository" was highlighted.

4 Data-Centric Visualization Architecture

In this section, we present our solution at an abstract level; in Sect. 5 we discuss its implementation in the field of geoscience for the oil and gas industry in particular. Using standard technologies, we propose a data-centric visualization architecture which stores users' interpretations back to the central database for reusability and knowledge sharing. We choose to build our data structure in a parallel relational database management system (RDBMS); we call this *spatially registered data structure* (SRDS) (Sect. 4.1). Since our data structure is stored in a relational database rather than in raw image files and geometric objects, we require an intermediate stage which builds in real time a volume in a format which can be directly rendered on the GPU. We call this a *feature-embedded spatial volume* (FESVo) (Sect. 4.2).

As illustrated in Fig. 1, the architecture links the SRDS on a database to one or more on-the-fly created FESVo through parallel feature-aware and global spatially-referenced queries which results in a parallel streaming of data units. FESVo, on the other side, is linked to a rendering engine. Users' interpretations are stored back to SRDS. In our current work, we directly store interpretations to SRDS and rebuild the intermediate local volume (FESVo) with the newly added data; in future work we can optimize this by caching users' interpretations in FESVo then later store it into SRDS.

4.1 Spatially Registered Data Structure (SRDS)

We follow the direction of a *data-centric* architecture as in MIDAS [29] (discussed in Sect. 2.2) to maximize data access by multiple users and thus provide efficient sharing. Our data structure is fine-grained so that each voxel-equivalent data unit is stored as a tuple, a row in a database. To support querying of tuple-based provenance [41], we restructure our data into a relational form. As a result, our visual analysis method shifts from the classic static raw file system into a central data structure built on a relational database. This also caters for the highly-structured relational modelling required by the integrated analytics paradigm of enterprise-scale business computing.

This structure principally features:

1. global (geographical) spatial reference on all data tuples,
2. interpretation tagging which accumulate users' interpretations into the database,
3. concurrent access allowing parallel multi-threading queries from multiple users.

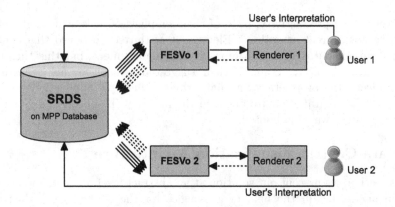

Fig. 1. This is a conceptual diagram of a data-centric visual analysis architecture which consists of three loosely coupled components. The (1) spatially registered data structure (SRDS) in a centrally located database is linked to multiple on-the-fly created (2) feature-embedded spatial volumes (FESVo) through parallel connections. FESVo is linked to (3) a renderer engine. Users' interpretations are directly fed back to SRDS; FESVo is then refreshed. Dashed arrows indicate SQL queries from FESVo to SRDS (0.2–0.7 KB each) or requests for a texture buffer from a renderer to FESVo. Full arrows indicate data units transfer from the database to FESVo (around 4 KB each), a texture buffer from FESVo to a renderer (multiple Mbytes) or updates from users to SRDS.

4.2   Feature-Embedded Spatial Volume (FESVo)

Since our data is restructured into a fine-grained form in SRDS, we need an intermediate stage to prepare data for rendering; we call this a *feature-embedded spatial volume* (FESVo). FESVo's primary roles are to (1) load data, (2) cache data, and (3) capture users' interpretations.

First, FESVo uses the indexing and parallel capabilities of the data structure to perform parallel queries, and an on-the-fly downsampling if a lower resolution is required, resulting in an intermediate volume which can be directly rendered on the GPU. Conventionally, data in 3D spatial volumes is stored in a firmed order and thus it can be read sequentially. However, an SRDS single datum is queried against its geographical location. Therefore, the loading mechanism in FESVo maps between the different coordinate systems: (1) geographical coordinate in SRDS, (2) intermediate volume coordinate in FESVo, and (3) texture coordinate in the GPU. We use a standard rendering technique to visualize this volume which is the data supplied by SRDS format. In Sect. 5.2, we explain how we implement this process for seismic imaging data.

Second, FESVo also acts as a data cache. The loaded dataset is cached and thus users can fully interact with and interpret it if the connection with SRDS is lost. Third, FESVo captures users' visual interpretations along with provenance metadata. The captured interpretation is mapped back from a local (FESVo) coordinate into a fine-grained geographical coordinate (SRDS).

5   Implementation for Geoscience

In this section, we first discuss the extension of SRDS and how data is initially prepared into it. Then, we discuss the continuous loading process by FESVo and how multiple users can contribute and track history. Finally, we briefly describe our rendering engine which renders the data loaded by FESVo.

5.1   SRDS for Geoscience

We extend SRDS from its abstract version to deal with geological and geophysical data. We implement SRDS on a Teradata database; thus we use standard SQL for queries and data updates. We use a hashing index in our tables; this allows an on-the-fly indexing using a hashing algorithm for direct access to data units and efficient update.

Using a hashing algorithm [42], the location of the required row can be determined through hashing functions without a construction or storage complexity; this is a feature offered by Teradata DBMS. This allows retrieval and writing back from and to the database at a complexity that is proportional only to the working dataset (the size of the dataset being retrieved or written back) and not to the total size of the tables.

In our case, the data is initially prepared from SEG-Y files and geometry. We start from post-stack 3D seismic conventional files (SEG-Y format) and extract *traces*, which are 1D vertical subsurface readings of amplitude values. The trace data is loaded into the database tagged with its geographical location, which is extracted from the trace header. Geological features, which were previously interpreted by users, are obtained in the form of geometry. This is converted into an *(x,y,z)* cloud of points and loaded into the database. Then, the on-going users' amendments to the features are directly stored in the same format, as a cloud of points with proper tagging and provenance metadata.

5.2   Data Loading

The data loading processes to render a dataset are as follows. A rendering engine requests a ready texture buffer to be directly rendered from FESVo based on a user request of desired datasets. Inside FESVo, a *data loader* calculates FESVo's current dimension and performs coordinate mapping between SRDS, FESVo and the GPU texture buffer. Upon the user's request, the *data loader* calculates the fine-grained data units (tuples) required to build the texture buffer. For each data unit, it first checks its internal cache. The data unit, if found, is placed in the texture buffer at the computed (mapped) position. If a data unit is, otherwise, not cached, the unit is added to one of a number of queues in a load balancing manner. Each queue is associated with a thread. After completing the search in the internal cache of FESVo, the threads start, each with its queue of data units (locations) to be concurrently fetched from the database. Each fetched data unit is loaded to FESVo and placed in the texture buffer at the computed position. As we are using a hashing algorithm to index the dataset, data retrieval is performed at a complexity of $O(k)$, where k is the number of data units being fetched from the database; this is independent of the total size of the table (dataset).

A GIS cell, in our seismic case, holds a subsurface dataset under a rectangular area (e.g. 12×12 m square) of the real-world. In texture world, this is mapped to a single 1D dataset. To search inside a GIS cell in the database, we can choose between two modes: (1) general discovery mode and (2) specific cached mode. We maintain both modes and perform one depending on the task.

In the first mode, we have no knowledge in advance about the exact coordinates of the data units; thus it is a discovery mode. To query the database for a dataset which lies in a GIS cell, we explicitly query every possible location (x,y) with a minimum step (e.g. 1 m). The reason why explicit values of x and y are provided in the query is to perform hash-based point-to-point queries and avoid a full table scan by the database. Due to the massive size of seismic datasets and because we place all raw seismic datasets in a single table for multi-datasets access, we always attempt to avoid a full table scan which leads to a decrease in performance.

In the specific mode, we pre-scan the tables for the required region and dataset source(s) and then cache all the (x,y) coordinates, using a sorted table. Starting with the texture coordinate we need to load, the mapped geographical coordinate is calculated: *(xInitial, yInitial)*. As all valid data coordinates are cached on the client, we can efficiently look for a point *(xTrue, yTrue)* which lies on the location of the current GIS cell. Having a valid and explicit coordinate, a point-to-point query is executed per required GIS cell. This mode overall performs at a complexity of $O(k)$, where k is the number of data units returned; this is regardless of the table size and number of datasets in the table.

5.3   Multi-user Input with History Tracking

Using the structure explained in Sect. 4.1, multiple users can interact by adding or changing others' interpretations while maintaining data provenance. For a user to insert some interpretations as an extension to another user's work, we do the following. We create a new entry in the grouping table linking the user's source ID to the source ID of the original interpretation to which the extension is applied. In this entry we insert a timestamp and the relation type of this grouping which is *insertion* in this case, since the user is inserting a new interpretation. Then, we insert the points which form the user's new interpretation into the features table with his/her user ID and the earlier timestamp inserted in the grouping table. In the case of deleting a previously created interpretation, the relation type would be *deletion* instead and we insert the points which the user wants to delete in the features table with his/her ID and the grouping timestamp. By doing so, we accumulate users' interpretations and do not physically delete but tag as deleted so users can roll back chronologically.

To retrieve a geological feature which involved multiple users in interpretation, we query the database such that we add points of an *insertion* relation type and subtract points of *deletion* relation type. Such points can be identified via the *source ID* and *timestamp*, linked to the *grouping* table. We can then retrieve a geological feature, which was interpreted by multiple users, by performing standard database queries on the metadata which is integrated into the database structure.

In this query, the *baseline* is the original interpretation which was first imported from an external source. We control the history tracking by manipulating the *timestamp* value.

5.4 Rendering Engine

At this stage, we adopt a back-to-front *textured slice mapping* rendering technique [39] along with a shader program, using OpenGL Shader Language (GLSL). Two texture objects (buffers) exist at any time: one for seismic raw data (volumetric datasets) and the other one is for all geological features.

6 Results and Case Studies

In this section, we first present a case study performed by geoscientists from a university geoscience department. Then, we present some performance measures. For these we used Teradata DBMS, running virtually on a 64-bit Windows Server 2003. Both FESVo and the renderer engine were deployed on laptop and desktop machines equipped with graphics cards of 256 MB to 1 GB of memory. Finally, we present and discuss a survey that we recently conducted on staff from the oil and gas industry as well as geoscientists from academia.

6.1 Case Study on Geoscientists

Six geoscientists from a university geoscience department participated in this case study. The participants were divided into three sessions. At each session, the participants were given an explication of the system and the purpose of the case study. Then each of the two participants in a session was given an access to the system, and were guided to load the same dataset at the same time from the centrally located database.

First, all participated geoscientists confirmed that the dataset was rendered correctly; their commercial software was considered as a guideline. We asked them to confirm this since the data in SRDS is completely restructured and thus this confirmation verifies our reconstruction (loading) method.

The participants were then asked to perform the following selected tasks using our system. For them, these cases are simple core tasks that may take part in their interpretation work flow. We selected these tasks only for the purpose of demonstrating the functionality of our architecture, such that provenance of users' interpretations is maintained using a two-way fine-grained visualization pipeline with a central relational database. In the following, we define a task then explain how it is technically achieved using our architecture.

Task 1: Multi-user Horizon Time Shifting. In this task, one user was asked to adjust a horizon by shifting its two-way-travel time (TWTT). Graphically, this *time* is the z axis of an early-stage seismic imaging data; it is later converted into real depth. The process of time shifting a horizon can be done in several ways in respect to selecting where the shift is applied. In our implementation, we allow the user to select the following:

1. the horizon to which the shift is applied
2. a seed point
3. a shift value $(+/-)$ (e.g. 50 ms)
4. a diameter value to which the shifting is applied (e.g. 400 m)

After setting these parameters, the time-shifting is implemented as follows. We start a *deletion* type grouping in SRDS linked to the original interpretation source ID and tagged with this user ID and a current timestamp. Then, all points lying within the selected diameter are inserted into the database in parallel threads, tagged with the user ID and the timestamp. Next, we end the *deletion* type grouping and start an *insertion* type grouping in SRDS with the same user ID but a new timestamp. Then, all points laying within the selected diameter are inserted into the database in parallel threads, with a new *time* value calculated in respect to the original value (this calculation is performed on the database) and tagged with the user ID and the new timestamp. Finally we end the grouping. Thus, user amendments are saved while the original interpretation is also maintained centrally with the original dataset. These steps are illustrated in Fig. 2.

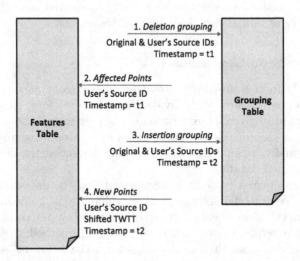

Fig. 2. This figure shows the steps taken in interaction with the database for a user to shift a previously interpreted horizon.

After the time-shift task was completed (took around 2–4 s), the second user in this session refreshed the application's view and was able to immediately visualize the changes on the horizon made by the first user. Figure 3 illustrates a similar view of this result.

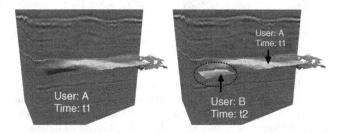

Fig. 3. The screenshot on the left shows an interpreted horizon by user A at time t1. The screenshot on the right shows a partially shifted horizon by user B at time t2. The change (shift) takes place only on the affect points since we implemented a fine-grained provenance.

In Fig. 3, the original horizon, as shown in the left screenshot, consisted of 160, 330 points (each is stored as a row in SRDS). The process of time-shift, as shown in the right screenshot, resulted in 1, 560 points tagged as deleted (from the original interpretation) and the same number of points (1, 560) tagged as inserted (forms the shifted area). Thus, a total number of 163, 450 points form this horizon including its provenance and new changes; this is a result of a fine-grained granularity model. In comparison to a coarse-grained (file-based) granularity, any small change to a feature object means a new whole object if we need to maintain its provenance. Thus, we end up with around double the number of original points, 320, 660 points.

Task 2: Deletion of an Interpreted Object with History Tracking. In this task, we assume that two users have previously added to an existing interpretation of a horizon from a particular source. A senior (more expert) user later visualized both interpretations and decided that one is more accurate than the other and therefore wanted to delete the less accurate interpretation.

One user of a session (acted as an expert) was asked to select a session with a data *insertion* tag to delete. Technically, this is achieved as follows. As in Task 1, we start a *deletion* type grouping in SRDS tagged with this user ID and a current timestamp. Then, a single update query containing the user ID and timestamp of the session to be deleted is executed. This results in re-inserting the points of this session but tagged with the expert user's ID and a timestamp of the created *deletion* type grouping. We then end this grouping. We then refresh FESVo and reload the latest version of interpretations which includes the original (previously existed) version and the additional interpretation by the more accurate user.

As we record a timestamp when starting a grouping between different interpretation sources, we can go back in history to visualize earlier versions. In this task, the second user of this session was able to track the history of this horizon; this is illustrated in Fig. 4.

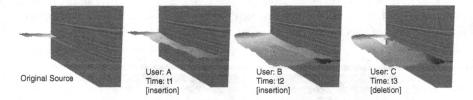

Fig. 4. From left, the second screenshot shows some interpretation added to the original one (first one from left) by User A at time t1. The third screenshot shows more contribution by User B at time t2. The fourth screenshot, an expert user (User C) decided to delete the interpretation by User A due to, for example, lack of accuracy. The interpretation of User A is in fact not deleted but tagged as deleted. Users, therefore, can go back in history and visualize previous versions of interpretations.

7   Extended Functionalities

Although the case studies presented in Sect. 6 illustrated basic functionalities to geoscientists, they have demonstrated the capability of the proposed architecture to capture provenance of interpretations from multiple users. Based on our results and our work with geoscientists, we now extend the functionality of our implementation in this section.

7.1   Richer User Interface

The fine-grained architecture allowed us to provide users to interact with the data on a per-point level. On mouse hovering, users can retrieve metadata on each point (e.g. location, interpreter, confidence level, etc.). Some of the metadata are inputted by users directly on the screen, where users can draw an area to, for example, assess; see Fig. 5.

Fig. 5. Left figure shows the ability of users to directly input metadata on the screen by drawing an area into which they wish to manipulate. Right figure shows per-point metadata of a geological feature being updated and shown as user move the mouse over.

In addition, users can highlight and filter the data loading on certain metadata values. The result of highlighting is illustrated in Fig. 5.

7.2 A Generic Provenance

Most importantly, we aim to support a generic provenance model. The data architecture presented in Sect. 4, the loading process, and the filtering mechanism are all designed and implemented to support multiple provenance properties.

At this stage, we have demonstrated the following provenance properties: (1) insertion, (2) deletion and (3) assessment. The first two were demonstrated in Sect. 6. The latter was later added to test the expandability of our architecture and to provide users with richer metadata. As discussed previously in this section, users are now able to directly assess their, or others', interpretations while visualizing the data. This provides other users, who might later user the data, with information about the provenance of the interpretation, allowing them to make their own assessment of its reliability.

8 Conclusion and Future Work

In this paper we have demonstrated a novel visualization architecture that can be used to manage different users' visual interpretations over multiple sessions. In this architecture, features in the data become first class objects stored alongside the original data. This also allows us to provide provenance of these interpretations so they become suitable for use in geographically distributed collaborations. We link this to a renderer through a loading mechanism involving a data structure that acts as an intermediary between the original data and the graphical objects used to visualize and render it. This intermediate structure allows users' amendments of interpretations expressed by manipulation of the visualization tool to flow back as new metadata to the data storage. In this paper, we have presented a case study on geologists testing our architecture and some performance results. We have also presented a survey which shows a need for an efficient data management in seismic visualization; we believe that our architecture is a step towards addressing this need.

Our plan for the future is to integrate our architecture with feature extraction techniques and algorithms, such as the work presented by Höllt et al. [8]. Also, we plan to provide more interactive functionalities available to users' interpretations of features. In addition, it is vital to test our methods on massive datasets due to the nature and demands of the oil and gas industry. Currently the original data and the metadata representing the interpreted features are stored in a relational database; we intend to test our method with other forms of data storage and to evaluate these in terms of their ability to supply data at a sufficiently high rate to allow interactive visualization of features from very large datasets. Finally, in this paper we have considered only seismic imaging datasets. However, our method can potentially be extended to support other types of spatial data, for example from oceanography, space physics, and medical imaging. This will involve creating new types of intermediate data structures compared to the FESVo we have created for seismic data. We believe that the efficient generation of such intermediate structures, which can allow visualization pipelines to be bi-directional, will form a very interesting new area in visualization research.

References

1. Robein, E.: Seismic Imaging: A Review of the Techniques, their Principles, Merits and Limitations. EAGE Publications bv, aaa (2010)
2. Plate, J., Tirtasana, M., Carmona, R., Fröhlich, B.: Octreemizer: a hierarchical approach for interactive roaming through very large volumes. In: Data Visualisation, pp. 53–60. Eurographics Association (2002)
3. Castanie, L., Levy, B., Bosquet, F.: VolumeExplorer: roaming large volumes to couple visualization and data processing for oil and gas exploration. In: IEEE Visualization, vol. im, pp. 247–254. IEEE (2005)
4. Lin, J.C.R., Hall, C.: Multiple oil and gas volumetric data visualization with GPU programming. In: Proceedings of SPIE 6495, pp. 64950U–64950U-8 (2007)
5. Plate, J., Holtkaemper, T., Froehlich, B.: A flexible multi-volume shader framework for arbitrarily intersecting multi-resolution datasets. IEEE Trans. Vis. Comput. Graph. **13**, 1584–1591 (2007)
6. Patel, D., Sture, O.Y., Hauser, H., Giertsen, C., EduardGröller, M.: Knowledge-assistedvisualization of seismic data. Comput. Graph. **33**, 585–596 (2009)
7. Patel, D., Bruckner, S., Viola, I.: Seismic volume visualization for horizon extraction. In: Proceedings of IEEE Pacific Visualization, vol. Vi, pp. 73–80. IEEE, Taipei (2010)
8. Höllt, T., Beyer, J., Gschwantner, F., Muigg, P., Doleisch, H., Heinemann, G., Hadwiger, M.: Interactive seismic interpretation with piecewise global energy minimization. In: 2011 IEEE Pacific Visualization Symposium (PacificVis), Hong Kong, pp. 59–66 (2011)
9. Visualization Sciences Group: Avizo Earth (2013). http://www.vsg3d.com/avizo/earth
10. Halliburton-Landmark: GeoProbe Volume Visualization (2013). https://www.landmarksoftware.com/Pages/GeoProbe.aspx
11. Schlumberger: Petrel Seismic to Simulation Software (2013). http://www.slb.com/services/software/geo/petrel.aspx
12. Bacon, M., Simm, R., Redshaw, T.: 3-D Seismic Interpretation. Cambridge University Press, Cambridge (2003)
13. Al-Naser, A., Rasheed, M., Irving, D., Brooke, J.: A data centric approach to data provenance in seismic imaging data. In: 75th EAGE Conference & Exhibition incorporating SPE EUROPEC. EAGE Publications bv, London (2013)
14. Al-Naser, A., Rasheed, M., Irving, D., Brooke, J.: A visualization architecture for collaborative analytical and data provenance activities. In: 2013 17th International Conference on Information Visualisation (IV), pp. 253–262 (2013)
15. Moreland, K.: A survey of visualization pipelines. IEEE Trans. Vis. Comput. Graph. **19**, 367–378 (2013)
16. Ahrens, J., Brislawn, K., Martin, K., Geveci, B., Law, C., Papka, M.: Large-scale data visualization using parallel data streaming. IEEE Comput. Graph. Appl. **21**, 34–41 (2001)
17. Biddiscombe, J., Geveci, B., Martin, K., Moreland, K., Thompson, D.: Time dependent processing in a parallel pipeline architecture. IEEE Trans, Vis. Comput. Graph. **13**, 1376–1383 (2007)
18. Stockinger, K., Shalf, J., Wu, K., Bethel, E.: Query-driven visualization of large data sets. In: IEEE Visualization, VIS 2005, pp. 167–174. IEEE (2005)
19. Gosink, L.J., Anderson, J.C., Bethel, E.W., Joy, K.I.: Query-driven visualization of time-varying adaptive mesh refinement data. IEEE Trans. Vis. Comput. Graph. **14**, 1715–1722 (2008)

20. Wu, K.: FastBit: an efficient indexing technology for accelerating data-intensive science. J. Phys. Conf. Ser. **16**, 556–560 (2005)
21. Wu, K., Ahern, S., Bethel, E.W., Chen, J., Childs, H., Cormier-Michel, E., Geddes, C., Gu, J., Hagen, H., Hamann, B., Koegler, W., Lauret, J., Meredith, J., Messmer, P., Otoo, E., Perevoztchikov, V., Poskanzer, A., Prabhat, Rübel, O., Shoshani, A., Sim, A., Stockinger, K., Weber, G., Zhang, W.M.: FastBit: interactively searching massive data. J. Phys. Conf. Ser. **180**, 012053 (2009)
22. Vo, H., Bronson, J., Summa, B., Comba, J., Freire, J., Howe, B., Pascucci, V., Silva, C.: Parallel visualization on large clusters using map reduce. In: 2011 IEEE Symposium on Large Data Analysis and Visualization (LDAV), pp. 81–88. IEEE (2011)
23. Pavlo, A., Paulson, E., Rasin, A., Abadi, D.J., DeWitt, D.J., Madden, S., Stonebraker, M.: A comparison of approaches to large-scale data analysis. In: Proceedings of the 2009 ACM SIGMOD International Conference on Management of Data, pp. 165–178. ACM (2009)
24. Simmhan, Y.L., Plale, B., Gannon, D.: A survey of data provenance in e-science. SIGMOD Rec. **34**, 31–36 (2005)
25. Ikeda, R., Widom, J.: Data lineage: a survey. Technical report, Stanford University (2009)
26. Bavoil, L., Callahan, S., Crossno, P., Freire, J., Scheidegger, C., Silva, C., Vo, H.: VisTrails: enabling interactive multiple-view visualizations. In: IEEE Visualization, VIS 2005, pp. 135–142. IEEE (2005)
27. Scheidegger, C.E., Vo, H., Koop, D., Freire, J., Silva, C.T.: Querying and creating visualizations by analogy. IEEE Trans. Vis. Comput. Graph. **13**, 1560–1567 (2007)
28. Hawtin, S., Lecore, D.: The business value case for data management - a study. Technical report, CDA & Schlumberger (2011)
29. Jomier, J., Aylward, S.R., Marion, C., Lee, J., Styner, M.: A digital archiving system and distributed server-side processing of large datasets. In: Siddiqui, K.M., Liu, B.J. (eds.) Proceedings of SPIE 7264, Medical Imaging 2009: Advanced PACS-based Imaging Informatics and Therapeutic Applications, vol. 7264, pp. 726413–726413-8 (2009)
30. Jomier, J., Jourdain, S., Marion, C.: Remote visualization of large datasets with MIDAS and ParaViewWeb. In: Proceedings of the 16th International Conference on 3D Web Technology, Paris, France, pp. 147–150. ACM (2011)
31. Alvarez, F., Dineen, P., Nimbalkar, M.: The Studio Environment: Driving Productivity for the E&P Workforce. White paper, Schlumberger (2013)
32. Ma, C., Rokne, J.: 3D seismic volume visualization. In: Zhang, D.D., Kamel, M., Baciu, G. (eds.) Integrated Image and Graphics Technologies, vol. 762, pp. 241–262. Springer, Netherlands (2004)
33. BP: BP statistical review of world energy June 2012. Technical report (2012)
34. BP: BP energy outlook 2030. Technical report, London (2012)
35. Society of exploration geophysicists: SEG Y rev 1 data exchange format. Technical Report, May 2002
36. Al-Naser, A., Rasheed, M., Brooke, J., Irving, D.: Enabling visualization of massive datasets through MPP database architecture. In: Carr, H., Grimstead, I. (eds.) Theory and Practice of Computer Graphics, pp. 109–112. Eurographics Association (2011)
37. Brooke, J.M., Marsh, J., Pettifer, S., Sastry, L.S.: The importance of locality in the visualization of large datasets. Concurrency Comput. Pract. Experience **19**, 195–205 (2007)

38. Zhang, S., Zhao, J.: Feature aware multiresolution animation models generation. J. Multimedia **5**, 622–628 (2010)
39. Mcreynolds, T., Hui, S.: Volume visualization with texture. In: SIGGRAPH, pp. 144–153 (1997)
40. Hadwiger, M., Ljung, P., Salama, C.R., Ropinski, T.: Advanced illumination techniques for GPU volume raycasting. In: ACM SIGGRAPH Courses Program, pp. 1–166. ACM, New York (2009)
41. Karvounarakis, G., Ives, Z.G., Tannen, V.: Querying data provenance. In: Proceedings of the 2010 ACM SIGMOD International Conference on Management of Data, SIGMOD 2010, pp. 951–962. ACM, New York (2010)
42. Rahimi, S.K., Haug, F.S.: Query Optimization. Wiley, New York (2010)

Generalized Pythagoras Trees: A Fractal Approach to Hierarchy Visualization

Fabian Beck[1]([⊠]), Michael Burch[1], Tanja Munz[1], Lorenzo Di Silvestro[2],
and Daniel Weiskopf[1]

[1] VISUS, University of Stuttgart, Stuttgart, Germany
fabian.beck@visus.uni-stuttgart.de
[2] Dipartimento di Matematica e Informatica, Università di Catania, Catania, Italy

Abstract. Through their recursive definition, many fractals have an inherent hierarchical structure. An example are binary branching Pythagoras Trees. By stopping the recursion in certain branches, a binary hierarchy can be encoded and visualized. But this binary encoding is an obstacle for representing general hierarchical data such as file systems or phylogenetic trees, which usually branch into more than two subhierarchies. We hence extend Pythagoras Trees to arbitrarily branching trees by adapting the geometry of the original fractal approach. Each vertex in the hierarchy is visualized as a rectangle sized according to a metric. We analyze several visual parameters such as length, width, order, and color of the nodes against the use of different metrics. Interactions help to zoom, browse, and filter the hierarchy. The usefulness of our technique is illustrated by two case studies visualizing directory structures and a large phylogenetic tree. We compare our approach with existing tree diagrams and discuss questions of geometry, perception, readability, and aesthetics.

Keywords: Hierarchy visualization · Fractals

1 Introduction

Hierarchical data (i.e., trees) occurs in many application domains, for instance, as results of a hierarchical clustering algorithm, as files organized in directory structures, or as species classified in a phylogenetic tree. Providing an overview of possibly large and deeply nested tree structures is one of the challenges in information visualization. An appropriate visualization technique should produce compact, readable, and comprehensive diagrams, which ideally also look aesthetically appealing and natural to the human eye.

A prominent visualization method are node-link diagrams, which are often simply denoted as *tree diagrams*; layout and aesthetic criteria have been discussed [24,31]. Although node-link diagrams are intuitive and easy to draw, visual scalability and labeling often is an issue. An alternative, in particular easing the labeling problem, are indented trees [7] depicting the hierarchical

© Springer International Publishing Switzerland 2015
S. Battiato et al. (Eds.): VISIGRAPP 2014, CCIS 550, pp. 115–135, 2015.
DOI: 10.1007/978-3-319-25117-2_8

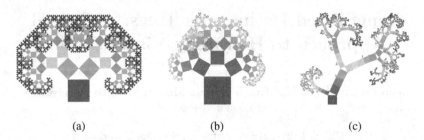

(a)   (b)   (c)

Fig. 1. Extending Pythagoras Trees for encoding information hierarchies: (a) traditional fractal approach; (b) Generalized Pythagoras Tree applied to an n-ary information hierarchy; (c) additionally visualizing the number of leaves by the size of the inner nodes.

structure by indentation. Further, layered icicle plots [17] stack boxes on top of each other for encoding a hierarchy, but waste space by assigning large areas to inner nodes on higher levels of the hierarchy. The Treemap approach [27], which is applying the concept of nested boxes, produces space-efficient diagrams but complicates interpreting the hierarchical structure.

In this paper, we introduce Generalized Pythagoras Trees as an alternative to the above hierarchy visualization techniques. It is based on Pythagoras Trees [5], a fractal technique showing a binary hierarchy as branching squares (Fig. 1(a)); the fractal approach is named after Pythagoras because every branch creates a right triangle and the Pythagorean theorem is applicable to the areas of the squares. We extend this approach to n-arily branching structures and use it for depicting information hierarchies (Fig. 1(b)). Instead of triangles, each recursive rendering step produces a convex polygonal shape where the corners are placed on a semi circle. The size of the created rectangles can be modified for encoding numeric information such as the number of leaf nodes of the respective subhierarchy (Fig. 1(c)).

We implemented the approach as an interactive tool and demonstrate its usefulness by applying it to large and deeply structured abstract hierarchy data from two application domains: a file system organized in directories and the NCBI taxonomy, a phylogentic tree that structures the living organisms on earth in a tree consisting of more than 300,000 vertices. Furthermore, a comparison to existing hierarchy visualization approaches provides first insights into the unique characteristic of Generalized Pythagoras Trees: a higher visual variety leads to more distinguishable visualizations, the fractal origin of the method supports identifying self-similar structures, and the specific layout seems to be particularly suitable for visualizing deep hierarchies. Finally, the created images are visually appealing as they show analogies to natural tree and branching structures.

2   Related Work

The visualization of hierarchical data is a central information visualization problem that has been studied for many years. Typical respresentations include node-link, stacking, nesting, indentation, or fractal concepts as surveyed by [13,26]. Many variants of the general concepts exist, for instance, radial [3,10] and bubble layouts [11,18] of node-link diagrams, circular approaches for stacking techniques [1,28,33], or nested visualizations based on Voronoi diagrams [2,22].

Although many tree visualizations were proposed in the past, none provides a generally applicable solution and solves all related issues. For example, node-link diagrams clearly show the hierarchical structure by using explicit links in a crossing-free layout. However, by showing the node-link diagram in the traditional fashion with the root vertex on top and leaves at the bottom, much screen space stays unused at the top while leaves densely agglomerate at the bottom. Transforming the layout into a radial one distributes the nodes more evenly, but makes comparisons of subtrees more difficult. Node-link layouts of hierarchies have been studied in greater detail, for instance, [6] investigated visual task solution strategies whereas [21] analyzed space-efficiency.

Indented representations of hierarchies are well-known from explorable lists of files in file browsers. Recently, [7] investigated a variant as a technique for representing large hierarchies as an overview representation. Such a diagram scales to very large and deep hierarchies and still shows the hierarchical organization but not as clear as in node-link diagrams. Layered icicle plots [17], in contrast, use the concept of stacking: the root vertex is placed on top and, analogous to node-link diagrams, consumes much horizontal space that is as large as all child nodes together.

Treemaps [27], a space-filling approach, are a prominent representative of nesting techniques for encoding hierarchies. While properties of leaf nodes can be easily observed, a limitation becomes apparent when one tries to explore the hierarchical structure because it is difficult to retrieve the exact hierarchical information from deeply nested boxes: representatives of inner vertices are (nearly) completely covered by descendants. Treemaps have been extended to other layout techniques such as Voronoi diagrams [2,22] producing aesthetic diagrams that, however, suffer from high runtime complexity.

Also, 3D approaches have been investigated, for instance, in Cone Trees [8], each hierarchy vertex is visually encoded as a cone with the apex placed on the circle circumference of the parent. Occlusion problems occur that are solved by interactive features such as rotation. Botanical Trees [14], a further 3D approach, imitate the aesthetics of natural trees but are restricted to binary hierarchies, that is, n-ary hierarchies are modeled as binary trees by the strand model of [12]; it becomes harder to detect the parent of a node.

The term fractal was coined by [20] and the class of those approaches has also been used for hierarchy visualization due to their self-similarity property [15,16]. With OneZoom [25], the authors propose a fractal-based technique for visualizing phylogenetic trees; however, n-ary branches need to be visually translated into binary splits. [9] visualize random binary hierarchies with a fractal approach as botanical trees; no additional metric value for the vertices is taken into account; instead, they investigate the Horton-Strahler number for computing the branch thicknesses.

The goal of our work is to extend a fractal approach, which is closer to natural tree structures, towards information visualization. This goal promises embedding the idea of self-similarity and aesthetics of fractals into hierarchy visualization. Central prerequisite—and in this, our approach differs from existing fractal approaches—is that n-ary branches should be possible. With respect to information visualization, the approach targets at combining advantages of several existing techniques: a readable and scalable representation, an efficient use of screen space, and the flexibility for encoding additional information. A downside of the approach, however, is that overlap may occur similar as in 3D techniques (though it is a 2D representation)—only varying the parameters of the visualization or using interaction alleviates this issue.

3 Visualization Technique

Our general hierarchy visualization approach extends the idea of Pythagoras Trees. Instead of basing the branching of subtrees on right triangles, we exploit convex polygons with edges on the circumference of a semi circle.

3.1 Data Model

We model a hierarchy as a directed graph $H = (V, E)$ where $V = \{v_1, \ldots, v_k\}$ denotes the finite set of k vertices and $E \subset V \times V$ the finite set of edges, i.e., parent–child relationships. One vertex is the designated root vertex and is the only vertex without an incoming edge; all other vertices have an in-degree of one. We allow arbitrary hierarchies, that is, the out-degree of the vertices is not restricted. A maximum branching factor $n \in \mathbb{N}$ of H can be computed as the maximum out-degree of all $v \in V$. For an arbitrary vertex $v \in V$, H_v denotes the subhierarchy having v as root vertex; $| H_v |$ is the number of vertices included in the H_v (including v). The depth of a vertex v' in H_v is the number of vertices on the path through the hierarchy from v to v'. We allow positive weights to be attached to each vertex of the hierarchy $v \in V$ representing metric values such as sizes. We model them as a function $w : V \rightarrow \mathbb{R}$. The weight $w(v) \in \mathbb{R}^+$ of an inner vertex v does not necessarily need to be the sum of its children, but can be.

3.2 Traditional Pythagoras Tree

The Pythagoras Tree is a fractal approach describing a recursive procedure of drawing squares. In that, it was initially not intended to encode information, but its tree structure easily allows representing binary hierarchies: each square represents a vertex of the hierarchy; the recursive generation follows the structure of the hierarchy and ends at the leaves.

Drawing a fractal Pythagoras Tree starts with drawing a square of side length c. Then, two smaller squares are attached at one side of the square—usually, at the top—according to the procedure illustrated in Fig. 2(a): Then, a

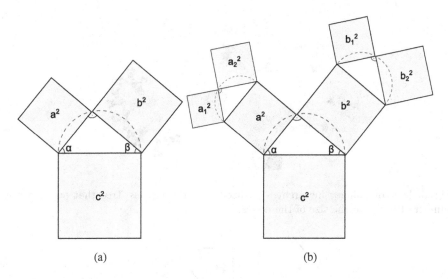

(a)   (b)

Fig. 2. Illustration of the traditional Pythagoras Tree approach: (a) a single binary branch; (b) recursively applied branching step.

right triangle with angles α and β where $\alpha + \beta = \frac{\pi}{2}$ is drawn using the side of the square as hypotenuse, which also becomes a diameter of the circumcircle of the triangle. The two legs of the triangle are completed to squares having side lengths a and b. In the right triangle, the Pythagorean theorem $a^2 + b^2 = c^2$ holds, i.e., the sum of the areas of the squares over the legs is equal to the area of the square over the hypotenuse. Applying this procedure recursively to the new squares as depicted for the next step in Fig. 2(b) creates a fractal Pythagoras Tree (the recursion is only stopped for practical reasons at some depth). The angles α and β can be set to a constant value or be varied according to some procedural pattern. Figure 1(a) provides an example of a fractal Pythagoras Tree where $\alpha = \beta = \frac{\pi}{4}$.

Transforming the fractal approach into an information visualization technique, the squares are interpreted as representatives of vertices of the hierarchy, called *nodes*. As a consequence, the fractal encodes a complete binary hierarchy, the recursion depth being the depth of the hierarchy. If the generated image should represent a binary hierarchy that is not completely filled to a certain depth, the recursion has to stop earlier for the respective subtrees. If the hierarchy is weighted as specified in the data model, the weights can be visually encoded by adjusting the sizes of the squares, i.e., the corresponding angles α and β.

Algorithm 1 describes in greater detail how an arbitrary binary hierarchy (i.e., a hierarchy where each vertex either has an out-degree of 2 or 0) can be recursively transformed into a Pythagoras Tree visualization. It is initiated by calling **PythagorasTree**(H_v, S): where $H_v = (V, E)$ is a binary hierarchy and $S = (c, \Delta s, \theta)$ is the initial square with center c, length of the sides Δs,

Fig. 3. Random binary hierarchy visualized as a Pythagoras Tree that encodes the number of leaves in the size of the nodes.

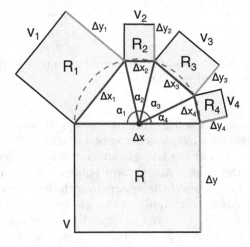

Fig. 4. Polygonal split of Generalized Pythagoras Trees creating an n-ary branch.

and slope angle θ. The recursive procedure first draws square S and proceeds if the current hierarchy still contains more than a single node. Then, encoding the node weights in the size of the squares, the angles α and β are computed according to the normalized weight of the node opposed to the angle. The angles form the basis for further computing the parameters of the two new rectangles S_1 and S_2. The drawing procedure is finally continued by recursively calling **PythagorasTree**(H_{v_1}, S_1) and **PythagorasTree**(H_{v_2}, S_2) for the two children v_1 and v_2 of the current root vertex.

When, for instance, using the number of leaf vertices as the weight of each vertex, the algorithm produces visualizations such as Fig. 3 that encodes a random binary hierarchy. Like the fractal approach, the visualization algorithm still produces overlap of subtrees that, however, becomes rarer through sparser hierarchies.

Algorithm 1. Pythagoras Tree.

PythagorasTree(H_v, S):

  // H_v: binary hierarchy

  // S: representative square $S = (c, \Delta s, \theta)$

    // $c = (x_c, y_c)$: center

    // Δs: length of a side

    // θ: slope angle

  drawSquare(S);   // draw square for current root vertex

  if $\mid H_v \mid > 1$ **then**

    // v_1 and v_2: children of H_v

    $\alpha := \frac{\pi}{2} \cdot \frac{w(v_2)}{w(v_1)+w(v_2)}$;

    $\beta := \frac{\pi}{2} \cdot \frac{w(v_1)}{w(v_1)+w(v_2)}$;

    $\Delta s_1 := \Delta s \cdot \sin \beta$;

    $\Delta s_2 := \Delta s \cdot \sin \alpha$;

    $c_1 := ComputeCenterLeft(c, \Delta s, \Delta s_1,)$;

    $c_2 := ComputeCenterRight(c, \Delta s, \Delta s_2)$;

    $S_1 := (c_1, \Delta s_1, \theta + \alpha)$;

    $S_2 := (c_2, \Delta s_2, \theta - \beta)$;

    PythagorasTree(H_{v_1}, S_1);   // draw subhierarchy H_{v_1}

    PythagorasTree(H_{v_2}, S_2);   // draw subhierarchy H_{v_2}

  end if

3.3 Generalized Pythagoras Tree

The Generalized Pythagoras Tree, as introduced in the following, can be used for visualizing arbitrary hierarchies, that are hierarchies allowing n-ary branches. Right triangles are replaced by convex polygons sharing the same circumcircle; the former hypotenuse of the triangle becomes the longest side of the triangle. For increasing the visual flexibility of the approach, squares are exchanged for general rectangles.

Figure 4 illustrates an n-ary branch, showing the polygon and its circumcircle. The polygon is split into a fan of isosceles triangles using the center of the circumcircle as splitting point. While the number of rectangles is specified by the degree of the represented branch, the angles and lengths can be modified to encode further information. In particular, we have two degrees of freedom:

– **Width function** $w_x : V \to \mathbb{R}^+$ **of rectangles**—Similar to binary hierarchies, the width Δx_i of a rectangle R_i can be changed, here, by modifying the corresponding angle α_i accordingly. The angle α_i should reflect weight $w_x(v_i)$ of a vertex v_i in relation to the weight of its siblings:

$$\alpha_i := \pi \cdot \frac{w_x(v_i)}{\sum_{j=1}^{n} w_x(v_j)}.$$

The width of the rectangle is $\Delta x_i := \Delta x \cdot \sin \frac{\alpha_i}{2}$ where Δx is the width of the parent node.

– **Length stretch function** w_y **of rectangles**—Analogously, the length Δy_i of the rectangle R_i can be varied. This length, in contrast to the width Δx_i, does not underly any restrictions such as the size of a cirumcircle. Nevertheless, we formulate the length dependent on the length of the parent Δy and the relative width $\sin\frac{\alpha_i}{2}$ in order to consider the visual context (otherwise, it would be difficult to define appropriate metrics not producing degenerated visualizations): the length of the rectangle is $\Delta y_i := w_y(v_i) \cdot \Delta y \cdot \sin\frac{\alpha_i}{2}$.

Algorithm 2 extends Algorithm 1 and describes the generation of Generalized Pythagoras Tree visualizations. Again, it is a recursive procedure and is initialized by calling **GeneralizedPythagorasTree**(H_v, R) where $H_v = (V, E)$ is an arbitrary hierarchy and $R = (c, \Delta x, \Delta y, \theta)$ represents the initial rectangle that, in contrast to the previous case, has a width Δx and a length Δy. For an n-ary branching hierarchy H_v with root vertex v, the algorithm first draws the respective rectangle before all children $v_1, \ldots, v_n$ are handled: for each child v_i, the computation of angle α_i forms the basis for deriving the width Δx_i and length Δy_i of the respective rectangle R_i as described above. Furthermore, the center and slope of the new rectangle need to be retrieved. Finally, **GeneralizedPythagorasTree**(H_{v_i}, R_i) can be recursively applied to subhierarchy H_{v_i} having rectangle R_i as root node.

Algorithm 2. Generalized Pythagoras Tree.

GeneralizedPythagorasTree(H_v, R):

   // H_v: hierarchy branching into $n \in \mathbb{N}_0$ subhierarchies $H_{v_1}, \ldots, H_{v_n}$

   // R: representative rectangle $R = (c, \Delta x, \Delta y, \theta)$

      // $c = (x_c, y_c)$: center

      // $\Delta x, \Delta y$: width and length

      // θ: slope angle

   drawRectangle(R);   // draw rectangle for parent vertex

   for all H_{v_i} **do**

      $\alpha_i := \pi \cdot \frac{w_x(v_i)}{\sum_{j=1}^{n} w_x(v_j)}$;

      $\Delta x_i := \Delta x \cdot \sin\frac{\alpha_i}{2}$;

      $\Delta y_i := w_y(v_i) \cdot \Delta y \cdot \sin\frac{\alpha_i}{2}$;

      $c_i :=$ComputeCenter$(c, \Delta x, \Delta y, (\alpha_1, \ldots, \alpha_{i-1}), \Delta x_i, \Delta y_i)$;

      $\theta_i :=$ComputeSlope$(\theta, (\alpha_1, \ldots, \alpha_i))$;

      $R_i := (c_i, \Delta x_i, \Delta y_i, \theta_i)$;

      GeneralizedPythagorasTree(H_{v_i}, R_i);

   end for

Figure 5 shows a sample visualization created with the algorithm. For this initial image width function w_x is set to a constant value and the length stretch function w_y is defined as 1. As a consequence, the nodes are squares again, equally sized for each branch but n-arily branching. An example with a similar

Fig. 5. Generalized Pythagoras Trees showing n-ary hierarchy using a constant width and length stretch function.

configuration can be found in Fig. 1(a); the same dataset is shown in Fig. 1(b) applying the number of leaf nodes as the width function w_x. Further configurations are discussed more systematically below. The discussion also includes the usage of color, which, in all figures referenced so far, visualizes the depth of the nodes. Furthermore, the order of rectangles can be modified and has an impact on the layout; in the generalized approach, we have a higher degree of freedom ($n!$ possibilities) than in the standard Pythagoras Trees where only a flipping between two angles can be applied.

3.4 Excursus: Fractal Dimension

The fractal dimension is typically used as a complexity measure for fractals. Looking back to the origin of the Generalized Pythagoras Tree visualization and interpreting it as a fractal approach, the extended fractal approach can be characterized by this dimension. To this end, however, not an information hierarchy can be encoded, but the approach needs to be applied for infinite n-arily branching structures; for simplification we do not consider scaling of rectangles. The following analysis shows that the fractal dimension, which is 2 for traditional Pythagoras Tree fractals, asymptotically decreases to 1 for a branching factor approaching infinity.

Any fractal can be characterized by its fractal dimension $D \in \mathbb{R}$ that is defined as a relation between the branching factor n and the scaling factor r given by $D = -\log_r n$. In our scenario, we have to first compute the scaling factor r depending on the branching factor n. Figure 6 illustrates the following formulas and shows an n-ary branch.

First of all, the n-ary branch creates a convex polygon, which is split into isosceles triangles as described before. Since all rectangles have the same width, the angle at the tip of the triangle is $\alpha = \frac{\pi}{n}$. The width of the rectangle then is

$$\Delta x' = \Delta x \cdot \sin \frac{\alpha}{2} = \Delta x \cdot \sin \frac{\pi}{2n}.$$

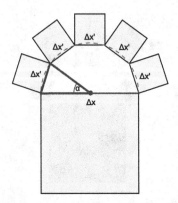

Fig. 6. Illustrating the fractal dimension of an n-ary branching hierarchy by showing the splitting into equally sized angles.

Relating the size of the square to the original square, the scaling factor can be derived as follows:

$$r = \frac{\Delta x'}{\Delta x} = \sin \frac{\pi}{2n}.$$

The fractal dimension finally is

$$D_n = -\frac{\log n}{\log \sin \frac{\pi}{2n}}.$$

This result confirms $D_2 = 2$ (traditional binary branches) and shows that the fractal dimension is approaching 1 for increasing n, i.e.,

$$\lim_{n \to \infty} D_n = 1.$$

3.5   Visual Parameters

The visualization approach has been described precisely but still has some degrees of freedom that shall be explored in the following. For example, the size of the rectangles can be varied, the order of the subhierarchies in a branch is not restricted, or the coloring of the nodes is open for variation. These parameters help optimizing the layout and support the visualization by extra information in form of weights assigned to each node. For illustrating the effect, Table 1 shows the same random hierarchy (75 nodes; maximum depth of 5) in different parameter settings. As a weight, the number of leaf nodes is applied; but the metric is interchangeable, for instance, by the number of subnodes, the depth of the subtree, or a domain-specific weight. One setting (Table 1, S1 = O1 = C1), which seemed to work most universally in our experience, is selected as default and applied in all following figures of the paper if not indicated otherwise.

Table 1. Exploring different parameter settings such as size, order, and color of rectangles for a sample dataset; framed images represent the default setting and are equivalent; the number of leaf nodes is applied as weight.

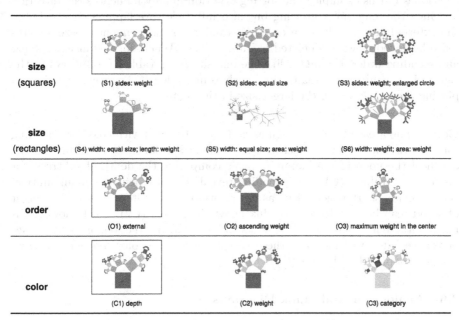

Size. Already for the traditional Pythagoras Tree approach, rectangles can be split in uniform size or non-uniform size. For the generalized approach, we define a width function as well as a length function (Sect. 3.3). When employing the same metric for both, all nodes are represented as squares. Table 1 (S1) uses the number of leaf nodes as the common metric, which seems to be a good default selection because more space is assigned to larger subtrees. In contrast, when all subnodes are assigned the same size (i.e. a constant function is employed), small subtrees become overrepresented as depicted in Table 1 (S2). A variant of the approach, which is shown in Table 1 (S3), extends the approach from using semi circles to larger sectors of a circle.

Inserting different functions for width and length further increases the flexibility—nodes are no longer squares, but differently shaped rectangles. For instance, Table 1 (S4) encodes the number of leaf nodes in the height and applies a constant value to the width. When defining the length function relative to the (constant) width so that the area of the rectangle is proportional to the number of leaves, those leaf nodes are emphasized as depicted in Table 1 (S5). A similar variant shown in Table 1 (S6) has a constant length and chooses the width accordingly for encoding the number of leaf nodes in the area.

Order. The subnodes of an inner node of a hierarchy are visualized as an ordered list. While, for some applications, there exist a specific, externally defined order, many other scenarios do not dictate a specific order. In case of the latter, the subnodes can be sorted according to a metric, which again is the number of leaf nodes in this example. The sorting criterion mainly influences the direction in which the diagram is growing but also influences overlapping effects. Often the external order, at least in case it is random or independent of size, creates quite balanced views as depicted in Table 1 (O1). When, for instance, applying an ascending order, the image like the one shown in Table 1 (O2) grows to the right. More symmetric visualizations such as in Table 1 (O3) are generated when placing the vertices with the larger size in the center.

Color. The areas of the rectangular nodes can be filled with color for encoding some extra information. Selecting the color on a color scale according to the depth of the node in the hierarchy helps comparing the depth of subtrees: for instance, in Table 1 (C1) this encoding reveals that the leftmost main subtree, though being shorter, is as deep as the rightmost one. Alternatively, the weight of a node can be encoded in color like shown in Table 1 (C2), which, however, is more suitable if the size of the node not already encodes the weight. If categories of vertices are available, also these categories can be color-coded by discrete colors as depicted in Table 1 (C3).

3.6   Analogy to Node-Link Diagrams

Though being derived from a fractal approach, Generalized Pythagoras Trees can be adapted—without changing the position of nodes—to become variants of node-link diagrams. An analogous diagram can be created as illustrated in Fig. 7 by connecting the circle centers of the semi circles of branches by lines. The circle centers become the nodes, the lines become the links of the resulting node-link diagram. Like the subtrees of a Generalized Pythagoras Tree might overlap, the analogous node-link drawing is not guaranteed to be free of edge crossings. We prefer the Pythagoras variant over the analogous node-link variant because it uses the available screen space more efficiently (which is important, for instance, for color coding) and shows the width of a node explicitly.

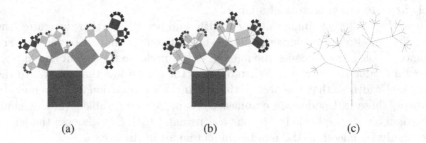

       (a)   (b)   (c)

Fig. 7. Relationship between Generalized Pythagoras Trees and node-link diagrams: (a) Generalized Pythagoras Tree; (b) Generalized Pythagoras Tree and analogous node-link diagram; (c) analogous node-link diagram.

4   Implementation

Our prototype implementation of Generalized Pythagoras Trees is written in
C++/Qt. It imports information hierarchies from text files in Newick format
or reads in directory trees from the file system. Each node can be assigned a
size that is specified in the imported file (or by the file size, in case of directory
structures). All parameters of the visualization presented can be adapted through
the user interface. For the width and length of nodes as well as for ordering and
coloring, the size metric can be employed, or alternatively, some node statistics
such as the number of subnodes or leaves. Additionally, the tool is capable of
reproducing the original fractal approach in different variants. All images of this
paper showing (Generalized) Pythagoras Trees—except for purely illustrating
figures—are generated with this tool.

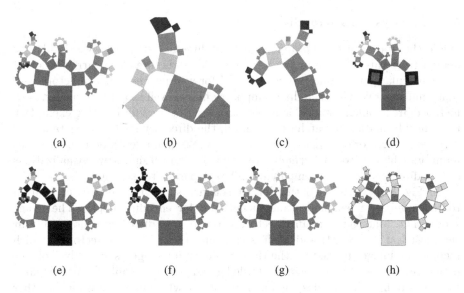

Fig. 8. Interaction in Generalized Pythagoras Trees: (a) sample hierarchy in default
representation, (b) geometric zooming, (c) selected subhierarchy, (d) collapsed nodes,
(e) highlighted path, (f) highlighted subhierarchy, (g) subhierary moved to front, and
(h) highlighting by search.

The tool does not only produce static images but is interactive (Fig. 8): Labels
can be activated for larger nodes and are retrievable for all nodes by hovering
or clicking. Selecting a node shows further statistics such as the number of children, subnodes, and leaves as well as its size value. Moreover, the tool provides
geometric zooming (Fig. 8(b)) as well as navigating through the hierarchy by
selecting a subhierarchy, which is then exclusively shown on screen (Fig. 8(c));
supplementary operations allow for moving a level up and jumping back to the
previous view. Nodes can be collapsed and expanded (Fig. 8(d)): a black bor-
derline indicates collapsed nodes; the thickness encodes the size of the collapsed

subtree. For improving the readability of the current view, paths (Fig. 8(e)) and subtrees (Fig. 8(f)) can be marked as well as, in case of overlap, a subtree can be moved to front (Fig. 8(g)). A search feature helps quickly finding specific nodes (Fig. 8(h)).

5   Case Studies

To illustrate the usefulness of our Generalized Pythagoras Tree visualization, we applied it to two datasets from different application domains—file systems with file sizes as well as the NCBI taxonomy that classifies species. In these case studies we demonstrate different parameter settings and also show how interactive features can be applied for exploration.

5.1   File System Hierarchy

While the approach can be applied to any directory structure, we decided to demonstrate this use case by reading in the file structure of an early version of this particular paper. Since we use LATEX for writing, the paper directory contains multiple text files including temporary files as well as a list of images. Also included are supplementary documents and a script used for creating exemplary random information hierarchies. All in all, the directory structure contains 139 vertices (7 directories and 132 files) having a maximum depth of 4 and a maximum branching factor of 38 (*figures* directory). Figure 9 shows two visualizations of this directory structure employing different parameter settings.

In Fig. 9(a), we applied the default settings sizing the vertices in relation to the number of leaf nodes and using color for encoding depth. The image shows that, among the main directories, the *figures* directory contains by far the most leaf nodes (94) and itself is split into three further directories, which include the images needed for the three more complex figures and tables of this paper: *canis* (Fig. 10), *parameters* (Table 1), and *samples* (Table 2). Additionally, *figures* also directly includes a number of images, which are needed for the other figures. The only other directory containing a reasonable number of files is the *hierarchy_generator* folder; besides the generator script it contains a number of generated sample datasets.

Customizing the parameters of the visualization for the use case of investigating file systems, we assigned the file size to the size of the vertices (directory sizes are the sum of the contained file sizes). Moreover, the file type is encoded in the color of the vertex (category coding) a legend providing the color–type assignments; directories are encoded in the color of the dominating file type of the contained files. The resulting visualization as depicted in Fig. 9(b) shows that the *figures* directory is also one of the largest main directories, but there exist other files and directories that also consume reasonable space such as the *additional material* directory. Comparing the size of the main PDF document to the *images* directory, it can be observed that not all image files contained in the directory are integrated into the paper because the paper is smaller than

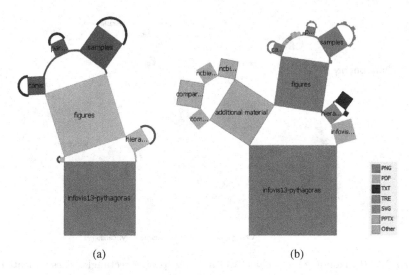

(a)   (b)

Fig. 9. Directory hierarchy of this paper on the file system: (a) size based on the number of leaf nodes with color-coded depth information; (b) size encoding the file and directory sizes with color-coded file types.

the *images* directory. The color-coded file types reveal that the most frequently occurring type are PNG files, not only in the *images* directory but also in general. The *hierarchy_generator* directory mostly includes TRE files (Newick format), but is dominated with respect to size by two TXT files (an alternative hierarchy format not as space-efficient).

5.2 Phylogenetic Tree

Moreover, our approach is tested on a hierarchical dataset commonly used by the biology and bioinformatics communities. The taxonomy here used has been developed by NCBI and contains the names of all organisms that are represented in its genetic database [4]. The specific dataset encoding the taxonomy contains 324,276 vertices (60,585 classes and 263,691 species) and has a maximum depth of 42. The Generalized Pythagoras Tree visualization applied to this dataset (Fig. 10(I)) creates a readable overview visualization of the very complex and large hierarchical structure. The vertices of the tree have different sizes according to the number of leaves of their subtrees. Each inner vertex represents a class of species and it is easy to point out the class that contains more species. The root node is an artificial class of the taxonomy that contains every species for which a DNA sequence or a protein is stored in the NCBI digital archive.

At the first level of the tree (see Fig. 10(I)), a big node represents *cellular organisms* and further nodes the *Viruses*, *Viroids*, unclassified species, and others (this information can be retrieved by using the geometric zoom). Selecting nodes and retrieving additional information facilitate the exploration of the tree. For instance, the biggest node at level 2 is the *Eukaryota* class, which includes all

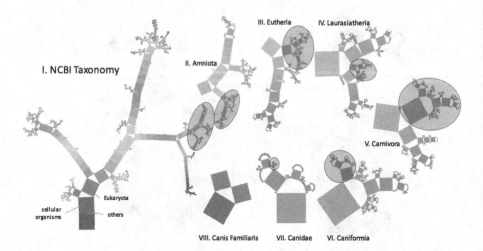

Fig. 10. NCBI taxonomy hierarchically classifying species; rectangles sizes indicate the number of species in a subtree, color encodes the depth; an example for exploring the taxonomy by semantic zooming is provided.

organisms whose cells contain a membrane-separated nucleus in which DNA is aggregated in chromosomes; it still contains 177,258 of the 263,691 species.

Besides gaining an overview of the main branches of the taxonomy, the visualization tool allows for analyzing subsets of the hierarchy down to the level of individual species by applying semantic zooming. As a concrete example, we demonstrate the exploration process in the right part of Fig. 10; in each step we selected the subtree of the highlighted node (red circle): Fig. 10(II) shows the *Amniota* class, which belongs to the *tetrapoda vertebrata* taxis (four-limbed animals with backbones or spinal columns). In the next steps (Fig. 10(III)–(V)), we followed interesting branches until we reach the *Carnivora* class in Fig. 10(V), which denotes meat-eating organisms; the subtree contains 301 species. From here, it is simple to proceed the exploration towards a well-known animal, such as the common dog, defined as *Canis Familiaris*, by zooming in the subtrees of *Caniformia*, literally "dog-shaped" (Fig. 10(VI)), then through *Canidae*, the family of dogs (Fig. 10(VII)) with 45 species, and finally *Canis Familiaris*.

6   Discussion

The introduced technique for representing hierarchical structures is discussed by taking existing other hierarchy visualization approaches into account. We applied different standard hierarchy visualization techniques to a number of randomly generated artificial datasets. The results are listed in Table 2. Each column represents a different data set with some characteristic feature: a binary hierarchy with a branching factor of 2, a deep hierarchy with many levels, a flat hierarchy with a high maximum branching factor, a degenerated hierarchy that grows

Table 2. Comparison of hierarchy visualization approaches for representatives of a selected set of hierarchy classes.

	binary hierarchy	deep hierarchy	flat hierarchy	degenerated hierarchy	symmetric hierarchy	self-similar hierarchy
	node degree of 2	high number of hierarchy levels (25)	high maximum node degree (20)	linearly growing depth	two equivalent subtrees	self similar tree structure
node-link						
indented tree						
icicle plot						
Treemap						
Generalized Pythagoras Tree						

linearly in depth with the number of nodes, a symmetric hierarchy having two identical subtrees, and a self-similar hierarchy following the same pattern at each level. The rows show standard visualization techniques in comparison to Generalized Pythagoras Trees. Though the graphics can only act as previews in a printed version of the paper, they are included in high resolution and are explorable in a digital version. The following analysis considers multiple levels of abstraction from geometry and perception to readability and aesthetics.

6.1 Geometry and Perception

Hierarchy visualizations aim at showing containment relationships between nodes and their descendants. Considering Gestalt theory [30], different approaches exist for visually encoding relationships: for instance, node-link diagrams use *connectedness* to express containment, while Treemaps are based on *common region* for showing that several nodes belong to the same parent. In contrast, Generalized Pythagoras Trees do neither directly draw a line between the nodes nor nest one node into the other, but they draw rectangles of decreasing size onto an imaginary curve. The human reader automatically connects the rectangles on the curve, which is denoted as the *law of continuation*. In all hierarchy visualization approaches shown in Table 2, *proximity* also plays a certain role (i.e., related nodes are placed next to each other) but should not be overinterpreted (i.e., nodes placed next to each other are not necessarily related).

In node-link diagrams, indented tree diagrams, or icicle plots, each level in the hierarchy creates another layer in the visualization. As a consequence, the amount of (vertical) space available for a layer is reduced when adding further levels. In Generalized Pythagoras Trees, however, there are no global layers for levels of nodes: adding a level only produces a kind of local layer that is arranged on a semi circle. With respect to this characteristic, Generalized Pythagoras Trees are similar to Treemaps, which neither have global layers but split the area of a node for introducing the next level.

Like in icicle plots and Treemaps, larger areas are used to encode the nodes in Generalized Pythagoras Trees. This makes it easier to use color for encoding some metrics (such as the hierarchy level) in the nodes because colors are easier to perceive for larger areas [29] (*Color for Labeling*). In contrast to Treemaps (and complete icicle plots), Generalized Pythogoras Trees do not create space-filling images. Areas, however, might overlap, which is discussed in detail below.

Comparing the images shown in Table 2 with respect to uniqueness, Generalized Pythagoras Trees show a high visual variety: not only the subtrees vary in size, they are also rotated. Only the splitting approach in Treemaps creates similarly varying images, however, just with respect to texture but not shape. A positive effect of a high visual variety is that the different datasets can be distinguished more easily—the visualization acts as a fingerprint. Together with the fractal roots of the approach, the uniqueness helps detect self-similar structures: Table 2 (last column) shows a tree having a self-similar structure, which is generated according to the same recursive, deterministic procedure for every node; the self-similar property of the hierarchy is best detectable in the Generalized Pythagoras Trees because every part of the tree is just a rotated version of the complete tree.

6.2   Readability and Scalability

A hierarchy visualization is readable if the users are able to efficiently retrieve the original hierarchical data from it and easily observe higher-level characteristics. However, readability is also related to visual scalability, which means preserving readability for larger datasets. While, for smaller datasets, the exact information is usually recognizable in any hierarchy visualization, the depicted information often becomes too detailed when increasing the scale of the dataset. The visualization approach, hence, needs to use the available screen space efficiently and has to focus on the most important information.

Generalized Pythagoras Trees clearly emphasize the higher-level nodes of the tree (i.e., the root node and its immediate descendants): most of the area that is filled by the visualization is consumed by these higher-level nodes, which can be easily perceived because surrounded by whitespace. Lower-level nodes and leaf nodes, however, become very small and are not visible. But the visualization allows for sizing the nodes according to their importance by using the number of leaf nodes as a metric as done in Table 2. Node-link diagrams, indented trees, and icicle plots are similar in their focus on the higher-level nodes; as well, lower-level nodes become difficult to discern because of lack of horizontal space. Since the vertical space assigned to each level does not become smaller in these visualizations, it is easier to retrieve the maximum depth of a subtree. Treemaps focus on leaf nodes and show largely different characteristics.

The ability of a visualization technique to display also large datasets in a readable way considerably widens its area of application. As shown in the case study, Generalized Pythagoras Trees can be used for browsing large hierarchies such as the NCBI taxonomy. While it is possible to interactively explore large hierarchies in a similar way with the other paradigms listed in Table 2, Generalized Pythagoras Trees show some characteristic scalability advantages: for specifically deep hierarchies such as the one in the second column of Table 2, it adaptively expands into the direction of the deepest subtree, here in spiral shape. Comparing it to the other approaches, deep subtrees are still readable in surprising detail. In contrast for flat hierarchies, which have a specifically high branching factor, Generalized Pythagoras Trees do not seem to be as suitable: the size of the nodes decreases too fast which constrains readability.

For a degenerated hierarchy (Table 2, fourth column), which grows linearly in depth with the number of nodes, Generalized Pythagoras Trees create an idiosyncratic but readable visualization, similar as it is the case for the other visualization approaches. Also a symmetry in a hierarchy such as two identical subtrees (Table 2, fifth column) can be detected: the identical tree creates the same image, which is rotated in contrast to the other approaches, where it is moved but not rotated.

A problem limiting the readability of Generalized Pythagoras Trees is that, depending on the visualized hierarchy, subtrees might overlap. The other visualization approaches do not share this problem; only Treemaps also employ a form of overplotting: inner nodes are overplotted by its direct descendants. While Treemaps use overplotting systematically, overlap only occurs occasionally in Generalized Pythagoras Trees and is unwanted. A simple way to circumvent the problem using the interactive tool is selecting the subset of the tree that is overdrawn by another. Also, reordering the nodes or adapting the parameters of the algorithm could alleviate the problem.

6.3   Aesthetics

Fractals often show similarities to natural structures such as trees, leaves, ferns, clouds, coastlines, or mountains [23]. Among the images shown in Table 2, the Generalized Pythagoras Trees clearly show the highest similarity to natural tree and branching structures. Since, according to the biophilia hypothesis, humans are drawn towards every form of life [32], this similarity suggests that Generalized Pythagoras Trees might be considered as being specifically aesthetic. Also the property of self-similarity that is partly preserved when generalizing Pythagoras Trees supports aesthetics: *"fractal images are usually complex, however, the propriety of self-similarity makes these images easier to process, which gives an explanation to why we usually find fractal images beautiful"* [19].

7   Conclusion and Future Work

In this paper, we introduced an extension of Pythagoras Tree fractals with the goal of using these for visualizing information hierarchies. Instead of depicting only binary trees, we generalize the approach to arbitrarily branching hierarchy structures. An algorithm for generating these Generalized Pythagoras Trees was introduced and the fractal characteristics of the new approach were reported. A set of parameters allows for customizing the approach and creating a variety of visualizations. In particular, metrics can be visualized for the nodes. The approach was implemented in an interactive tool. A case study demonstrates the utility of the approach for analyzing large hierarchy datasets. The theoretical comparison of Generalized Pythagoras Trees to other hierarchy visualization paradigms, on the one hand, suggested that the novel approach is capable of visualizing various features of hierarchies in a readable way comparably to previous approaches and, on the other hand, might reveal unique characteristics of the approach such as an increased distinguishability of the generated images and detectabiltiy of self-similar structures. Further, the approach may have advantages for visualizing deep hierarchies and provides natural aesthetics.

An open research questions is how the overplotting problem of the approach can be solved efficiently and how the assumed advantages can be leveraged in practical application. Moreover, formal user studies have to be conducted to further explore the characteristics of the approach.

Acknowledgements. We would like to thank Kay Nieselt, University of Tübingen, for providing the NCBI taxonomy dataset.

References

1. Andrews, K., Heidegger, H.: Information slices: visualising and exploring large hierarchies using cascading, semicircular disks. In: Proceedings of IEEE Symposium on Information Visualization, pp. 9–11 (1998)
2. Balzer, M., Deussen, O., Lewerentz, C.: Voronoi treemaps for the visualization of software metrics. In: Proceedings of Software Visualization, pp. 165–172 (2005)
3. Battista, G.D., Eades, P., Tamassia, R., Tollis, I.G.: Graph Drawing: Algorithms for the Visualization of Graphs. Prentice-Hall, Englewood Cliffs (1999)
4. Benson, D.A., Karsch-Mizrachi, I., Lipman, D.J., Ostell, J., Sayers, E.W.: Genbank. Nucleic Acids Res. **38**(suppl 1), D46–D51 (2010)
5. Bosman, A.E.: Het wondere onderzoekingsveld der vlakke meetkunde. N.V. Uitgeversmaatschappij Parcival, Breda (1957)
6. Burch, M., Konevtsova, N., Heinrich, J., Höferlin, M., Weiskopf, D.: Evaluation of traditional, orthogonal, and radial tree diagrams by an eye tracking study. IEEE Trans. Vis. Comput. Graph. **17**(12), 2440–2448 (2011)
7. Burch, M., Raschke, M., Weiskopf, D.: Indented pixel tree plots. In: Proceedings of International Symposium on Visual Computing, pp. 338–349 (2010)
8. Carrière, S.J., Kazman, R.: Research report: interacting with huge hierarchies: beyond cone trees. In: Proceedings of Information Visualization, pp. 74–81 (1995)

9. Devroye, L., Kruszewski, P.: The botanical beauty of random binary trees. In: Proceedings of Graph Drawing, pp. 166–177 (1995)
10. Eades, P.: Drawing free trees. Bull. Inst. Comb. Appl. **5**, 10–36 (1992)
11. Grivet, S., Auber, D., Domenger, J., Melançon, G.: Bubble tree drawing algorithm. In: Proceedings of International Conference on Computer Vision and Graphics, pp. 633–641 (2004)
12. Holton, M.: Strands, gravity, and botanical tree imaginery. Comput. Graph. Forum **13**(1), 57–67 (1994)
13. Jürgensmann, S., Schulz, H.J.: A visual survey of tree visualization. In: IEEE Visweek 2010 Posters (2010)
14. Kleiberg, E., van de Wetering, H., van Wijk, J.J.: Botanical visualization of huge hierarchies. In: Proceedings of Information Visualization, pp. 87–94 (2001)
15. Koike, H.: Generalized fractal views: a fractal-based method for controlling information display. ACM Trans. Inf. Syst. **13**(3), 305–324 (1995)
16. Koike, H., Yoshihara, H.: Fractal approaches for visualizing huge hierarchies. In: Proceedings of Visual Languages, pp. 55–60 (1993)
17. Kruskal, J., Landwehr, J.: Icicle plots: better displays for hierarchical clustering. Am. Stat. **37**(2), 162–168 (1983)
18. Lin, C.C., Yen, H.C.: On balloon drawings of rooted trees. Graph Algorithms Appl. **11**(2), 431–452 (2007)
19. Machado, P., Cardoso, A.: Computing aesthetics. In: de Oliveira, F.M. (ed.) SBIA 1998. LNCS (LNAI), vol. 1515, pp. 219–228. Springer, Heidelberg (1998)
20. Mandelbrot, B.: The Fractal Geometry of Nature. W.H. Freeman and Company, New York (1982)
21. McGuffin, M., Robert, J.: Quantifying the space-efficiency of 2D graphical representations of trees. Inf. Vis. **9**(2), 115–140 (2009)
22. Nocaj, A., Brandes, U.: Computing voronoi treemaps: faster, simpler, and resolution-independent. Comput. Graph. Forum **31**(3), 855–864 (2012)
23. Peitgen, H.O., Saupe, D. (eds.): Science of Fractal Images. Springer, New York (1988)
24. Reingold, E., Tilford, J.: Tidier drawings of trees. IEEE Trans. Softw. Eng. **7**, 223–228 (1981)
25. Rosindell, J., Harmon, L.: OneZoom: a fractal explorer for the tree of life. PLOS Biol. **10**(10), e1001406 (2012)
26. Schulz, H.J.: Treevis.net: a tree visualization reference. IEEE Comput. Graph. Appl. **31**(6), 11–15 (2011)
27. Shneiderman, B.: Tree visualization with tree-maps: 2-D space-filling approach. ACM Trans. Graphics **11**(1), 92–99 (1992)
28. Stasko, J.T., Zhang, E.: Focus+context display and navigation techniques for enhancing radial, space-filling hierarchy visualizations. In: Proceedings of the IEEE Symposium on Information Visualization, pp. 57–65 (2000)
29. Ware, C.: Information Visualization, Second Edition: Perception for Design (Interactive Technologies), 2nd edn. Morgan Kaufmann, Burlington (2004)
30. Wertheimer, M.: Untersuchungen zur Lehre von der Gestalt. II. Psychol. Res. **4**(1), 301–350 (1923)
31. Wetherell, C., Shannon, A.: Tidy drawings of trees. IEEE Trans. Soft. Eng. **5**(5), 514–520 (1979)
32. Wilson, E.O.: Biophilia. Harvard University Press, Cambridge (1984)
33. Yang, J., Ward, M.O., Rundensteiner, E.A., Patro, A.: InterRing: a visual interface for navigating and manipulating hierarchies. Inf. Vis. **2**(1), 16–30 (2003)

A Potential Field Function for Overlapping Point Set and Graph Cluster Visualization

Jevgēnijs Vihrovs[✉], Krišjānis Prūsis, Kārlis Freivalds, Pēteris Ručevskis, and Valdis Krebs

Institute of Mathematics and Computer Science, University of Latvia,
Raiņa bulvāris 29, Riga LV-1459, Latvia
jevgenijs.vihrovs@lumii.lv

Abstract. In this paper we address the problem of visualizing overlapping sets of points with a fixed positioning in a comprehensible way. A standard visualization technique is to enclose the point sets in isocontours generated by bounding a potential field function. The most commonly used functions are various approximations of the Gaussian distribution. Such an approach produces smooth and appealing shapes, however it may produce an incorrect point nesting in generated regions, e.g. some point is contained inside a foreign set region. We introduce a different potential field function that keeps the desired properties of Gaussian distribution, and in addition guarantees that every point belongs to all its sets' regions and no others, and that regions of two sets with no common points have no overlaps.

The presented function works well if the sets intersect each other, a situation that often arises in social network graphs, producing regions that reveal the structure of their clustering. It performs best when the graphs are positioned by force-directed layout algorithms. The function can also be used to depict hierarchical clustering of the graphs. We study the performance of the method on various real-world graph examples.

Keywords: Information visualization · Implicit surfaces

1 Introduction

Point sets emerge from the study of many real-world data structures, from social networks to geographical maps. The sets identify important groups of objects inside the structure. We focus on the visualization of graph clustering, for example, scientific article co-authorship [17] or social circles in organizations [13]. Overlaps often occur naturally in such graphs, with sets sharing common intermediary vertices. Such graphs are also often hierarchical. For example, human and biological networks rarely cluster in clean ways, and one vertex may belong to many groups.

Supported by ERAF project 2010/0318/2DP/2.1.1.1.0/10/APIA/VIAA/104.

S. Battiato et al. (Eds.): VISIGRAPP 2014, CCIS 550, pp. 136–152, 2015.
DOI: 10.1007/978-3-319-25117-2_9

After the identification of different point sets in the given data, there is still the problem of presenting this information in a quickly and easily comprehensible way. In order for the visualization of overlapping point sets to be effective, it should adhere to various criteria. Firstly, it should be unambiguous: the user should be able to identify the sets and their points, as well as their overlaps, without any misunderstandings. It should also represent the geometrical layout of the points themselves as closely as possible. In this paper we deal with visualizing point sets that have an arbitrary, but fixed positioning.

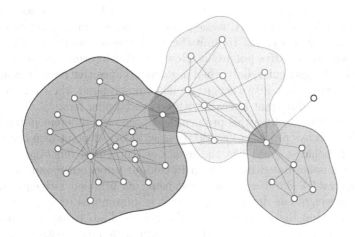

Fig. 1. An example of the proposed visualization. There are three overlapping point sets, which correspond to vertex clusters in the given graph. In this example edges do not affect the visualization.

The common approach in visualization is to enclose the points of each set in a region that represents this set, forming an Euler diagram. There are several works that list the desirable properties of visualizations that produce such regions [6,16]. Here we present a brief summary.

P1. Each point of a set is contained inside the region of this set.
P2. Each point not present in a set is not contained inside the region of this set.
P3. The regions of two different sets must have overlaps only where there is a point that belongs to both of the sets.
P4. The regions must have a strict boundary that separates the included and excluded space.
P5. The boundary of the region should stay close to the points of the set.
P6. The boundaries of the regions should be sufficiently smooth.
P7. The region of one set should be connected.

A widely used visualization method is to compute the regions by bounding a potential field function. The advantages are the smooth shape borders and the flexibility of the result, as the function parameters can be adjusted to obtain a better visualization. The main problems are ensuring the correct membership of the points (P1, P2) and region overlapping (P3), as well as maintaining each region connected (P7). Traditionally, most potential functions are based on the Gaussian distribution function [2,10,19] or quadratic polinomial functions [4,21]. These functions naturally satisfy properties P4–P6.

The main problem with such functions is maintaining the membership properties P1–P3. If each set region is calculated ignoring the influence of the foreign points, then P1 holds, but P2 and P3 are completely ignored. To mitigate this, the foreign points are assigned a negative potential, thus repelling the influence of the points of the given set. This improves P2 and P3 (though does not fully satisfy either), but violates P1, as negative potentials can overwhelm positives ones even directly at a point. To limit the effect this has on P1, smaller weights are often used for negative potentials. However, this weakens the improvements to P2 and P3. In particular, only when negative and positive potentials are equal does the following hold: if two sets share no common point, their regions do not overlap, which is important in cases with no overlapping sets.

In this work we address this problem and propose a different potential field function that always produces visualization with correct point membership, and visually still behaves much like the mentioned potential functions. It is very similar to Gaussian-based methods in isocontour appearance and satisfies the same properties P4–P6. In addition, when using our function with equal negative and positive weights as described earlier, it is guaranteed to satisfy properties P1, P2 concerning the membership of the points. Thus it also guarantees the property P3 for every two sets with no common points. We show the results of the visualization on different artificial and real-world examples and compare them to other potential functions. The function has also two parameters that allow to adjust the size and appearance of the produced regions.

Maintaining property P7 is a topological problem, and cannot be solved by a specific potential function design. Still, it can be ensured using standalone techniques such as including the edges of each set's spanning tree when computing the potential function [4], but usually at the cost of ignoring correct overlap properties P2 and P3. We show how this method works with the proposed function. However, as the proposed function is designed to improve properties P1–P3, this leads to unusual visual behavior. Without this property the region of each set can consist of multiple disconnected shapes. In our examples, we use color-coding to differentiate between set regions.

2   Related Work

There have been a substantial number of works that deal with overlapping point set visualization. The problem is highly relevant in visualizing graph vertex clusters in graph drawing, and many approaches are described in the context of graph clusters [1,8,11,20].

The main tasks pursued in most visualizations are maintaining connectivity of the regions of a set (P7) and point region membership (P1–P3). The connectivity is very important in the sense that it allows the user to quickly identify a particular set. The correctness of the sets' overlaps may be crucial in understanding the membership of the points and the relationships of the sets. Both these properties may be conflicting: if the positioning of the points is fixed, it is frequently not possible to ensure connectivity and eliminate all unnecessary region overlaps. It is also important whether the layout of the points can be changed or not; various algorithms of point positioning may greatly improve the readability of the visualization.

The most straightforward approach for obtaining the region of a set is to obtain its convex hull, used, for example, in [11,17]. Here, the boundaries of the convex hulls are smoothed by using Bézier curves. However, the convex hull approach ignores the P2 and P3 properties, and many points may happen to reside inside foreign set regions. This is addressed in [3], which describes visualization of UML diagrams. The areas where foreign elements overlap the convex hull are excluded from the region heuristically, and afterwards the resulting polygon is shrunk inwards to better represent the layout of the elements in the set. The borders of the regions are also softened which achieves a more natural look of the shapes.

A more sophisticated method that also computes discrete hulls is proposed in [18]. The intersection graph of the sets (which is planar) is built and positioned using a force-directed algorithm [5]. Then a polygonal skeleton is built around the graph vertices, and the points are inserted inside the corresponding polygons; these polygons are then expanded by applying the force-directed layout algorithm several more times. The resulting visualization has correct point membership, and each set region is connected, thus all the constraints P1–P3, and P7 are fully satisfied. Still, this visualization can be applied only when the positioning of the points is not important.

Another method is to use the Voronoi diagram of the given points to create the space partitioning for the visualization. A simple application of such approach for visualizing self-organizing maps can be found in [14]. The result fully ensures all the membership constraints P1–P3, and is very simple to implement. However, the regions have many sharp edges and all borders are shared (P6 does not hold). Another problem is the long and sharp region parts that do not contain any set vertices, which are common in Voronoi diagrams, as these parts distort the regions and lessen readability, violating P5. GMap [8] mitigates this by inserting many artificial points in the diagram around the areas of interest. A particular technique for smoothing the boundaries of the regions is described in [16], which also uses Voronoi diagrams. They create the region boundaries by using the Voronoi diagram to obtain a hull of the set and then drawing a distance field around this hull. This does not, however, work on boundaries between adjacent or intersecting regions. A problem with all Voronoi-based methods is that they do not ensure region connectivity.

The approach used in this work is to compute the regions for point sets by using a potential field function. The value of influence of a point in the given space is defined by the potential function, and the total influence of a point set in some location is the sum of influences for all the points in the set. Then the region for a set is obtained by thresholding the total influence of this set.

The properties of such visualization depend on the choice of the potential field function. Most standard potential functions are based on the Gaussian distribution. Its main advantage is the contour smoothness of the regions, they look similar to hand-drawn. One of the first analyses of such a potential function appears in [2]. It also describes how the "blobbiness" of the visualization can be adjusted using the parameters of the function.

Another type of potential functions are polynomial approximations of Gaussians that are faster to compute and produce nearly identical results, e.g. [1]. Also used are bounded quadratic functions [4,21], they generate similar results.

This approach has been widely applied in point set visualization. Some works visualize and cluster point sets in 3D [1,19]. The latter also focuses on hierarchical depiction of a clustered graph. To avoid incorrect overlaps of different set regions, one can assign negative potential to point of foreign sets, which is noted in [2], and used in [4,10,21].

Bubble Sets [4] directly focus on 2D point set visualization; while it also uses a potential field function and negative influences for foreign set points, it maintains P7 by using the edges of a spanning tree of each set in addition to the vertices when computing the potential functions. However, these edges from different sets can overlap, violating P3.

Other novel methods include Euler diagrams with connected regions [15], as well as Kelp diagrams [6], which focus primarily on region connectivity in visualizing points with fixed geometrical layout.

3   The Potential Function

We are given a set of points and the subsets we have to visualize, along with a geometrical positioning of the points in an Euclidean space. For clarity, we will call the given points the *vertices* (from the use cases of representing social graphs), to avoid confusion with spatial points. For each set, we independently compute a function which assigns a real number to each point denoting the set's influence on it. Vertices belonging to the set positively influence the value of the function, whereas other vertices influence it negatively. The amount a vertex influences a point is determined by their distance.

After calculating the function for a set, the shape for this set is extracted using a fixed threshold. We describe the potential function in Sect. 3.1. In Sect. 3.3, we discuss the specifics of the implementation.

3.1   Function Description

Suppose we are given N non-empty sets $S_1 \ldots S_N$ where $\forall i : S_i \subseteq V$, and V is the set of all the vertices. Some vertices can also belong to no set. These vertices in any case negatively influence the value of the function for all sets.

For the visualization, we need the vertices to be geometrically positioned. The positioning can be expressed by a function $P : V \to \mathbb{R}^n$ that assigns a position to each vertex in a n-dimensional Euclidean space. Clearly, in practice, the visualization can be meaningful to users when $n \leq 3$. For use case without a predefined layout, constructing the function P is itself a separate, well-studied problem [5]. In this work, we do not propose any new positioning methods, and the visualization itself does not require any special positioning. For clustered graphs, we will often use a positioning based on a force-directed layout algorithm.

The field function approach works as follows: first, define the *influence* of a vertex v with respect to a point p, which is a function $I(v, p)$. Then the *influence* of a vertex set S on a point p is defined as the sum of the influence of its vertices:

$$I(S, p) = \sum_{v \in S} I(v, p). \tag{1}$$

The potential field function for set S_i at a point p is defined as the difference between the influence of this set and the total influence of all other vertices:

$$F_i(p) = I(S_i, p) - w \cdot I(V \setminus S_i, p) \tag{2}$$

All the vertices that belong to the i-th set influence the value of the function positively, whilst all other vertices influence it negatively. The nonnegative parameter w is the weight of negative influence, which is usually less than 1 (i.e. the weight of the positive influence).

To obtain a shape that represents the i-th set, use a nonnegative threshold t. The region of the vertex set is defined to be the set of points such that for any point p in the set the inequality $F_i(p) > t$ holds. By adjusting the value of t the shapes can be made smaller or larger. In practice, it can be difficult to estimate the shape size from t, therefore we adjust the desired radius of the shape of a single vertex R_t and set t to be the influence of a single vertex at distance R_t.

Since the influence of the vertices belonging to S_i is taken with a positive sign and the influence of other vertices with a negative sign, vertices not belonging to the set repel its shape.

As the function $I(v, p)$ we use the following:

$$I(v, p) = \max(\|P(v) - p\|^{-b} - m, 0) \tag{3}$$

where b and m are nonnegative constants (parameters). We also use the weight $w = 1$ (we show later this contributes to an important property of the function). When $\|P(v) - p\| = 0$, influence $I(v, p)$ is defined as $\lim_{x \to 0} \frac{1}{x} = +\infty$.

There can also be a situation when two or more vertices have equal positions, where by (3) it is not clear what value should F_i take. For example, there can be one vertex v_1 from set i and two other v_2, v_3 from set j in the same point p. For all three vertices the influence in this point is equal to infinity, so F_i formally is undefined. In this case F_i should be equal to $-\infty$, since for all other points q the value of F_i is $-I(v_3, q) = I(v_1, q) - I(v_2, q) - I(v_3, q)$. So in such situations, if the number of set i vertices in point p is less than the number of foreign set vertices in this point, then $F_i(p)$ should be equal to $-\infty$, and $+\infty$ or 0 if it is greater or equal respectively. This may not be very important as it affects only individual points but should be noted during implementation.

Parameter b regulates the slope of the function. When b grows larger, the function decreases much faster when distance is greater than 1. Visually this adjusts how strongly the blobs of the vertices interact with each other. Figure 2 shows the graphs of the function with different b values ($m = 0$). The threshold b was chosen so that for each function the radius of a single blob would be constant. When m is small, the blobs of nearby vertices meld together. When b is large, they appear as overlapping circles.

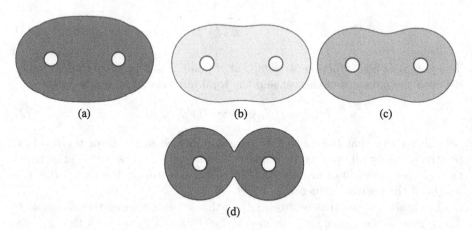

(a)   (b)   (c)

(d)

Fig. 2. Graphs of the influence function with different b values. This example shows how b affects blob interaction: **(a)** $b = 0.2$; **(b)** $b = 1$; **(c)** $b = 2$; **(d)** $b = 20$. The radius of a single blob R_t is the same in all examples.

The value m is subtracted to set function values to 0 starting from a certain distance. The purpose of this is to limit the maximum effect radius of a single vertex, preventing large concentrations of vertices from accumulating a large radius of the corresponding region with respect to other regions that contain less vertices. This is shown in Fig. 3: m normalizes the radiuses of the small and large set regions. Again, as with threshold, we regulate m with the use of the value R_m, which is defined as the distance at which the value of influence of a single vertex is equal to 0. It is clear that R_m should be greater than R_t; we mostly set the value of R_m to be a multiple of R_t.

3.2 Properties and Comparison

In this section we discuss how the presented function deals with properties P1–P3 and how it compares to already known functions. As influence function $I(v, p)$ we also consider the Gaussian distribution function $e^{-\frac{x^2}{2\sigma^2}}$ and quadratic polynomial function $(\max(r - x, 0))^2$, where $x = \|P(v) - p\|$. For Gaussian distribution function, standard deviation parameter σ also adjusts blobbiness. Overall the

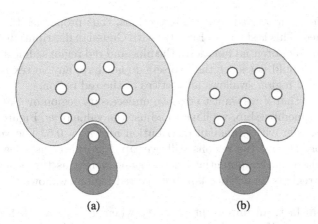

Fig. 3. The effect of parameter m on the consistency of region radiuses: **(a)** R_m is unbounded ($m = 0$), the larger set region has a much larger radius then the smaller set region; **(b)** $R_m = 2.5R_t$, the radiuses of the regions are almost the same.

proposed function keeps the desired properties of these functions, and in addition guarantees correct point region membership because the potential function diverges to plus infinity at $x = 0$.

A simple example in Fig. 4 shows that the proposed function maintains the visual appearance of Gaussian distribution and quadratic polynomial potential functions. Of course, each function has its parameters that adjust the blobbiness and size of the regions, so each can be used to produce different visualizations with the same example case. However, the figure shows that all three functions can be adjusted to produce similar-looking results.

Further we examine how these functions satisfy the membership properties P1–P3. In contrast to other functions, the proposed function guarantees properties P1, P2. It also ensures property P3 for sets with no common points.

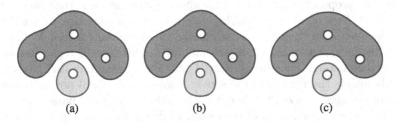

Fig. 4. A simple example of how the proposed function maintains the appearance of common potential functions: **(a)** Gaussian distribution; **(b)** quadratic polynomial; **(c)** proposed function.

The property P1 is easy to establish for any potential function by taking $w = 0$ in (2), i.e. ignoring foreign vertex influence for the current set. This introduces many unwanted region intersections, thus properties P2 and P3 are not maintained. This is shown in Fig. 5(a) with Gaussian distribution (of course it happens with any potential function): the blue and red regions almost completely overlap (they should not since the two sets do not have any vertex in common), besides the blue region swallows all vertices of the red region.

This means that if we want to remove unnecessary region overlaps or ensure correct vertex membership, we have to adjust the value of w. Figure 5(b) shows the same example with Gaussian distribution and $w = 0.5$. The vertex membership is correct, but the regions still overlap. Figure 5(c) uses Gaussian with $w = 1$, i.e. the negative potential has the same weight as the positive. In this case the two regions do not overlap, but the red region swallows the single blue vertex.

That is, in fact, an important property when $w = 1$, which holds for any potential function. If two sets have no vertices in common, then their regions do not overlap (this is actually a special case of the P3 property). The proof is trivial: suppose the total positive influences of the two sets at some spatial point are I_1 and I_2. Only one of the values $I_1 - I_2$ and $I_2 - I_1$ can be positive, thus only one can be greater than threshold t. Other set vertices do not change this property, since for each spatial point the influence from them is subtracted from both of the given set total influences.

This property is very important if we need to visualize sets with no intersections: then no two regions will have any overlaps. Therefore we would want to keep it while also ensuring correct vertex membership. We examine this property for a small example with a fixed R_t.

Using a Gaussian function with a fixed parameter σ, as it is shown in Fig. 5(a)–(c), all of the mentioned properties at once are not achievable. In the given example, we can change σ (from 2.5 to 3.5), and obtain a correct visualization, see Fig. 5(d); however, this is an unstable improvement and slightly adjusted vertex positions again produce an incorrect visualization even with the new σ, see Fig. 5(e). In addition, σ also regulates blobbiness and it is not desirable to adjust it in order to maintain correctness.

With the proposed function all of the above is guaranteed in the visualization, see Fig. 5(f). Firstly, we keep weight $w = 1$, so non-intersecting sets will have no overlaps between their regions. Secondly, the potential function is based on inverse distance and reaches positive infinity at distance 0, regardless of the function parameters. Thus properties P1 and P2 are ensured: each vertex will always be inside the regions of the sets it belongs to and will not be inside the regions of the sets it doesn't belong to. This allows using the parameters b and m to adjust the visual quality of the result, while the membership of the vertices will always be correct.

The only case when P1 and P2 do not hold is in the same case if several different set vertices are in the same position p. However, it is obviously impossible to satisfy these properties with any visualization in this case.

Using a quadratic polynomial function produces results similar to those of using a Gaussian function, see Fig. 6. Parameter r can be adjusted to regulate the

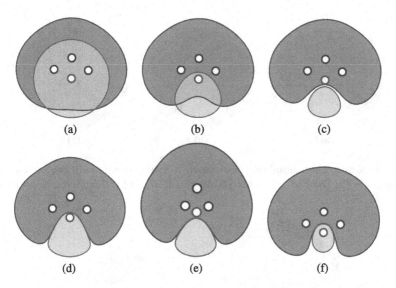

Fig. 5. Comparison with the Gaussian distribution function. The example shows two non-intersecting sets: 1 vertex (blue); 3 vertices (red). In this example the threshold radius R_t is the same for all six cases. (**a**) Gaussian, $\sigma = R_t/2.5$, $w = 0$. (**b**) Gaussian, $\sigma = R_t/2.5$, $w = 0.5$. (**c**) Gaussian, $\sigma = R_t/2.5$, $w = 1$. (**d**) Gaussian, $\sigma = R_t/3.5$, $w = 1$. (**e**) Different vertex positioning, Gaussian, $\sigma = R_t/3.5$, $w = 1$. (**f**) The proposed function, $b = 2$, $R_m = 2R_t$ (Color figure online).

blobbiness. The red region swallows the blue vertex in Fig. 6(a), when $r = 1.4R_t$. Again, it can be solved by adjusting r to $1.02R_t$ (see Fig. 6(b)), and again this solution is unstable and doesn't work when the vertex positions are changed slightly (see Fig. 6(c)).

The example uses the same R_t in all cases, but a vertex positioning producing similar results can be obtained for any R_t. In practice, changing the radius of the blobs is also unwanted if we need to obtain shapes of particular size. With the proposed function there is no need for fine-tuning to obtain a correct vertex membership visualization.

It should be noted that the situation with P3 is different if two sets do have some vertices in common. As it was shown that any vertex belonging to a set will reside inside the region of this set, a vertex belonging to two sets will also always reside in some intersection of their regions. However, among all overlaps of the regions, there can be some that contain no vertices from any of the given sets. These overlaps may be unwanted since they do not hold any semantic meaning of set relationship, see Fig. 7(a). Still, such overlaps can also contribute to the smooth region borders, see Fig. 7(b); without them, the image would not look natural. Generally, this problem has a topological nature and most probably cannot be solved solely through potential function design.

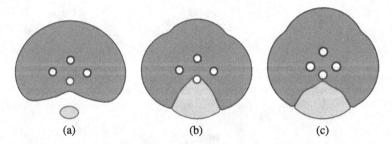

Fig. 6. Quadratic polynomial function $(\max(r-x,0))^2$ results with the same example and the same R_t: **(a)** $r = 1.4R_t$; textbf(b) $r = 1.02R_t$; **(c)** different vertex positioning, $r = 1.02R_t$.

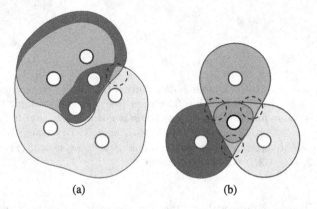

Fig. 7. Unnecessary region overlaps (with the proposed potential function): **(a)** the highlighted overlap is unwanted; **(b)** the highlighted overlaps are not necessary, but mandatory for smooth shape borders. Vertices that have black borders in each example belong to all sets.

3.3 Implementation

In this section we discuss the implementation specifics of the proposed potential function. We use a radial sweep algorithm to find the region boundaries for each set (marching squares could be used as well), which are simplified using the Douglas-Peucker algorithm [7]. For high-quality images (as in this paper) we also interpolate these polygons with splines [12].

One aspect of the implementation of our potential function is the possibility of infinity values. In fact, due to datatype restrictions (for example, floating point) such values could be achievable not only exactly at a vertex, but also within some ϵ-distance from the vertex. This problem can be solved by adjusting the used datatype for our function specifics:

- if the operation would cause the value of a variable to overflow, it should take the special ∞-value with an appropriate sign (the floating point standard guarantees this);

- if an ∞-value is added a value which is not an ∞-value with the opposite sign, the result should remain the same ∞-value (the floating point standard guarantees this);
- if an ∞-value is added to an ∞-value with the opposite sign, the result should be 0 (this is different from the floating point standard).

Still, in our implementation we used the standard floating point datatype, as it is fully sufficient in practice and produces no visible errors, because such ϵ-regions are negligible in size.

Another aspect of the implementation is choosing appropriate values for b, m and t. Recall the following auxiliary parameters we used in the previous section: R_t — the desired radius of the shape of a single vertex; R_m — the distance where the influence of a single vertex reaches 0. These parameters can be used in the implementation to adjust the look of the visualization. We advise to set R_m to be some multiple of R_t, for example, $R_m = 2R_t$. We use the following rules to compute the actual parameters for the function:

- $b = 2$, but adjust in real-time if needed.
- $m = R_m^{-b}$.
- $t = R_t^{-b} - m$, provided $R_m > R_t$.

The running time of the implementation is mainly dependent on the particulars of the radial sweep or marching squares algorithm used. The only time-consuming part connected with the function is the calculation of the influence on the given scalar field, which is essentially $O((\text{set count}) \cdot (\text{vertex count}) \cdot (\text{field size}))$. There are a few worthy optimizations to this:

- since $F_i(p) = I(S_i, p) - I(V \setminus S_i, p)$, first calculate only $I(S_i, p)$: if it is less than the used threshold t, there is no need to calculate $I(V \setminus S_i, p)$;
- if the distance $\|P(v) - p\|$ is greater than R_m, then the influence is 0 and there is no need to take v into account for point p;
- to avoid expensive exponentiation in $\|P(v) - p\|^{-b}$, the value of $b = 2$ can be used, leaving only the inverse of the easily computable squared distance between v and p.

4   Case Studies

In this section we demonstrate our visualization on several real-world examples, mostly clustered graphs. We compare it to the Gaussian distribution-based visualization. We show that in cases where the Gaussian distribution works well, so does the proposed function. In addition, we show that our function can be used to achieve a good result where that is not possible using the Gaussian distribution. We focus on the Gaussian distribution function, as quadratic polynomial potential functions produce essentially the same results.

The first example shows that the proposed function can be used to obtain similar results to those of the Gaussian distribution function, see Fig. 8. The latter works well in this example, and we have produced essentially the same

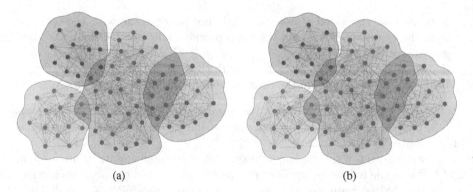

(a)   (b)

Fig. 8. This figure illustrates a visualization of a simple real-world overlapping set example: **(a)** Gaussian distribution; **(b)** the proposed function. There are no difficulties in ensuring correct point membership for both functions in this case. The example shows that our function is able to produce results that are very similar to the visualization with Gaussian distribution function. Both cases illustrate that this visualization method works well with graph clustering.

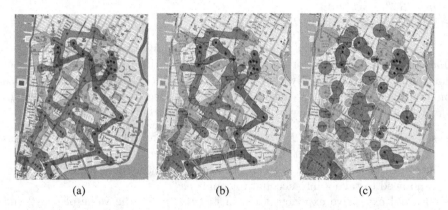

(a)   (b)   (c)

Fig. 9. Hotels (orange), subway stations (brown), and medical clinics (purple) in Manhattan: **(a)** Bubble Sets [4]; **(b)** our function with spanning tree edges added to the visualization; **(c)** our function without any modifications. This example shows how region connectivity can be ensured while using our function (Color figure online).

result using our function. The example itself demonstrates a small company with four work locations. The central red cluster corresponds to the company headquarters. Each vertex in the graph represents an employee, colored according to the location they work at. Graph edges denote frequent, work-related communications between employees. Cluster overlaps reveal which employees frequently interact with other locations. Besides the comparison of the functions, this example shows how this visualization can be used to depict graph overlapping clustering [13].

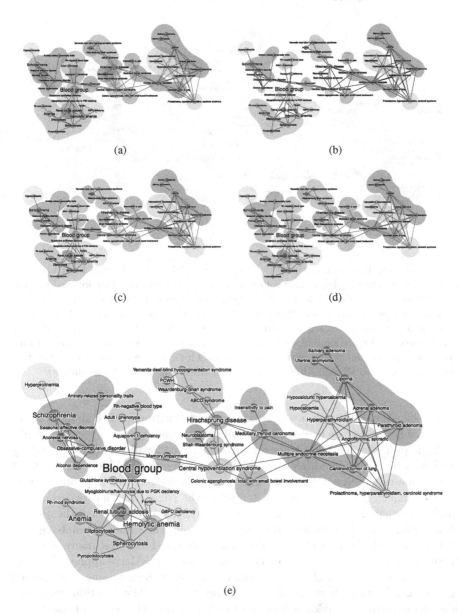

Fig. 10. A part of the human disease network [9]. Case **(a)** is recreated from the Diseasome poster and uses a Gaussian distribution function with no negative weights. In this case there are many unnecessary region overlaps. Cases **(b)** and **(c)** show the Gaussian distribution function with negative weights. In **(b)**, we tried to preserve the blobbiness of the shapes in **(a)**, however this results in some regions disappearing completely. In **(c)** this is remedied at the cost of the blobbiness of the regions. Cases **(d)** and **(e)** show the results of the proposed method. **(d)** closely resembles the results of the Gaussian function in (c). However, **(e)** combines the smooth regions of (a) with correct overlaps, a result not achievable using a Gaussian function.

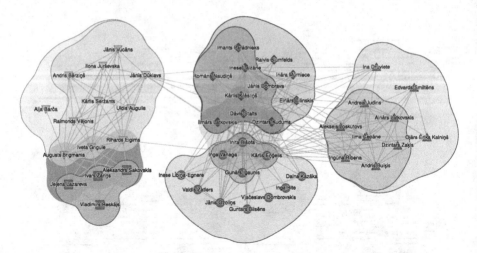

Fig. 11. Voting blocks of Saeima, the parliament of Latvia. The top level bordered clusters (as well as the shapes and colors of the vertices) correspond to political parties, with the two center parties forming the main coallition and the two left-most parties forming the opposition. The darker inner clusters denote the core blocks of each party with the most similar voting patterns (for Saskaņas Centrs, the lower left party, the whole party has highly similar voting patterns). The two borderless clusters (light blue and light green) denote cross-party voting blocks. This example shows that the function can be also used to visualize hierarchical graph clustering (Color figure online).

In the second example we compare our visualization with Bubble Sets [4], see Fig. 9 and illustrate how their method of ensuring region connectivity works with the proposed method. The example depicts the locations of hotels, subway stations and clinics in Manhattan. Case (a) shows the Bubble Sets result. It uses a quadratic polynomial function as the potential field function, and ensures region connectivity by assigning potential to the edges of a (not necessary the minimum) spanning tree of the set vertices in addition to the vertices themselves. When computing the influence of an edge on a point the standard point-segment distance is used. As a result the regions have unwanted overlaps, many of which, however, contribute to region connectivity. In case (b) we have applied the same spanning tree method with our function. In addition to maintaining connectivity, in our visualization there are also no region overlaps. The function properties lead to interesting behavior at edge intersection points, where each of the regions is connected by a single point. We leave it to the reader to decide whether such a result is aesthetically pleasing. In case (c) we apply our visualization without any modifications.

In the third example we visualize a part of the human disease network described in [9], see Fig. 10. In this graph, diseases are linked by common genetic associations, with the sets denoting various types of disease. This graph contains no set overlaps, so using equal weights for positive and negative influence ensures that there are no region overlaps. We show that our visualization works better than the Gaussian in this example. We were also interested in improving the visualization performed by Mathieu Bastian and Sébastien Heymann from Gephi.

Finally, we analyse voting blocks in Saeima (the parliament of Latvia), see Fig. 11. In this graph, each vertex corresponds to a member of Saeima, with edges denoting that those two members frequently vote the same on significant legislation. The clusters in this graph show both the polical parties and unofficial voting blocks. In this example, we can see that our method works well with hierarchically clustered graphs. It also allows densly but easily displaying multiple layers of information in a single image.

5   Conclusion

We presented a new potential field function for overlapping point set visualization in the form of Euler diagrams. In contrast to the most widely used potential functions, the proposed function ensures correct point membership in the set regions. Moreover, it retains all desired Gaussian-based potential field function properties, e.g., the set region shapes are smooth and visually pleasing. Set regions are easily identifiable and closely match the layout of the points. The smoothness and size of the regions can be also adjusted using the parameters of our function.

We have applied our function on different real-world examples and compared the result to the earlier methods. The proposed function is very effective in cases with no intersecting sets, since then the regions are guaranteed not to overlap. It also works well with overlapping sets, with regions creating an easily comprehensible Euler diagram, retaining correct point membership. We have demonstrated that our function works well in cases where it is not possible to obtain a good result using the Gaussian potential function. We have also illustrated how the overall approach can be successfully used to visualize overlapping graph clustering.

References

1. Balzer, M., Deussen, O.: Level-of-detail visualization of clustered graph layouts. In: 2007 6th International AsiaPacific Symposium on Visualization, pp. 133–140 (2007)
2. Blinn, J.F.: A generalization of algebraic surface drawing. ACM Trans. Graph. 1(3), 235–256 (1982)
3. Byelas, H., Telea, A.: Towards realism in drawing areas of interest on architecture diagrams. J. Vis. Lang. Comput. 20(2), 110–128 (2009)
4. Collins, C., Penn, G., Carpendale, S.: Bubble sets: revealing set relations with isocontours over existing visualizations. IEEE Trans. Vis. Comput. Graph. 15(6), 1009–1016 (2009)
5. Di Battista, G., Eades, P., Tamassia, R., Tollis, I.G.: Graph Drawing: Algorithms for the Visualization of Graphs. Prentice Hall PTR, Upper Saddle River (1998)

6. Dinkla, K., van Kreveld, M.J., Speckmann, B., Westenberg, M.A.: Kelp diagrams: point set membership visualization. Comput. Graph. Forum **31**(3pt1), 875–884 (2012)
7. Douglas, D.H., Peucker, T.K.: Algorithms for the reduction of the number of points required to represent a digitized line or its caricature. Cartographica Int. J. Geogr. Inf. Geovisualization **10**(2), 112–122 (1973)
8. Gansner, E.R., Hu, Y., Kobourov, S.: GMap: Visualizing graphs and clusters as maps. In: 2010 IEEE Pacific Visualization Symposium PacificVis, pp. 201–208 (2010)
9. Goh, K.-I., Cusick, M.E., Valle, D., Childs, B., Vidal, M., Barabsi, A.-L.: The human disease network. Proc. Natl. Acad. Sci. USA **104**(21), 8685–8690 (2007)
10. Gross, M.H., Sprenger, T.C., Finger, J.: Visualizing information on a sphere. In: Proceedings of the 1997 Conference on Information Visualization, pp. 11–16 (1997)
11. Heer, J., Boyd, D.: Vizster: visualizing online social networks. In: IEEE Symposium on Information Visualization 2005 INFOVIS 2005, vol. 5, pp. 32–39 (2003)
12. Hobby, J.D.: Smooth, easy to compute interpolating splines. Discrete Comput. Geom. **1**(2), 123–140 (1986)
13. Krebs, V.: Managing the 21st century organization. IHRIM J. **XI**(4), 2–8 (2007)
14. Matsumoto, Y., Umano, M., Inuiguchi, M.: Visualization with Voronoi tessellation and moving output units in self-organizing map of the real-number system. Neural Netw. **1**, 3428–3434 (2008)
15. Riche, N.H., Dwyer, T.: Untangling Euler diagrams. IEEE Trans. Vis. Comput. Graph. **16**(6), 1090–1099 (2010)
16. Rosenthal, P., Linsen, L.: Enclosing surfaces for point clusters using 3D discrete Voronoi diagrams. Comput. Graph. Forum **28**(3), 999–1006 (2009)
17. Santamaría, R., Therón, R.: Overlapping clustered graphs: co-authorship networks visualization. In: Butz, A., Fisher, B., Krüger, A., Olivier, P., Christie, M. (eds.) SG 2008. LNCS, vol. 5166, pp. 190–199. Springer, Heidelberg (2008)
18. Simonetto, P., Auber, D., Archambault, D.: Fully automatic visualisation of over-lapping sets. Comput. Graph. Forum **28**(3), 967–974 (2009)
19. Sprenger, T.C., Brunella, R., Gross, M.H.: H-BLOB: a hierarchical visual clustering method using implicit surfaces. In: Visualization 2000. Proceedings, pp. 61–68, October 2000
20. Van Ham, F., Van Wijk, J.J.: Interactive visualization of small world graphs. In: IEEE Symposium on Information Visualization, pp. 199–206 (2004)
21. Watanabe, N., Washida, M., Igarashi, T.: Bubble clusters: an interface for manipu-lating spatial aggregation of graphical objects. In: Proceedings of the 20th Annual ACM Symposium on User Interface Software and Technology, UIST 2007, pp. 173–182. ACM, New York (2007)

Designing Close and Distant Reading Visualizations for Text Re-use

Stefan Jänicke[1]([✉]), Thomas Efer[2], Marco Büchler[3], and Gerik Scheuermann[1]

[1] Image and Signal Processing Group, Leipzig University, Leipzig, Germany
{stjaenicke,scheuermann}@informatik.uni-leipzig.de
[2] Natural Language Processing Group, Leipzig University, Leipzig, Germany
efer@informatik.uni-leipzig.de
[3] Göttingen Centre for Digital Humanities,
University of Göttingen, Göttingen, Germany
mbuechler@gcdh.de

Abstract. We present various visualizations for the Text Re-use found among texts of a collection to support answering a broad palette of research questions in the humanities. When juxtaposing all texts of a corpus in form of tuples, we propose the *Text Re-use Grid* as a distant reading method that emphasizes text tuples with systematic or repetitive Text Re-use. The *Text Re-use Browser* provides a closer look on the Text Re-use between the two texts of a tuple. Additionally, we present *Text Re-use Alignment Visualizations* to improve the readability of Text Variant Graphs that are used to compare various text editions to each other. Finally, we illustrate the benefit of the proposed visualizations with four usage scenarios for various topics in literary criticism.

Keywords: Text Re-use · Text visualization · Text variant graph · Literary criticism · Digital humanities

1 Introduction

The conscientious analyzation and interpretation of small text passages, so called *Close Reading*, is a major technique for researches in literary criticism. But the digital age with algorithms that automatically retrieve vast amounts of data expedite *Distant Reading* methods [17] that give the observer an impression about the data distribution. The Information Seeking Mantra "Overview first, zoom and filter, details-on-demand" [22] is accomplished, when distant reading views are interactively used to switch to close reading views. The general task is to provide a visualization that shows an overview of the data, so that patterns potentially interesting for the observer are salient. A drill down on these patterns for further exploration is the bridge between distant and close reading.

In this paper, we want to outline the process of designing close and distant reading visualizations for the Text Re-use determined between all texts of a given collection. *Text Re-use* is defined as the oral or the written reproduction

© Springer International Publishing Switzerland 2015
S. Battiato et al. (Eds.): VISIGRAPP 2014, CCIS 550, pp. 153–171, 2015.
DOI: 10.1007/978-3-319-25117-2_10

of textual content and is roughly divided into two categories [3]. On the one hand, a text passage is re-used deliberately, like direct quotes and phrases like winged words and wisdom sayings. Translations of a text into other languages also count to this category and are called interlingual Text Re-use. A very popular form of deliberate Text Re-use is plagiarism, which has gained major attention in the recent years, mainly driven by plagiarism allegations in politics. On the other hand, a Text Re-use may be unintended, like boilerplates, e-mail headers or the repetition of news agency texts when writing daily newspapers [6]. Further examples are idioms, battle cries and so called multi word units.

The analysis of Text Re-use among historic texts with the goal to explore known and discover unforeseen relationships in cultural heritage has become an important task within various Digital Humanities projects. Thereby, the humanities scholars are interested which texts share patterns of consecutive similar units (systematic Text Re-use) and how specific phrases are used (repetitive Text Re-use). Furthermore, the analysis of these similar text units regarding structure, context and used expressions is of special interest. This is also a substantial task in literary criticism, when various editions of a text are cautiously compared to each other.

This paper shows, how intuitive, interactive visualizations help humanities scholars in understanding and interpreting the Text Re-use occurrences. In particular, we present the following visualizations that support close and distant reading for Text Re-use:

- **Text Re-use Grid:** a chart that juxtaposes all texts of a collection in relation to amount and type of Text Re-use,
- **Text Re-use Browser:** a user interface consistent of an interactive Dot Plot View and a Text Re-use Reader that allows for the inspection and browsing through all Text Re-uses between two texts,
- **Text Re-use Alignment Visualization:** a visualization for aligned text units that improves the readability for *Text Variant Graphs*.

2   Related Work

The discovery of relationships between different texts and the alignment and visualization of the findings has been a challenging task in various works. Xanadu, founded in 1960, can be seen as one of the pioneer projects that attend to this matter [19]. The current prototype shows the dedicated text in the center of the screen, related texts are positioned on both sides and shared patterns are aligned and highlighted using various colors. John et al. [15] propose a focus and context approach for the visualization of texts sharing similar patterns. A vertical ribbon for each text shows the distribution of these patterns, and interactively, the user can drill down to regions of interest. Cheesman offers a visualization for the alignment of multilingual text passages in Shakespeare's Othello, where the user can interactively browse through the texts of two editions [5]. Likewise, related text entities are visually linked to each other. To illustrate the computationally determined Text Re-uses in ancient Greek texts, Büchler suggests a graph to show the results for a certain author by number, citing authors, years of citing authors and passages of the book [4]. Additionally, the user can inspect individual text snippets with highlighted re-used passages. For plotting Text Re-use between Bible books [16], Lee uses a static Dot Plot View, which was originally designed for bioinformatics to compare two genome sequences to each other [11]. A single dot marks a correlation between the genomes and multiple dots form patterns that indicate similar genomic segments. Lee utilizes this approach to highlight patterns of systematic Text Re-use. Various visualization methods also exist to highlight plagiarized passages of a given source text [12,20]. A complete overview of the whole text is given and each page, chapter or plagiarized text passage receives its own block. Coloring is used to show the amount of re-used text or to indicate potential sources.

All the above visualization techniques focus on displaying relationships between a limited number of texts. Mostly, a certain source text is given and its correlations to other texts can be analyzed. A comparative overview between all texts of an arbitrary text collection is not provided. For this purpose, we propose the *Text Re-use Grid* as a distant reading solution for Text Re-use (Sect. 4). Moreover, we present the *Text Re-use Browser* that supports close reading of the Text Re-use between two selected texts (Sect. 5).

Text Variant Graphs are data structures that represent various editions of a text [21]. CollateX is one of the standard tools in the Digital Humanities that computes a static directed acyclic graph with vertices showing the various text fragments and edges labeled with edition identifiers connecting subsequent text fragments [8]. The plain design makes it hard for the user to follow how an edition disseminates in the graph. Furthermore, the vertices do not reflect the amount of occurrences and synonyms are not properly aligned to each other. Although extensions for user-driven annotation and modification exist [1], there are only few works that attend to the matter of designing Text Variant Graphs. A visualization, which allows for weighted nodes is the *Word Tree* [25]. It cannot be directly applied to Text Variant Graphs, since it only aligns shared beginnings of sentences in form of a tree. Each variation splits a node of the tree into several leaves. The font size of a node label reflects the number of occurrences. A plain solution to align Text Re-use is given in [4]. The original text snippet is drawn as a main branch and variations of Text Re-use candidates are sub-branches with a certain color. This approach works fine for small examples with minor variations, but it fails for major differences, especially, when multiple Text Re-uses share the same sub-branches. A similar visualization for the uncertainty in lattice graphs also supports various sub-branches [7]. But merging of multiple nodes of the same kind is not provided, although the metaphor for uncertainty could be used for this purpose.

In Sect. 6, we propose the *Text Re-use Alignment Visualization* that utilizes some of the presented ideas with the goal to design a well readable layout for Text Variant Graphs.

3 Theoretical Basis of Text Re-use

Let $A_1, \ldots, A_n$ denote a corpus of n texts. After splitting each text into units (e.g. sentences), the Text Re-uses between all text unit tuples are determined. Each detected Text Re-use $\{a_i, b_j\}$ consists of two corresponding Text Re-use units a_i (e.g. i-th sentence of text A) and b_j (e.g. j-th sentence of text B). The *Scoring value* $t(a_i, b_j)$ defines a weight for $\{a_i, b_j\}$ dependent on the text unit lengths of a_i and b_j and their *Re-use Overlap*, which is the proportion of matching to non-matching tokens. t is ranged in the interval [0,1]; 0 means no similarity between the two units, 1 means that a_i and b_j are equal. The complete Text Re-use result list contains only relevant Text Re-uses above a certain threshold for t. A more detailed description of the underlying algorithms for Text Re-use detection and the computation of t can be found in Büchler's dissertation [3].

Researchers working with Text Re-use have various research questions that require a definition of the following Text Re-use types:

Systematic Text Re-use. The consecutive occurence of the same pattern of Text Re-use is of particular interest for researchers when comparing different texts to each other. Such type of Text Re-use could be an indication for plagiarism. For instance, the pattern $\{a_i, b_j\}, \{a_{i+1}, b_{j+1}\}, \{a_{i+2}, b_{j+2}\}$ is a *Systematic Text Re-use* of three consecutive phrases.

Repetitive Text Re-use. This type of Text Re-use appears, when the researcher is interested in analyzing a phrase that is frequently used in a certain text. The goals in this use case are to explore the contexts, in which a phrase appears as well as to what extent a specific phrase is spread in the text. *Repetitive Text Re-use* for a phrase a exists for a set of Text Re-use pairs in the form $\{a, b_1\}, \{a, b_2\}, \{a, b_3\}, \ldots$.

Isolated Text Re-use. We classify a Text Re-use $\{a_i, b_j\}$ as isolated if it does not occur within a certain pattern, more precisely, if it is neither systematic nor repetitive. As systematic Text Re-use doesn't necessarily need to be consecutive in both texts due to potential insertions, deletions or changes in the ordering of the textual entities, we need to discriminate isolated from systematic Text Re-use. We define $\{a_i, b_j\}$ as isolated if there is no Text Re-use $\{a_u, b_v\}$ within a certain neighborhood ε, so that:

$$\varepsilon = \sqrt{\frac{|i - u| + |j - v|}{2}} < 10$$

Empirically, we determined 10 as the best value to separate systematic ($\varepsilon <$ 10) from isolated Text Re-use ($\varepsilon \geq 10$).

4 Text Re-Use Grid

The intention of this visualization is to give the researcher an overview of the Text Re-use distribution among all texts of a corpus. We transform the result of the Text Re-use detection algorithm into an intuitive, readable visual interface that immediately (1) reflects the amount of Text Re-uses between each pair of texts, and (2) provides evidence for the type of Text Re-use. For this purpose, we define three parameters:

1. **Text Re-use Amount** σ. σ is the number of Text Re-uses detected between two texts A and B.
2. **Systematic Text Re-use Index** λ. λ is an assessment for structures of systematic Text Re-use between two texts A and B with an ordered list of text units, so that $A = \{a_{first}, \ldots, a_i, \ldots, a_{last}\}$ and $B = \{b_{first}, \ldots, b_j, \ldots, b_{last}\}$. To detect these structures, we preliminary filter the Text Re-use results by removing all repetitive and isolated Text Re-uses. This filter process results in a decomposition of the remaining n Text Re-uses into m clusters $C = \{c_1, \ldots, c_h, \ldots, c_m\}$ containing more than one Text Re-use $\{a_i, b_j\}$ each. For each of these clusters c_h with $|c_h|$ Text Re-uses in total, we compute a correlation coefficient $\rho(c_h)$ as

$$\rho(c_h) = \frac{\sum\limits_{\{a_i,b_j\} \in c_h} (i - \bar{i}_h)(j - \bar{j}_h)}{\sqrt{\sum\limits_{\{a_i,b_j\} \in c_h} (i - \bar{i}_h)^2 \sum\limits_{\{a_i,b_j\} \in c_h} (j - \bar{j}_h)^2}}$$

$$\text{with } \bar{i}_h = \sum\limits_{\{a_i,b_j\} \in c_h} \frac{i}{|c_h|} \text{ and } \bar{j}_h = \sum\limits_{\{a_i,b_j\} \in c_h} \frac{j}{|c_h|}$$

to estimate the strength of the linear relationship between the Text Re-uses in c_h. Finally, the Systematic Text Re-use Index is defined as:

$$\lambda = \sum_{h=0}^{m} \frac{|c_h|}{n} \rho(c_h)$$

λ ranges in the interval $[0,1]$, whereas high values indicate that patterns of systematic Text Re-uses are contained.
3. **Repetitive Text Re-use Index** ω. ω is a measure for the amount of repetitive Text Re-use. Let N denote the number of Text Re-uses found between two texts A and B. To define ω, we remove each Text Re-use $\{a_i, b_j\}$, if both text units a_i and b_j occur only once within all Text Re-uses. Finally, we define ω in the interval $[0,1]$ with the remaining n Text Re-uses as

$$\omega = \frac{n}{N}$$

Grid Visualization. For the visual mapping, we construct a grid with each cell representing the Text Re-uses found between two texts of a corpus. For each cell, we compute σ, λ and ω for the corresponding two texts. The cells are displayed in form of rectangles with bounds proportional to the lengths of the corresponding texts. Interactively, the user can change the display to equal-sized squares, so that even cells representing short texts are properly visible.

Because of the importance for the researchers to detect and analyze texts with extensive systematic or repetitive Text Re-use, we use a specific coloring for the grid cells, so that the type of Text Re-use (represented by λ and ω) and the amount of Text Re-use (σ) can be easily recognized. As the human's ability to discriminate colors is limited, we chose a class based approach to compute a limited number of cell colors. As proposed by Slocum et al., we use an optimal classification method [23] to group the cells into two sets of classes in dependency of σ, λ and ω. With the Jenks-Caspall-Algorithm [14] using reiterative cycling, we compute a configurable number of classes. We receive n classes $\alpha_1, \ldots, \alpha_n$ for the type of Text Re-use (systematic or repetitive), so that α_1 contains the cells with the smallest λ (or ω) and α_n contains the cells with the largest λ (or ω). Furthermore, we compute m classes $\beta_1, \ldots, \beta_m$ for the amount of Text Re-use with β_1 containing the cells with smallest σ and β_m containing the cells with the largest σ. We determine the color for a grid cell in the HSV color space dependent on these classes as follows:

$$H = 240 + \frac{i-1}{n-1} \cdot 120 \qquad S = \frac{j}{m} \cdot 100 \qquad V = 100$$

The type of Text Re-use defines the hue on the "Cold-Hot" color scale Diehl proposes [9] for the *EpoSee* tool from blue (cold) to red (hot). Thus, we receive cold hues for cell colors with less, and hot hues for cell colors with a lot of systematic (or repetitive) Text Re-use between the corresponding texts. The amount of Text Re-use defines the saturation, so that cells with a high number of Text Re-use receive highly saturated, and cells with less Text Re-uses lightly saturated colors. For the examples in this paper we used $n = m = 4$.

In Fig. 1, the resultant Text Re-use Grids for the Bible books of the American Standard Version compared to each other highlighting systematic (Fig. 1(a)) and repetitive Text Re-use (Fig. 1(b)) can be seen. With the help of a legend, the user is able to immediately categorize type and amount of Text Re-use between two Bible books. Interactively, the user can change from highlighting systematic to highlighting repetitive Text Re-use. By mouse clicking onto a cell, the user has the ability to switch from the distant reading grid view to the closer Text Re-use Browser view that is explained in the next section.

5   Text Re-use Browser

Whereas, the Text Re-use Grid allows for distant reading of all Text Re-uses occurring within a text collection, the Text Re-use Browser provides a closer look at the Text Re-uses found between two texts $A = \{a_{first}, \ldots, a_i, \ldots, a_{last}\}$ and $B = \{b_{first}, \ldots, b_j, \ldots, b_{last}\}$. This still complies to the characteristics of distant

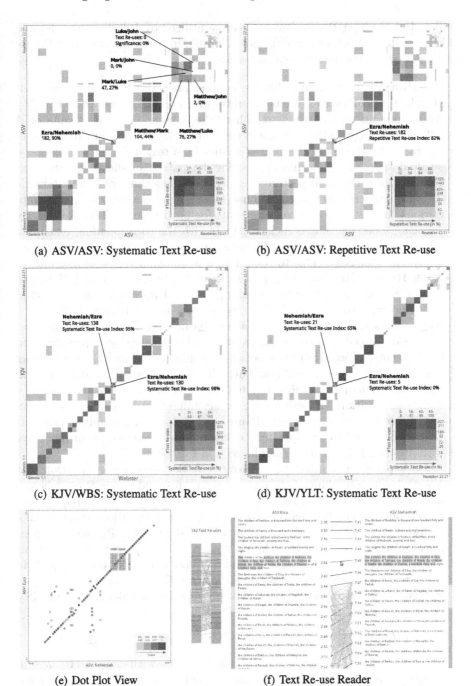

(a) ASV/ASV: Systematic Text Re-use

(b) ASV/ASV: Repetitive Text Re-use

(c) KJV/WBS: Systematic Text Re-use

(d) KJV/YLT: Systematic Text Re-use

(e) Dot Plot View

(f) Text Re-use Reader

Fig. 1. Text Re-use Grids for the juxtaposition of various Bible editions (a–d) and panels of the Text Re-use Browser showing systematic and repetitive Text Re-use patterns detected between the books *Ezra* and *Nehemiah* (e–f).

reading, but the Text Re-use Browser can be utilized to drill down to a limited set of Text Re-uses, which supports close reading. In particular, the Text Re-use Browser provides two panels for this purpose:

1. **Dot Plot View.** We also utilize the approach of a Dot Plot View to emphasize the types of Text Re-use between the given texts. In contrast to Lee [16], we provide an interactive chart, where the number $|A|$ of text units of A defines the range for the x-axis, and the number $|B|$ of text units of B defines the range for the y-axis. Each Text Re-use for a text unit pair is drawn as a single dot. As in bioinformatics, specific dot patterns indicate specific Text Re-use types. Diagonal patterns highlight sections that contain systematic Text Re-use, whereas vertical and horizontal patterns appear for phrase repetitions. In Fig. 1(e) we detect patterns for both types of Text Re-use. By selecting a dot via mouse click, a popup with the corresponding text units and a Text Re-use Alignment Visualization (see Sect. 7) is shown. Interactively, the user is also able to zoom into a rectangular region of interest.

2. **Text Re-use Reader.** This panel allows for browsing A and B in two opposite windows. Whenever a re-used text unit appears in the viewport of one window, a connection to the opposite text unit is drawn. A click on a connection scrolls both texts, so that the text units of the corresponding Text Re-use are placed on the same horizontal level and a step-by-step exploration of consecutive Text Re-use is possible. A mouseover highlights matching tokens in both units. An additional overview for the texts gives an impression about all occurring Text Re-uses, and can be utilized to directly jump to a dedicated position. In both views, an accumulation of parallel lines is an indication for systematic Text Re-use, and *hubs* (a single unit of one text that is connected to multiple units of the opposite text) occur for repetitive Text Re-use. These features can be seen in Fig. 1(f).

Both panels are linked to each other. A dot selection in the Dot Plot View triggers a scrolling of the texts to the corresponding positions, whereas a connection selection in the Text Re-use Reader opens the popup for the corresponding dot. For coloring the Text Re-use glyphs (dots, connections), we use again a class based approach. We group the Text Re-uses in dependency of their scoring value t into p classes $\gamma_1, \ldots, \gamma_p$, so that γ_1 contains Text Re-uses with the smallest t, and γ_p these ones with the largest t. In order to avoid misinterpretations, we chose a different color scheme in comparison to the Text Re-use Grid. The glyph colors are defined in the HSV color space as:

$$H = 60 + \frac{k-1}{p-1} \cdot 60 \qquad S = 100 \qquad V = 100 - \frac{k-1}{p-1} \cdot 50$$

Thus, the hue of a glyph color for a Text Re-use with class γ_k ($1 \leq k \leq p$) ranges from yellow to green. To gain visually distinctive colors, the color value ranges between 100 and 50. For the examples in this paper we used $p = 4$.

Some text juxtapositions contain lots of Text Re-uses that form various patterns. To visually filter for specific Text Re-uses, we allow to hide glyphs

King James Version (KJV)	In the beginning God created the heaven and the earth.
American Standard Version (ASV)	In the beginning God created the heavens and the earth.
Bible in Basic English (BBE)	At the first God made the heaven and the earth.
Young's Literal Translation (YLT)	In the beginning of God's preparing the heavens and the earth –
Smith's Literal Translation (SLT)	In the beginning God formed the heavens and the earth.

(a) Five English translations of *Genesis 1:1*

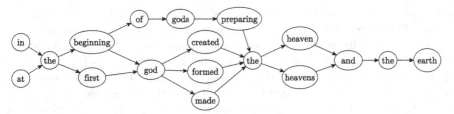

(b) Resultant Text Variant Graph

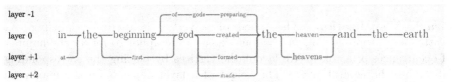

(c) Visualization using the KJV as main branch (layer 0)

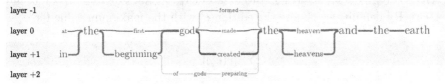

(d) Visualization using the BBE as main branch (layer 0)

Fig. 2. Text Re-use Alignment Visualizations for five editions of *Genesis 1:1*.

of repetitive or systematic Text Re-use. Additionally, Text Re-uses with low scoring values can be hidden and a slider can be used to hide isolated Text Re-uses without adjacent Text Re-uses in a certain neighborhood.

6   Text Re-use Alignment Visualization

One of the substantial tasks in the field of textual criticism is called collation, which is the cautious comparison of various editions of a text. Since it is an extremely laborious approach for humanities scholars to do this manually, few projects investigate methods that compute alignments for the digitized texts and visualize the resulting directed acyclic Text Variant Graph in a simple manner (see Sect. 2). In this section, we present an intuitive, for the humanities scholars easily readable and comprehensible layout for Text Variant Graphs.

A text edition is a specific kind of Text Re-use as it was derived from a specific source text, but also occurrences of repetitive Text Re-use can be aligned and visualized with this approach. Let $\{e^1, \ldots, e^n\}$ denote a set of editions for a re-used text unit. Various alignment algorithms exist (e.g. [2,18]), we use a brute force approach that performs well for small text units. After tokenization and normalization, we insert each edition $e^i = e^i_1 \ldots e^i_{|e^i|}$ in form of a directed path $v(e^i_1) \ldots v(e^i_{|e^i|})$ with vertices representing tokens in the initial Text Variant Graph. Then, we iteratively merge vertices of different paths with equal tokens and choose this alignment that reaches a maximum number of merge iterations while keeping the Text Variant Graph acyclic. Each vertex $v = \{e^i_s, e^j_t, e^k_u, \ldots\}$ of the graph is an alignment of the tokens $\{e^i_s, e^j_t, e^k_u, \ldots\}$. The *token degree* $|v|$ is the number of tokens assigned to v and $v(e^i_s)$ is the corresponding vertex for the s-th token of edition i. Figure 2(b) shows such a graph for five editions of the first Bible verse (Fig. 2(a)).

Graph Visualization. The token of a vertex is used for labeling. As Wattenberg proposes [25], we use font size as a metaphor to reflect the number of occurrences of a token. We layout the vertices by placing the corresponding labels onto horizontal layers. The height of a layer depends on the maximum height of the labels placed on it. We start by placing the labels for the vertices $v(e^i_1), \ldots, v(e^i_{|e^i|})$ of an arbitrary edition e^i in left-to-right order on layer 0 (main branch). By default, we choose the edition e^i with the maximum value for

$$\sum_{s=1}^{|e^i|} |v(e^i_s)|$$

which means e^i has lots of tokens assigned to vertices with large token degrees. Afterwards, we iteratively determine the subpaths of the edition with most vertices already assigned to layers. Each subpath $\{v_1, \ldots, v_n\}$ has assigned layers for v_1 and v_n and the layer for the vertices of $p = \{v_2, \ldots, v_{n-1}\}$ needs to be determined. Let i denote the layer of v_1 and j the layer of v_n. We aim to place p as close as possible to its adjacent vertices v_1 and v_n. Starting with layer $k = \lfloor (i+j)/2 \rfloor$, we iteratively search for a layer with enough free space for the labels of the vertices of p in the order $k, k+1, k-1, k+2, k-2$, and so on. If the total width of the labels of p is larger than the space between v_1 and v_n, we stretch the distance between v_1 and v_n. After the proper layer is found, we move all vertices of the graph horizontally, so that (1) the labels do not overlap each other, (2) a minimum space of configurable width between all adjacent vertices is given, and (3) each vertex is placed in the barycenter of its neighbors. We perform this process for all subpaths of all editions to complete the layout for the Text Variant Graph. We draw undirected edges (for the user the direction is obvious) between adjacent vertices of the same layer in form of horizontal lines. To ensure a good readability of the graph, we use horizontal and vertical links connected with bends for edges between vertices of different layers. More details about the edge routing algorithm can be found in [13]. The edges of the

Text Variant Graph are drawn in gray, and to identify different edition flows, we use colors of the 12-color palette for categorial usage suggested by Ware [24] to facilitate maximal visual differentiation by the user.

We provide several means of interaction for the analysis of the resultant alignment visualization. The user is able to decide between various methods to display the edges of the Text Variant Graph. This includes an edge overview like shown for the resultant Text Re-use Alignment Visualization for the five editions of the first Bible verse in Fig. 2(c), and a display of all edges drawn in the corresponding edition colors (e.g. Fig. 6). Furthermore, thick grey majority edges that are passed by a minimum number of editions can be drawn (Fig. 2(d)), so that only varying paths are highlighted. Hovering a token removes all edges from editions not passing it and selecting a token via mouse click shows the information about its corresponding editions in a popup window (Fig. 6(b)).

7   Usage Scenarios

The Text Re-use Visualizations presented in this paper are utilized in various Digital Humanities projects. In this section, we take a look at four usage scenarios from these projects.

7.1   Various English Translations of the Bible

The interdisciplinary Digital Humanities project *eTRACES*[1] wants to discover, analyze and evaluate intertextual similarities in form of Text Re-use among historical texts of a given corpus. Since the Bible is known as one of the most often read and studied books, and therefore, potential findings are easily evaluable, it was chosen as a proof of concept for the project. The text corpus contains 23 different English translations of the Bible covering a time period from the 14th (Wycliffe Bible) to the 21th century (World English Bible). Since each translation was driven by a specific motivation, the involved humanities scholars had a great variety of research questions to be answered. In this section, we present some of their findings.

Of particular interest for the humanities scholars was the comparison of Bible books of the same edition regarding systematic Text Re-use. The Text Re-use Grid shows for the three evangelists *Matthew*, *Mark* and *Luke* strong interdependencies, whilst *John* has few or no Text Re-use at all with those three – confirming a well known fact by visualizing it. These interdependencies were detected for the ASV (Fig. 1(b)) and for various other editions the visualization showed a similar pattern. The visualization reveals further insights by highlighting other cells of the grid. For example, there is an indication for vast systematic Text Re-use between the books *Ezra* and *Nehemiah*. Also, there is evidence for repetitive Text Re-use given (Fig. 1(b)). Picking the corresponding cell in the Text Re-use Browser allows for close reading and reveals a rectangular cluster of repeatedly

---

[1] http://etraces.e-humanities.net/.

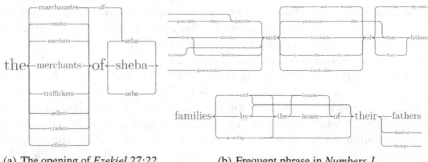

(a) The opening of *Ezekiel 27:22*.   (b) Frequent phrase in *Numbers 1*.

Fig. 3. Text Re-use Visualizations show the variability of English Bible translations.

used phrases to be compared using the Text Re-use Alignment Visualization, and a large systematic Text Re-use pattern between the beginning of *Ezra* and the middle section of *Nehemiah* (*Ezra 2:1/Nehemiah 7:6 - Ezra 2:70/Nehemiah 7:73*). When juxtaposing the KJV and its revision by Webster (Fig. 1(c)) the systematic Text Re-use pattern for *Ezra* and *Nehemiah* is still highlighted. For the juxtaposition of the KJV and YLT, which uses Hebrew syntax, the overall number of Text Re-uses strongly decreases and a systematic Text Re-use pattern for *KJV:Ezra* and *YLT:Nehemiah* is not detected (Fig. 1(d)). Interestingly, for the juxtaposition of *KJV:Nehemiah* and *YLT:Ezra* a small part of the systematic Text Re-use pattern is still preserved. Those are results causing the user to analyze Text Re-use further and to gain knowledge that wasn't expected or even looked for.

The Text Re-use Alignment Visualization turned out to be very useful for philological matters since syntactic similarities and differences between repetitive Text Re-use occurrences and verses of different editions can be analyzed easily. Variations are easy to detect, for example many synonyms for "merchants" as seen in Fig. 3(a) for the opening of *Ezekiel 27:22*. Most of the early Bible versions translated into Middle English used the token "marchauntes" except the Wycliffe Bible, which used "silleris". Most often, the early Middle English translations vary strongly from modern English translations which approves a separate analysis. For a repeatedly used phrase in *Numbers 1*, we detect a stronger variance for the 6 early translations compared to the 17 modern translations. The number of uses in different translations of the Bible implies that the long, possibly more precise and most often used phrase "families by the house of their fathers" (8 times) in the modern translations could be the most literal translation of the original text, an impression that can now be researched and verified or falsified (Fig. 3(b)).

The humanities scholars also stated that the Text Re-use visualizations can help to determine, whether English versions of the Bible that claim to translate the Hebrew and ancient Greek original very literally, do this in a similar way or not and which one could be considered the most literal one.

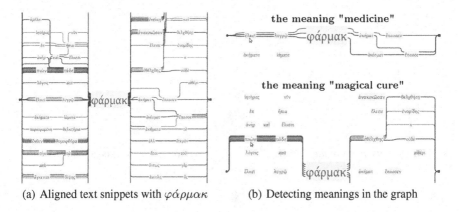

(a) Aligned text snippets with φάρμακ   (b) Detecting meanings in the graph

Fig. 4. Utilizing Text Re-use Alignments to discover concepts in ancient Greek texts.

7.2   The Meaning of Terms in Ancient Greek Texts

The purpose of the Digital Humanities project *eXChange*[2] is to explore the various meanings of a term, how these meanings changed over the years and how they were transferred in the past. Based upon ancient Greek texts, the meanings of a term are described by a set of various words.

One of the examples interesting for the collaborating classical philologists is the various meanings of φάρμακον. Utilizing the Text Re-use Alignment Visualization, text snippets containing the truncated form φάρμακ can be analyzed (Fig. 4(a)). Immediately, the known two meanings of φάρμακ to be a "medicine" or a "magical cure" can be perceived. Interactively, one is able to highlight the corresponding branches in the visualization (Fig. 4(b)). The first meaning can be identified by the token ἕλκει (wound) and various variants of ἀκήματα (cure). The token ἐθέλχθης (cast a spell) clusters text snippets for the second meaning. An important role for the given meanings plays the administration of a drug. If a drug is applied (ἔπασσε), it is a medicine, whereas if it is drunk (πίων), it is a magical cure. In contrast to the traditional methods of analyzing the contexts of the given terms' occurrences, the visualization facilitates a rapid comprehension of its opposing meanings by clustering related tokens that define the meaning of a term.

7.3   Text Re-use in Historic Arabic Texts

To explore the Text Re-use among historic arabic texts for the first time digitally, historians from the Aga Khan University utilized Büchler's algorithm for detecting Text Re-use and the Text Re-use Browser for visualizing and exploring the results. Predominantly, the focus was on the analysis of systematic Text Re-use. On the one hand, known facts were confirmed to assess the reliability

---

[2] http://exchange-projekt.de/.

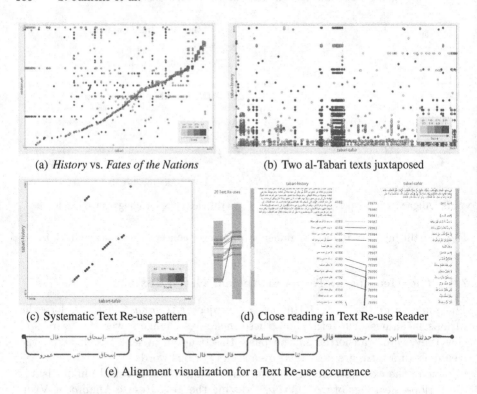

(a) *History* vs. *Fates of the Nations*   (b) Two al-Tabari texts juxtaposed

(c) Systematic Text Re-use pattern   (d) Close reading in Text Re-use Reader

(e) Alignment visualization for a Text Re-use occurrence

Fig. 5. Text Re-use Browser to discover systematic Text Re-use in Arabic texts.

of the visualization, on the other hand, unexpected and unknown patterns were analyzed further.

Figure 5(a) shows evidence for systematic Text Re-use in form of a diagonal pattern between two chronologies: *History*, called *Ta'rīkh al-rusul wa-l-mulūk* by al-Tabari (839–923), and *Fates of the Nations*, called *Tajārib al-umam* by Miskawayh (932–1030). Modern historians have often argued that Miskawayh relied heavily on al-Tabari's text. The analysis with the visualization suggests a more complex picture, namely, that Miskawayh more selectively copied al-Tabari's text. With the interaction capabilities of the Text Re-use Browser, the historians discovered that Miskawayh copied al-Tabari's text directly for the Umayyad (661–750) and Abbasid (750–1517) periods but copied very little of it for the period up to 651, which includes Iran's pre-Islamic history and the history of the early Muslim community. It seems possible that Miskawayh wanted a fresh reading. To examine this judgment, the historians plan a further Text Re-use analysis, including a comparison of Miskawayh's text against a larger pool of digitzed Arabic texts.

Another research question tries to discover what conclusions can be drawn from common passages in a single author's works. In Fig. 5(b), the Text Re-use between al-Tabari's *History* and his *Commentary on the Qur'an*, called

Jāmiʿ al-bayān ʿan taʾwīl āy al-Qurʾān, is shown. After removing vast occurrences of repetitive Text Re-use, especially stock phrases, the remaining systematic Text Re-use patterns can be analyzed. An example is given in Fig. 5(c). The pattern begins with the statement *"According to what someone with knowledge claimed ..."* (Fig. 5(d)). Read on its own, one might think this was al-Tabari's introduction to a topic or report. This might be the case, with al-Tabari repeating himself. It seems at least as likely, however, that this small bit of introduction derives from an original source, which al-Tabari (or perhaps a member of his editorial workshop) copied into both of his texts. Detecting chunks like this across his text, and comparing them to textual units in other classical Arabic texts, might give a sense of the size of units that passed through the tradition.

For both the examples, the Text Re-use Visualization Alignment was an effective method to investigate syntactic differences of a detected Text Re-use. An example, taken from the latter use case, is given in Fig. 5(e).

7.4   The Comparison of German Shakespeare Translations

In the occasion of the 450th anniversary of William Shakespeare's birth and the 300th anniversary of the German Shakespeare Society[3], researchers from the Max Planck Institute for Mathematics in the Sciences, Leipzig and the Natural Language Processing Group of the Leipzig University developed novel language analysis methods and applications on Shakespearean works [10]. The contribution described rather formal, block entropy based measures for language complexity, but also had a focus on making structural aspects of the dramatic progression in Shakespeare's plays and the used language visible to the user in an interactive environment. As part of the latter, the Text Re-use Aligment Visualization was used to point out the differences and similarities of several German translations. Figure 6(a) shows the parallel text of over 20 different German versions of an Othello scene. Even this small segment shows the variety of visible aspects of the language. Regarding word position and sequence, the so called syntagmatic level bears fixed and grammar-related expressions, such as singular "mein brief" (my letter) versus plural "meine briefe" (my letters) or active constructions "nennt mir" (tell me) versus passive "mir wird gemeldet" (is reported to me), which can be tracked visually, mainly along the horizontal axis. The vertical axis yields word substitutions and chosen alternatives, the so called paradigmatic level, for example the nouns "brief" (letter) and "schreiben" (writing). The visualization lets the user estimate the relative popularity of sequences and alternatives and the longer-ranging effects of the choice of words. As a very interesting find, the original "hundred and seven galleys" are quite often changed to 106 in translations. Most probably, this is done for its shorter syllable count and the resulting softer pronunciation with "ga-lee-ren". However, other (as valid) variations (like 108) were never chosen, hinting at a high level of influence within the concerned translations.

---

[3] Deutsche Shakespeare-Gesellschaft, http://shakespeare-gesellschaft.de/.

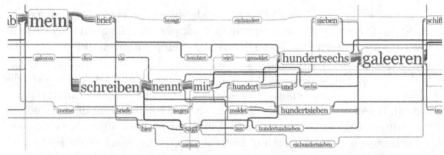

(a) Part of Shakespeare's *Othello, Act 1, Scene 3*

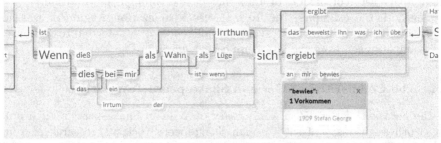

(b) Part of Shakespeare's *Sonnett 116*

Fig. 6. Text Re-use Alignments for various German translations of Shakespeare texts.

The reception of this intuitive form of text variant representation was very positive. This lead to Text Re-use Aligment Visualizations being included as a central part in a follow-up cooperation with the Department of British Culture at the University of Bamberg focusing on Shakespeare's Sonnets. To compare works of poetry, it has proven useful to add a special line break marker that helps segmenting the verses visually and that also provides "incentives" for the alignment algorithm, to adhere to the verse structure, as can be seen in Fig. 6(b). The interactive features of the Text Re-use Alignment Visualization were reported as very useful for quick inquiries into the data. The hover-based edge filtering clears the visual representation to show only the set of relevant parallel versions while retaining the broader context on all other used tokens. The click-induced popup box provides the details for all crossing versions unambiguously. Both interactions are of great use in scenarios showing ten, twenty and more editions, which are common in the translation analysis of very popular authors, like Shakespeare.

8   Conclusion and Future Work

To support humanities scholars in exploring and analyzing the Text Re-use among historic text collections, we designed various close and distant reading visualizations. The *Text Re-use Grid* is a novel distant reading approach to discover type and amount of Text Re-use between each pair of texts of a given text corpus. At the researcher's convenience, one is able to highlight either grid cells with systematic or repetitive Text Re-use. The *Text Re-use Browser* facilitates a further exploration of such Text Re-use patterns between two texts and allows for close reading of individual text passages. This bridge between both perspectives, which fulfills Shneiderman's Information Seeking Mantra, turned out to be an important aspect for the collaborating humanities scholars. The *Text Re-use Alignment Visualization*, a further close reading visualization, allows for exploring similarities and differences between various editions of a given text. Thereby, we focused on improving the design and the readability for so called Text Variant Graphs. Instead of vertices, we place the vertices' labels with variable font size that reflect the number of occurences on horizontal layers. We attached great importance to the vertical alignment of variations to allow an easy detection of synonyms.

During the development phase, the collaborating humanities scholars steadily evaluated the design of the Text Re-use visualizations. We wanted to ensure creating intuitive and flexible interfaces to be able to help answering a broad palette of research questions. Four usage scenarios for various English Bible editions, the meaning of terms in ancient Greek texts, the Text Re-use among historic Arabic texts and the comparison of German Shakespeare translations confirm the benefit of this iterative process and the adaptability of the visualizations independent on the language of the texts in the given corpus.

In the future, we will direct our attention on the development of distant reading visualizations for Text Re-use Alignments. This would allow for comparing the various editions of a text on a different level and for detecting global patterns among those editions.

Acknowledgements. The authors like to thank Sarah Bowen (Aga Khan University), who utilized the presented Text Re-use Visualizations for historic Arabic texts, Eva Wöckener-Gade (Leipzig University), who worked with the Text Re-use Alignment Visualization to analyze the various meanings of ancient Greek terms and Annette Geßner (Göttingen Centre for Digital Humanities) for the collaboration when designing the Text Re-use Visualizations for English Bible translations. This research was funded by the German Federal Ministry of Education and Research.

References

1. Andrews, T.L., Macé, C.: Beyond the tree of texts: Building an empirical model of scribal variation through graph analysis of texts and stemmata. Literary and Linguistic Computing (2013)
2. Bourdaillet, J., Ganascia, J.G.: Practical block sequence alignment with moves. In: Loos, R., Fazekas, S. Z., Martn-Vide, C. (eds.) LATA. vol. Report 35/07, pp. 199–210. Research Group on Mathematical Linguistics, Universitat Rovira i Virgili, Tarragona (2007)
3. Büchler, M.: Informationstechnische Aspekte des Historical Text Re-use (2013)
4. Büchler, M., Geßner, A., Eckart, T., Heyer, G.: Unsupervised detection and visualisation of textual reuse on ancient Greek texts. J. Chicago Colloquium Digit. Humanit. Comput. Sci. 1(2) (2010). https://letterpress.uchicago.edu/index.php/jdhcs/article/view/60/71

5. Cheesman, T., Flanagan, K., Rybicki, J., Thiel, S.: Six maps of translations of Shakespeare. In: Wiggin, B., Macleod, C., DiMassa, D., Theis, N. (eds.) Un/Translatables: New Maps for Germanic Literatures. Northwestern University Press, Evanston (2014)
6. Clough, P., Gaizauskas, R., Piao, S.S.L., Wilks, Y.: METER: MEasuring TExt Reuse. In: Proceedings of the 40th Annual Meeting on Association for Computational Linguistics, pp. 152–159, ACL 2002. Association for Computational Linguistics, Stroudsburg, PA, USA (2002)
7. Collins, C., Carpendale, S., Penn, G.: Visualization of uncertainty in lattices to support decision-making. In: Proceedings of the 9th Joint Eurographics/IEEE VGTC Conference on Visualization, pp. 51–58, EUROVIS 2007. Eurographics Association, Aire-la-Ville, Switzerland (2007)
8. Dekker, R. H., Middell, G.: Computer-Supported Collation with CollateX: Managing Textual Variance in an Environment with Varying Requirements. Supporting Digital Humanities (2011)
9. Diehl, S.: Software Visualization: Visualizing the Structure, Behaviour, and Evolution of Software. Springer, Secaucus (2007)
10. Efer, T., Heyer, G., Jost, J.: Text Mining am Beispiel der Dramen Shakespeares. In: Jansohn, C., (ed.) Proceedings of the Symposium "Shakespeare unter den Deutschen" (2014)
11. Gibbs, A.J., McIntyre, G.A.: The diagram, a method for comparing sequences. Its use with amino acid and nucleotide sequences. Eur. J. Biochem. **16**(1), 1–11 (1970)
12. GuttenPlag: GuttenPlag Wiki Visualizations (2013). http://de.guttenplag.wikia.com/wiki/Visualisierungen. Accessed 10 June 2013
13. Jänicke, S., Büchler, M., Scheuermann, G.: Improving the layout for text variant graphs. In: VisLR: Visualization as Added Value in the Development, Use and Evaluation of Language Resources, pp. 41–48 (2014)
14. Jenks, G.F., Caspall, F.C.: Error on choroplethic maps: definition, measurement, reduction. Ann. Assoc. Am. Geogr. **61**(2), 217–244 (1971)
15. John, M., Heimerl, F., Müller, A., Koch, S.: A visual focus+context approach for text comparison tasks. In: VisLR: Visualization as Added Value in the Development, Use and Evaluation of Language Resources, pp. 29–32 (2014)
16. Lee, J.: A computational model of text reuse in ancient literary texts. In: Association for Computational Linguistics, Proceedings of the 45th Annual Meeting of the Association of Computational Linguistics, pp. 472–479 (2007)
17. Moretti, F.: Distant Reading. Verso (2013)
18. Needleman, S.B., Wunsch, C.D.: A general method applicable to the search for similarities in the amino acid sequence of two proteins. J. Mol. Biol. **48**(3), 443–453 (1970)
19. Nelson, T.H.: Xanalogical structure, needed now more than ever: parallel documents, deep links to content, deep versioning, and deep re-use. ACM Comput. Surv. (CSUR) **31**(4es), 33 (1999)
20. Ribler, R.L., Abrams, M.: Using visualization to detect plagiarism in computer science classes. In: Proceedings of the IEEE Symposium on Information Visualization, pp. 173–178, INFOVIS 2000. IEEE Computer Society, Washington, DC, USA (2000)
21. Schmidt, D., Colomb, R.: A data structure for representing multi-version texts online. Int. J. Hum.-Comput. Stud. **67**(6), 497–514 (2009)
22. Shneiderman, B.: The eyes have it: a task by data type taxonomy for information visualizations. In: Visual Languages, Proceedings, pp. 336–343 (1996)

23. Slocum, T.A., McMaster, R.B., Kessler, F.C., Howard, H.H.: Thematic Cartography and Geovisualization. Prentice Hall Series in Geographic Information Science, 3, international edn. Prentice Hall, Englewood Cliffs (2009)
24. Ware, C.: Information Visualization: Perception for Design. Morgan Kaufmann Publishers Inc., San Francisco (2004)
25. Wattenberg, M., Viégas, F.B.: The word tree, an interactive visual concordance. IEEE Trans. Vis. Comput. Graph. **14**(6), 1221–1228 (2008)

Computer Vision Theory and Applications

Age Estimation Using 3D Shape of the Face

Baiqiang Xia[⊠], Boulbaba Ben Amor, and Mohamed Daoudi

Télécom Lille/LIFL (UMR CNRS 8022), Cité Scientifique, Villeneuve D'ascq, France
xia.baigiang@telecom-Lilie.fr
http://www-rech.telecom-lille1.eu/miire/

Abstract. The 3D shape of human faces deform with time and contain rich aging information. However, in the literature, no work has been done with the 3D shape of face for age estimation. Thus, we propose in this paper to explore the 3D facial surface in age estimation. Based on Riemannian shape analysis of facial curves, we extract four types of *Dense Scalar Field* (DSF) descriptions from the 3D facial surface, which reflect the face *Averageness*, face *Symmetry*, and the *Spatial* and *Gradient* deviations of the face. Experiments are carried out following the Leave-One-Person-Out (LOPO) cross-validation on the earliest 466 scans in FRGCv2 dataset, using the Random Forest Regressor. With the DSF features, the proposed approach achieves 3.29 years Mean Absolute Error (MAE) in the gender-general experiments, and 3.15 years MAE in the gender-specific experiments. Results confirm the idea that the face aging differs with gender. To address the high dimensionality of DSF features and the imbalance in training instances, we propose to use a *weighted PCA* method. In both the gender-general and gender-specific experiments, the age estimation performances using *weighted PCA* are comparable to the performances using the DSF descriptions. While, the size of feature is significantly smaller with the *weighted PCA*.

Keywords: Age estimation · 3D face · Dense scalar field · Principal component analysis · Random forest regression

1 Introduction

Face age estimation performs important social roles in human-to-human communication. Studies in cognitive psychology, presented as a review by [4], have discovered that human beings develop the ability of face age estimation naturally in early life, and can be fairly accurate in deciding the age or age group with a given face. These studies, based on subjective age estimation given to face image from human participants, have also found that multiple cues contribute to age estimation, including the holistic face features (like the outline of the face, face shape and texture, etc.), local face features (like the eyes, nose, the forehead, etc.) and their configuration (like the bilateral symmetry of the face [20]). Whereas, claims has also been given that individuals are not sufficiently reliable to make fine-grained age distinctions, and individuals age estimation suffers from the subjective individual factors and contextual social factors.

© Springer International Publishing Switzerland 2015
S. Battiato et al. (Eds.): VISIGRAPP 2014, CCIS 550, pp. 175–190, 2015.
DOI: 10.1007/978-3-319-25117-2_11

The aging process is a cumulative, uncontrollable and personalized slow process, influenced by intrinsic factors like the gene and gender, and extrinsic factors like lifestyle, expression, environment and sociality [3,6]. The appearance and anatomy of human faces changes remarkably with the progress of aging [12]. The general pattern of the aging process differs in faces of different person (personalized or identity-specific), in faces of different age (age-specific), in faces of different gender (gender-specific), and in different facial components [3–5,16,17]. Typically, the craniofacial growth (bone movement and growth) takes place during childhood, and stops around the age of 20, which leads to the re-sizing and re-distribution of facial regions, such as the forehead, eyes, nose, cheeks, lips, and the chin. From adulthood to old age, face changes mainly in the skin, such as the color changes (usually darker and with more color changes) and the texture changes (appearance of wrinkles). The shape changes of faces continues from adulthood to old age. With the droops and sags of facial muscle and skin, the faces are tend to be more a shape of trapezoid or rectangle in old faces, while the typical adult faces are more of a U-shaped or upside-down-triangle [4].

Automatic face age estimation is to label a face image with the exact age or age group objectively by machine. With the rapid advances in computer vision and machine learning, recently, automatic face age estimation have become particularly prevalent because of its explosive emerging and promising real-world applications, such as electronic customer relationship management, age-specific human-computer-interaction, age-specific access control and surveillance, law enforcement (e.g., detecting child-pornography, forensic), biometrics (e.g., age-invariant person identification [16]), entertainment (e.g., cartoon film production, automatic album management), and cosmetology. Compared with human age estimation, automatic age estimation yields better performance as demonstrated in [6]. The performance of age estimation is typically measured by the mean absolute error (MAE) and the cumulative score (CS). The MAE is defined as the average of the absolute errors between the estimated age and the ground truth age, while the CS, proposed firstly by [13] in age estimation, shows the percentage of cases among the test set where the absolute age estimation error is less than a threshold.

As pointed in [4,14], the earliest age estimation works used the mathematical cardioidal strain model, derived from face anthropometry that measures directly the sizes and proportions in human face, to describe the craniofacial growth. These approaches are useful for young ages, but not appropriate for adults. After this, abundant works exploiting 2D images have been published in the literature with more complex approaches. Different with the comprehensive surveys given by [4,14], which categorized the literature concerning different aging modeling techniques, we represented the literature with the different ideas underlying these technical solutions. Based on the previous statements, we describe the face appearance as a function of multiple factors, including the age, the intrinsic factors (permanent factors like gene, gender, ethnicity, identity, etc.), and the extrinsic factors (temporary factors like lifestyle, health, sociality, expression, pose, illumination, etc.).

A. General aging patterns in face appearance. Essentially, face age estimation is to estimate the age of a subject by the aging patterns shown visually in the appearance. To analyze the appearance given in the face image is the basic ways to estimate the age. In the literature of age estimation, works were carried out with several different perceptions of the general aging patterns in face appearance. As aging exhibits similar patterns among different person, several approaches have been designed to learn the general public-level aging patterns in face appearance for age estimation. The most representative ones are the Active-Appearance-Model (AAM) based approaches, the manifold embedding approaches, and the Biologically-Inspired-Feature (BIF) based approaches. The common idea underlying these approaches is to project a face (linearly or non-linearly) into a subspace, to have a low dimensional representation. Respectively, *(i)* [10,12] use an Active Appearance Model (AAM) based scheme for projecting face images linearly into a low dimensional space. The AAM was initially proposed by [21], in which each face is represented by its shape and texture deviations to the mean face with a set of model parameters. Age estimation results with a quadratic regressor showed that the generic aging patterns work well for age estimation. Moreover,[10] illustrated that different face parameters obtained from training are responsible for different changes in lighting, pose, expression, and individual appearance. Considering that these parameters work well for age estimation, we can conclude that these face co-variants are influential in age estimation. *(ii)* The goal of manifold embedding approaches is to embed the original high dimensional face data in a lower-dimensional subspace by linear or non-linear projection, and take the embedding parameters as face representation. In the work of [17,18], the authors extracted age related features from 2D images with a linear manifold embedding method, named Orthogonal Locality Preserving Projections (OLPP). [26] learned age manifold with both local preserving requirements and ordinal requirements to enhance age estimation performance [25] projected each face as a point on the Grassmann Manifold with the standard SVD method, then the tangent vector on these points of the manifold were taken as features for age estimation. *(iii)* Inspired by a feed-forward path theory in cortex for visual processing, [5] introduced the biologically inspired features (BIF) for face age estimation. After filtering a image with a Gabor filter and a standard deviation based filter consecutively, the obtained features are processed with PCA to generate lower-dimension BIF features. The results demonstrated the effectiveness and robustness of bio-inspired features in encoding the generic aging patterns. Beyond the public-level aging patterns, there could be some less generic aging patterns when dealing with a subset of faces, such as a group of faces with high similarity, or a temporal sequence of face images for the same person. Based on the observation that similar faces tend to age similarly, [10,12] presented an appearance-specific strategy for age estimation. Faces are firstly clustered into groups considering their inter similarity, then training is performed on each group separately to learn a set of appearance-specific age estimators. Given a previously unseen face, the first step is to assign it to the most appropriate group, then the corresponding age estimator makes the age estimation. Experimental results showed that the group-level aging patterns are more accurate in age estimation compared with the generic-aging patterns. In case there is no similar enough face image for a testing face image in the database, [12] presented a weighted-appearance-specific which also yield fine performance. As different individual ages differently, [13,19] proposed the Aging-Pattern-Subspace (AGES), which studies the individual-level aging patterns from a temporal sequence of images of an individual ordered by time. For a test face, the aging pattern and the age is determined by the projection in the subspace that has the least reconstruction error. Experiments confirm that individual aging patterns contributes to age estimation. As different face components age differently, the component-level aging patterns are studied for age estimation. [24] represented faces with a hierarchical And-Or Graph. Face aging is then modeled as a Markov process on the graphs and the learned parameters of the model are used for age estimation. They found that the forehead and eye regions are the most informative for age estimation, which is also supported by discoveries of [6] using the BIF features.

B. *Considering the intrinsic/extrinsic factors in facial aging.* As stated at the beginning of this introduction, the appearance of face is influenced by intrinsic factors like the gene, gender, and extrinsic factors like lifestyle, expressions, environment and sociality [3,6]. Several studies have given consideration of the influences of these factors in age estimation with enhanced age estimation performance reported. Specifically, thinking that faces age differently in different age, age-specific approaches are adopted by [10], where age estimation is obtained by using a global age classifier first, then adjusted the estimated age by a local classifier which operates within a specific age range. Similarly, [17,18] proposed a Locally Adjusted Robust Regressor (LARR) for age estimation, which begins with a SVR-based global age regression, then followed by a local SVM-based classification that adjusts the age estimation in a local age range. All of these age-specific approaches have achieved better performance compared with their corresponding approaches without local adjustment. Considering that different gender ages differently with age [14,18], [11,14,18,22] carried out age estimation on male and female groups separately. Considering the individual lifestyle, [12] encoded this information together with facial appearance in age estimation, and demonstrated that the importance of lifestyle in determining the most appropriate aging function of a new individual. [11] gave weights to different lighting conditions for illumination-robust face age estimation. [26] gave consideration of the feature redundancy and used feature selection to enhance age estimation.

As stated before, in the childhood, face deformation mainly takes the form of craniofacial growth with facial features re-sized and re-distributed. From adulthood to old age, with the droops and sags of facial muscle and skin, the old faces usually deform to a trapezoid or rectangle shape from a typically

U-shaped or upside-down-triangle in adult face [4]. Another significant shape deformation is the introduction of facial wrinkles with aging. While, given the fact that face shape deforms significantly with age in three dimensions, and given the robustness of 3D face scans to illumination and poses compared with 2D face images, all the previous works in the literature used 2D face datasets for age estimation, no work has been done concerning the 3D face. Thus, in this work, we introduce the investigation of age estimation with 3D face scans. The rest of the paper is organized as follows: in Sect. 2, we present an overview of our methodology and summarize the contributions; in Sect. 3, we explain our methodology of features extraction from the 3D faces based on Riemannian shap analysis; in Sect. 4, we detail the regression strategy for age estimation using Random Forest; experimental results and discussions are presented in Sects. 5 and 6. Section 7 makes the conclusion.

2   Methodology and Contribution

From the analysis above, it emerges that most of the existing works study age estimation with aging patterns chosen at a specified level and some aging factors enrolled for enhancement. As far as we concern, all these works are based on 2D images, no work concerning 3D face scans has been attached to age estimation. Thus, we introduce in the present work a new study of 3D-base face age estimation to the domain. In our approach, we consider the public-level aging patterns and gender factor for age estimation. First, we extract four types of Dense Scalar Field (DSF) features from each pre-processed face, namely the Average DSF, the Symmetry DSF, the Spatial DSF and the Gradient DSF. These DSFs are derived from different face perception ideas and their computation is grounding on Riemannian shape analysis of facial curves. Then we perform age estimation using Random Forest Regression on each type of DSFs with two protocols: one experiment on DSFs of the whole dataset directly and the other experiment on male and female DSFs separately. We have also designed a simple result-level fusion with different type of the DSFs, to see if the performance improves with all these face perception ideas combined. In summary, the main contributions of this work are as follows. First, as far as we know, this is the first work in 3D-based age estimation. Although 3D face growth has been notice for a long time [27,28], no work has been reported to 3D face age estimation. Secondly, in this work, we introduce four different perspectives of faces perception for face representation. With the Dense Scalar Field features, we have obtained significant accuracy with each of the perspectives, compared with typical 2D-based age estimation performance. Thirdly, we have enhanced the age estimation performance by experimenting on the scans of each gender separately, which confirms that different gender ages differently. Forthly, we have proposed a *weighted PCA* method for addressing the high dimensionality of the original DSF features, and the imbalance of age distribution in training instances. With the *weighted PCA*, we have achieved comparable results than with the original DSF features, but using significantly less features.

3   Feature Extraction

As mentioned earlier, we adopt the Dense Scalar Field features in our approach. Based on pair-wise shape comparison of curves, the Dense Scalar Field (DSF) grounding on Riemannian shape analysis [9] is capable for capturing the local shape deformation between corresponding feature points. Formally, for any curve in the space of $\mathbb{R}^3$, $\beta: I \to \mathbb{R}^3$, where $I = [0, 1]$, it is first represented mathematically by the *square-root velocity function* $q(t)$, according to: $q(t) = \frac{\dot{\beta}(t)}{\sqrt{\|\dot{\beta}(t)\|}}$ [9]. With the $\mathbb{L}^2$ norm $\| \cdot \|$ scaled to 1, the space of such functions: $\mathcal{C} = \{q : I \to \mathbb{R}^3, \|q\| = 1\} \subset \mathbb{L}^2(I, \mathbb{R}^3)$ becomes a Riemannian manifold with the $\mathbb{L}^2$ metric on its tangent spaces. Since $\|q\| = 1$, $\mathcal{C}$ is a also a Hypersphere in the Hilbert space $\mathbb{L}^2(I, \mathbb{R}^3)$. Given two curves β_1 and β_2, they are first represented by the *square-root velocity function*, then unified to q_1 and q_2 with $\|q\| = 1$. The geodesic path ψ^* between q_1, q_2 on the manifold $\mathcal{C}$ is given by the minor arc of great circle connecting them on this Hypersphere, $\psi^* : [0, 1] \to \mathcal{C}$ an given by given by (1),

$$\psi^*(\tau) = \frac{1}{\sin(\theta)} \left(\sin((1 - \tau)\theta)q_1 + \sin(\theta\tau)q_2 \right) \tag{1}$$

where $\theta = d_{\mathcal{C}}(q_1, q_2) = cos^{-1}(\langle q_1, q_2 \rangle)$ is the angle between q_1 and q_2. The tangent vector field on this geodesic $\psi^* : [0, 1] \to T_\psi(\mathcal{C})$ is then given by (2):

$$\dot{\psi}^* = \frac{d\psi^*}{d\tau} = \frac{-\theta}{\sin(\theta)} \left(\cos((1 - \tau)\theta)q_1 - \cos(\theta\tau)q_2 \right) \tag{2}$$

Knowing that on a geodesic path, the covariant derivative of its tangent vector field is equal to 0. Thus, $\dot{\psi}^*|_{\tau=0}$ is sufficient to represent this vector field. Accordingly, (2) becomes:

$$\dot{\psi}^*|_{\tau=0} = \frac{\theta}{\sin(\theta)} (q_2 - \cos(\theta)q_1) \tag{3}$$

With the magnitude of $\dot{\psi}_\alpha^*$ at each all the N indexed points of the curve, we build a *Dense Scalar Field* (DSF) , $V = \{\|\dot{\psi}^*|_{(\tau=0)}(k)\|, k = 1, 2, 3, .., N\}$, which quantifies the shape difference between two curves.

In our approach, the raw 3D face scans are first pre-processed for hole-filling, cropping, smoothing and pose normalization, and then represented by a set of parameterized radial curves emanating from the nose tip of the preprocessed face denoted with S. The radial curve that makes an clockwise angle of α with the radial curve which passes through the forehead (β_0) is denoted as β_α, and the neighbor curve of β_α that has an angle increase of $\Delta\alpha$ is denoted as $\beta_{\alpha+\Delta\alpha}$. Such representation can be seen as a approximation of the preprocessed face S. To extract the DSF features, one need to first define the correspondence of curves in pair-wise shape comparison. With four different perspectives from face perception, we define four different types of correspondence in pair-wise shape comparison, which results into four different types of DSF features with all the radial curves considered in a face, namely the Symmetry DSF, the Averageness DSF, the Spatial DSF and the Gradient DSF. Figure 1 gives an illustration of these DSF features. The Symmetry DSF shown in sub-figure (a) captures the deformation between a pair of bilateral symmetrical radial curves (β_α^S and $\beta_{2\pi-\alpha}^S$) in a preprocessed face S. The Symmetry DSF conveys the idea that the bilateral facial symmetry loses with age. The Averageness DSF shown in sub-figure (b) compares a pair of curves with the same angle index from a preprocessed face β_α^S and an average face template β_α^T. The average face template T (as presented in sub-figure (b)) is defined as the middle point of geodesic deformation path from a representative male scan to a representative female scan. The Averageness DSF represents the idea that faces become more personalized and thus deviates more from the average face shape with age. The Spatial DSF shown in sub-figure (c) captures the deformation of a curve β_α to one reference radial curve β_0 in the forehead in a preprocessed face S. As β_0 is the most rigid curve in the face, the Spatial DSF can be perceived as the cumulative deformation from the most rigid part of the face. The Gradient DSF shown in sub-figure (d) captures the deformation between a pair of neighbor curves (β_α^S and $\beta_{\alpha+\Delta\alpha}^S$) in a preprocessed face S. In contrast with the Spatial DSF, the Gradient DSF can be viewed as a representation of local deformation on the face. In each sub-figure of Fig. 1, the left part shows the extracted radial curves in the face and correspondence for curve comparison, the right part shows the corresponding DSF features as color-map on the face, where on each face point, the hotter the color, the lower the DSF magnitude.

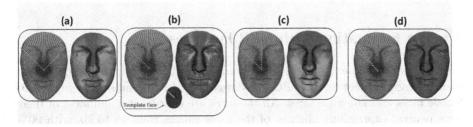

Fig. 1. Illustrations of different DSFs on preprocessed face S. (a) Symmetry DSF: the DSF from radial curve β_α^S to its bilateral symmetrical curve $\beta_{2\pi-\alpha}^S$; (b) Averageness DSF: DSF from radial curve β_α^S in a preprocessed face to radial curve β_α^T in an average face template (with the same angle index α); (c) Spatial DSF: DSF from radial curve β_α^S to the middle radial curve β_0^S in the forehead; (d) Gradient DSF: DSF from radial curve β_α^S to its neighbor curve $\beta_{\alpha+\Delta\alpha}^S$.

4 Random Forest Regression

Age estimation can be considered as a classification problem, when each age is taken as a class label. On the other hand, age estimation can also be considered as a regression problem, since the age could be interpreted as continuous value. Note that there are only 15 subjects of more than 40 years old in FRGCv2, the number of faces is too small to train classifiers for those ages. Thus, in our approach, we view the age estimation as a regression problem. Similar reason has been used by [5] for choosing the regression strategy for age estimation on the FG-net dataset, where the images from old subjects is also very rare. As summarized by [1], the regression task is, given a labeled set of training data, learning a general mapping which associates previously unseen, independent test data points with their dependent continuous output prediction. In [7], Random Forest regression has demonstrated good age estimation performance (3.43 MAE) in LOPO experiments for the young age subset of the FG-net dataset. As far as we concern, no studies have investigated the age estimation performance of Random Forest with the overall age distribution. Thus, we adopt the Random Forest in our regression experiments to demonstrate its capability in age estimation. Technically, Random Forest is an ensemble learning method that grows many classification trees $t \in \{t_1, .., t_T\}$. To estimate age from a new face from an input vector (DSF-based feature vector $v = V_\alpha^k$), each tree gives a regression result and the forest take the average of estimated ages as the final result. In the growing of each tree, two types of randomness are introduced consecutively. Firstly, a number of N instances are sampled randomly with replacement from the original data, to make the training set. Then, if each instance comprises of M input variables, a constant number m $(m<<M)$ is specified. At each node of the tree, m variables are randomly selected out of the M and the best split on these m variables is used to split the node. The process goes on until the tree grows to the largest possible extent without pruning, where the resulted subsets of the node are totally purified in label.

5   Age Estimation with DSF Features

We carry out the experiments with Random Forest Regression, on the 466 earliest scans of each subject in the FRGCv2 dataset [8]. In this subset, there are 203 female faces and 263 male faces. All the faces are near-frontal and most of them have neutral expression. The age of the faces ranges from 18 to 70, with 90 % less than 30 years old. With this subset, we design two experiment protocols. The first protocol, named **Gender-General-Protocol** (GGP), experiments on the 466 scans directly. Each time one scan is picked out for testing and the rest 465 scans are used for training. The second protocol, named **Gender-Specific-Protocol** (GSP), separates the 466 scans into male group and female group first, and then performs experiments on each group separately. This protocol carries the idea that face aging effect differs with gender. In [2], *Samal et al.* have shown that the degree of dimorphism changes as a function of age (e.g., the average age at which the sexual dimorphism becomes more significant is around 13). For all the two protocols, experimental results are generated using the **Leave-One-Person-Out** (LOPO) cross-validation strategy, where each time one scan of the concerning data (all 466 scans or scans of each gender) is used as testing face

Table 1. Age estimation results with DSF features (MAE: Mean Absolute Error; AVR: Averageness; SYM: symmetry; GRA: gradient; SPA: spatial; FUS: fusion).

Age	Gender General					Gender Specific				
	AVR	SYM	GRA	SPA	FUS	AVR	SYM	GRA	SPA	FUS
≤ 20	3.48	3.43	3.77	3.30	2.20	3.25	3.38	3.46	3.19	2.14
(20, 30]	2.18	2.58	2.32	2.38	1.98	2.03	2.16	2.14	2.18	2.04
(30, 40]	9.99	7.60	10.05	8.92	9.18	8.97	8.52	9.18	8.81	10.43
> 40	24.82	23.66	24.56	25.36	25.75	20.81	22.59	21.32	22.22	24.05
Overall	**3.76**	**3.79**	**3.94**	**3.76**	**3.29**	**3.42**	**3.57**	**3.58**	**3.51**	**3.15**

once, with the remaining scans used in training. In the cross-validation, each scan is tested equally only once. The age estimation results for each description are shown in Table 1 as the Mean Absolute Error (MAE), under both the GGP and GSP protocols. As shown in this table, we have also designed a fusion method. It takes the minimum estimated age given by the four descriptions as the fusion result.

As shown in the left part of Table 1, under the *Gender-General-Protocol*, the Averageness and Spatial DSFs achieve about 3.7 years MAE in age estimation. For the other two descriptions, the MAEs are higher, while both of them are under 4 years. These results demonstrate that our approach with all the four descriptions are effective in age estimation. Moreover, when examing the results in each age group, we find that the age estimation performance declines significantly with age. We assume that the big decrease of the number of scans in aged groups (from about 200 to about 20) accounts largely for this. At the same time, we observe that the fusion method yields a better overall mean absolute error of 3.29 years. It means that the age related cues in these descriptions are different and complimentary in age estimation. Also, we find the enhancement of overall performance is mainly coming from young age groups. It is probably due to the fact that for young age groups, more scans are available in training for each description. Thus the estimation results from each description for young age groups are less biased for making the fusion decision. The corresponding cumulative scores (CS) of age estimation under the *Gender-General-Protocol* are shown in Fig. 2. The x-axis is the level of Mean Absolute Error. The y-axis show the cumulative score of accuracy by percentage of acceptance. Thus, a point *(a,b)* on the curve shows, with a Mean Absolute Error tolerance of *a* years, it achieves an acceptance of *b* percent. As shown in Fig. 2, with an Error Level of 5 years, we achieve an acceptance of more than 75 % over the 466 scans; when the Error Level is 10 years, the acceptance increases to ≥ 90 %. The fusion result is significantly higher than the result of each individual description. Thus, in terms of the cumulative score (CS), we confirm that our age estimation approaches with all the four descriptions are effective under the *Gender-General-Protocol*, and the result-level fusion can enhance the age estimation performance.

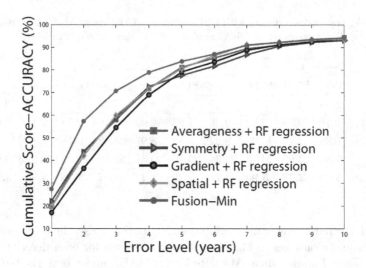

Fig. 2. Cumulative score in the GGP experiments with DSF features.

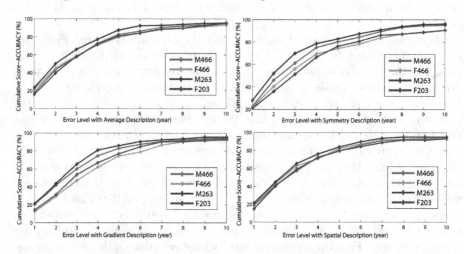

Fig. 3. Results from the Gender-Specific-Protocol (GSP) and the Gender-General-Protocol (GGP) for each gender. (M466: male group in the GGP experiments; F466: female group in the GGP experiments; M263: male group in the GSP experiments; 203: female group in the GSP experiments).

In parallel, for the **Gender-Specific-Protocol**, as shown in the right part of Table 1, the overall MAE in each description is always lower than the MAE in the Gender-General-Protocol. We also achieve better results in each age group with all these descriptions, except for the symmetry description in the (30,40] age group. These results demonstrate that the **Gender-Specific-Protocol** outperforms the Gender-General-Protocol in age estimation. It confirms the claims in [2], that faces of different gender convey different morphology of aging. Again, the overall fusion result outperforms the result of each description in the GSP experiments, and also the overall fusion result in the GGP experiments. It shows again the benifit of the fusion. The corresponding cumulative scores of age estimation under the *Gender-Specific-Protocol* are shown in Fig. 3 by gender. For comparison purpose, the results from the Gender-General-Protocol are also shown in each subplot. From Fig. 3, we observe that, only except for the beginning part of result with the female group in symmetry description, the experimental results are always significantly higher for both male and female groups in the GSP experiments for all the descriptions. That is to say, although trained with less data, the GSP experiments have the advantage of giving better age regression results. One probable reason for that is, in the GSP experiments, the regression do not suffer the influence from the scans in the other gender, which conveys a significantly different aging morphology. With Fig. 3, we further confirm that the aging effect differs with gender, and **Gender-Specific-Protocol** outperforms the Gender-General-Protocol in age estimation.

6 Age Estimation After PCA-Based Feature Dimensionality Reduction

With the original DSF features, we have achieved effective age estimation results, and have confirmed that the *gender-specific-protocol* works better for age estimation. However, the original DSF features have 20,000 dimensions, which has put big inconvenience in training and testing. Thus, we propose to use the Principle Component Analysis (PCA) method for feature dimensionality reduction. The PCA method is a well-known non-supervised dimensionality reduction method. It first centers the instances by subtracting the mean of all instances, then extracts the eigenvectors from the covariance matrix of the centered instances. The elements of the covariance matrix represent the differences between two dimensions of the features. Thus the eigenvectors of the covariance matrix sumarize the changes in different feature dimensions. For each eigenvector, its eigenvalue signifies the rank of the changes in the direction of the eigenvector. The higher the eighenvalue, the more changes in the underlying direction. By taking a small number of eigenvectors as projection matrix, the features of the original instances are transformed into a lower dimensional subspace. To retain the main changes in the instances, only the eigenvectors with high eigenvalues are chosen for projection.

According to the methodology of PCA, it can be infered that the distribution of the instances plays a crucial role in the process. It influences the centering of the instances and the building of the covariance matrix for the centered instances. If the distribution of instances is significantly imbalance to some data group, during the PCA process, the instances would be mis-centered, and the covariance matrix would represent more changes in specific data group. For example, in our case, the age distribution of the 466 instances (scans) is significantly imbalanced. 47 of the 466 instances are in the age range of [30,70] (only 10 %), and the rest 419 instances (near 90 %) are focused in the age range of [18,29]. If we apply PCA directly, the mean of the 466 instances would be significantly baised to young age group, and the covariance matrix would also represent mainly the changes in young group. Consequentially, the PCA-base features would have little discriminating power in old age group. To handle this, we propose to first balance the age distribution of the training set in LOPO experiment, by repeating the scans of $\geqslant 30$ year old. In Table 2, we illustrate this procedure using the 466 scans as example. While in reality, as 1 of the 466 is selected for testing in each round of LOPO experiments, this precedure only works for the remaining 465 training scans. As illustrated in Table 2, for the training instances of [30,35) years old, we repeat the scans 16 times. For training instances of [35,40) years old, we repeat the scans 31 times. For training instances of [40,45) years old, we repeat the scans 46 times. Finally, training instances of [45,70] years old are repeated 61 times. After repeating the scans, the age distribution of the instances is much more balanced, than without repetition. We then apply PCA on the testing instance and the training instances (after repeating) for feature dimensionality reduction. Actually, the precedure of repeating the training scans is equal to give weights to the training instances when using PCA. Thus in our work, we name our method as **weighted PCA**.

Table 2. Illustration of balancing the age distribution on the 466 instances.

Age group	[18,30)	[30,35)	[35,40)	[40,45)	[45,70]	All
♮ of scans (pre)	419	22	8	6	11	466
repeating times	1	16	31	46	61	—
♮ of scans (post)	419	352	248	276	671	1966

Following the previous study, we perform the Gender-Specific and Gender-General experiments with the PCA-based features. For comparison, we have also performed the LOPO experiments with PCA working directly on the 466 scans. The experimental results are shown in Fig. 4. In each subplot of Fig. 4, the x-axis shows the dimensionality of PCA-based feature space, and the y-axis shows the MAE of age estimation. We observe that, in the gender general protocol, the weighted-PCA (green line) generally achieves lower MAE than using PCA directly on the 466 scans (blue line). It means that the weighted PCA we proposed captures better the age-related changes in the imbalanced instances. From Fig. 4, we perceive also that with the weighted PCA, the MAEs are again lower in the Gender-Specific-Protocol (red line), than with the Gender-General-Protocol (green line). It confirms the previous finding that faces of different gender convey significantly different aging morphology, and the Gender-Specific-Protocol works better in age estimation.

In Table 3, we detail the best results achieved by each description in both the protocols. We observe that except in the (20,30] age group, the resulted MAEs in GSP experiments are generally lower than in the GGP experiments. Considering the overall MAE, the GSP experiments outperform obviously the results from

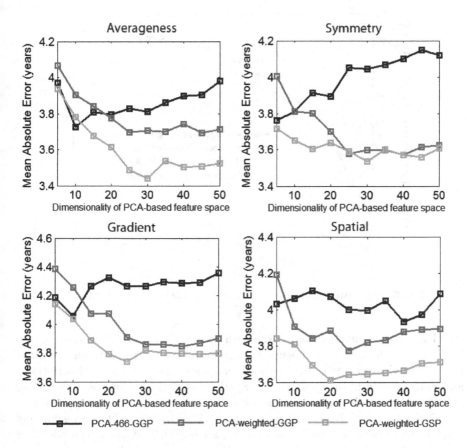

Fig. 4. Age estimation MAEs with different size of PCA-based features.

Table 3. Age estimation results using PCA-based feature dimensionality reduction (MAE: Mean Absolute Error; AVR: Averageness; SYM: symmetry; GRA: gradient; SPA: spatial; FUS: fusion).

Age	Gender General				Gender Specific			
	AVR	SYM	GRA	SPA	AVR	SYM	GRA	SPA
≤ 20	3.48	3.12	3.39	3.32	3.04	2.88	3.03	2.94
(20, 30]	2.04	2.03	2.34	2.34	2.02	2.22	2.43	2.41
(30, 40]	10.27	9.67	10.88	9.73	9.85	9.66	10.41	10.25
> 40	24.68	26.41	24.74	24.86	22.99	24.81	24.81	22.73
Overall	**3.69**	**3.57**	**3.84**	**3.77**	**3.44**	**3.53**	**3.73**	**3.61**
♯ of pc	25	25	40	25	30	30	25	20

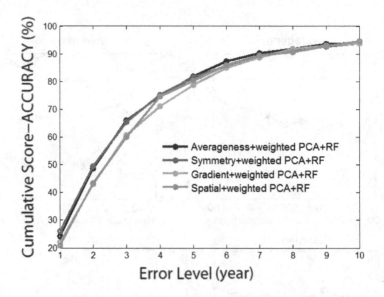

Fig. 5. Cumulative score in the GSP experiments with weighted PCA.

the GGP experiments. In comparison with the results from DSF features shown in Table 1, in the GSP experiments, the age estimation MAEs with our weighted PCA method are comparable in the Averageness and Symmetry descriptions, and a litter higher in the Gradient and Spatial descriptions. However, here we use only 20−30 principle components as features, as shown in the last row of Table 1. This is far less than 20000 dimensions in the DSF features. The corresponding Cumulative Scores of these results are shown in Fig. 5. With an error level of 5 years, all the descriptions achieve ∼80 % acceptance. When the error level increases to 10 years, more than 93 % acceptance is achieved in each description. Thus, with the results in Figs. 4, 5 and Table 3, we demonstrate that our weighted PCA method performs effectively in age estimation.

7   Conclusions

In this paper, we have proposed the first work in the 3D-based face age estimation domain. We have explored four types of the DSF features from 3D face which reflect four different perspectives of face perception, and performed the Leave-One-Person-Out age estimation with the Random Forest Regressor. With all the four descriptions, we have achieved effective age estimation results. By investigating the age estimation separately on Female and Male subsets, we have achieved better age estimation results, which justifies that the face aging differs considerably with gender. For addressing the high dimensionality of features, and the imbalance of training instances, we have proposed the *weighted PCA* method, which has produced comparable age estimation performance than with the original DSF feature, but with significantly smaller size of features.

References

1. Criminisi, A., Shotton, J.: Regression forests. In: Criminisi, A., Shotton, J. (eds.) Decision Forests for Computer Vision and Medical Image Analysis, pp. 49–58. Springer, Berlin (2013)
2. Samal, A., Subramani, V., Marx, D.: Analysis of sexual dimorphism in the human face. J. Vis. Comp. Imag. Rep. **18**, 453–463 (2007)
3. Fu, Y., Guo, G., Huang, T.S.: Age synthesis and estimation via faces: a survey. J. Pattern Anal. Mach. Intell. **11**, 1955–1976 (2010). IEEE
4. Rhodes, M.G.: Age estimation of faces: a review. J. Appl. Cogn. Psyc. **23**, 1–12 (2009)
5. Guo, G., Mu, G., Fu, Y., Huang, T.S.: Human age estimation using bio-inspired features. In: IEEE Conference on Computer Vision and Pattern Recognition, pp. 112–119. IEEE (2009)
6. Han, H., Otto, C., Jain, A.K.: Age estimation from face images: human vs. machine performance. In: 6th International Conference on Biometrics, pp. 1–8 (2013)
7. Montillo, A., Ling, H.: Age regression from faces using random forests. In: 16th IEEE International Conference on Image Processing, pp. 2465–2468. IEEE (2009)
8. Phillips, P.J., Flynn, P.J., Scruggs, T., Bowyer, K.W., Chang, J., Hoffman, K., Marques, J., Min, J., Worek, W.: Overview of the face recognition grand challenge. In: IEEE Conference on Computer Vision and Pattern Recognition, vol. 1, pp. 947–954. IEEE (2005)
9. Srivastava, A., Klassen, E., Joshi, S.H., Jermyn, I.H.: Shape analysis of elastic curves in euclidean spaces. J. Pattern Anal. Mach. Intell. **33**, 1415–1428 (2011). IEEE
10. Lanitis, A., Draganova, C., Christodoulou, C.: Comparing different classifiers for automatic age estimation. J. Syst. Man Cybern. Part B: Cybern. **34**, 621–628 (2004). IEEE
11. Ueki, K., Sugiyama, M., Ihara, Y.: Perceived age estimation under lighting condition change by covariate shift adaptation. In: 20th International Conference on Pattern Recognition, pp. 3400–3403. IEEE (2010)
12. Lanitis, A., Taylor, C.J., Cootes, T.F.: Toward automatic simulation of aging effects on face images. J. Pattern Anal. Mach. Intell. **24**, 442–455 (2002). IEEE
13. Geng, X., Zhou, Z.H., Smith-Miles, K.: Automatic age estimation based on facial aging patterns. J. Pattern Anal. Mach. Intell. **29**, 2234–2240 (2007). IEEE
14. Ramanathan, N., Chellappa, R., Biswas, S.: Computational methods for modeling facial aging: a survey. J. Vis. Lang. Comp. **20**, 131–144 (2009). Elsevier
15. Albert, A.M., Ricanek, K., Patterson, E.: A review of the literature on the aging adult skull and face: implications for forensic science research and applications. J. Forensic Sci Int. **172**, 1–9 (2007). Elsevier
16. Park, U., Tong, Y., Jain, A.K.: Age-invariant face recognition. J. Pattern Anal. Mach. Intell. **32**, 947–954 (2010). IEEE
17. Guo, G., Fu, Y., Huang, T.S., Dyer, C.R.: Locally adjusted robust regression for human age estimation. In: IEEE Workshop on Applications of Computer Vision, pp. 1–6. IEEE (2008)
18. Guo, G., Fu, Y., Dyer, C.R., Huang, T.S.: Image-based human age estimation by manifold learning and locally adjusted robust regression. J. Imag. Process. **17**, 1178–1188 (2008). IEEE
19. Geng, X., Zhou, Z.H., Zhang, Y., Li, G., Dai, H.H.: Learning from facial aging patterns for automatic age estimation. In: 14th Annual ACM International Conference on Multimedia, pp. 307–316. ACM (2006)

20. The Relationship Between Age and Facial Asymmetry. http://meeting.nesps.org/2011/80.cgi

21. Cootes, T.F., Edwards, G.J., Taylor, C.J.: Active appearance models. In: Burkhardt, H., Neumann, B. (eds.) ECCV'98. LNCS, vol. 1407, pp. 484–498. Springer, Heidelberg (1998)

22. N.S., L., J., B., Majumder, S.: Age estimation using gender information. In: Venugopal, K.R., Patnaik, L.M. (eds.) ICIP 2011. CCIS, vol. 157, pp. 211–216. Springer, Heidelberg (2011)

23. Chao, W.L. Liu, J.Z., Ding, J.J.: Facial age estimation based on label-sensitive learning and age-specific local regression. In: IEEE International Conference on Acoustics, Speech and Signal Processing, pp. 1941–1944. IEEE (2012)

24. Suo, J.L., Zhu, S.C., Shan, S.G., Chen, X.L.: A compositional and dynamic model for face aging. J. Pattern Anal. Mach. Intell. **32**, 385–401 (2010). IEEE

25. Tao, W., Turaga, P., Chellappa, R.: Age estimation and face verification across aging using landmarks. J. Info. Foren. Sec. **7**, 1780–1788 (2012)

26. Li, C.S., Liu, Q.S., Liu, J., Lu, H.Q.: Learning ordinal discriminative features for age estimation. In: IEEE Conference on Computer Vision and Pattern Recognition, pp. 2570–2577. IEEE (2012)

27. Mark, L.S., Todd, J.T.: The perception of growth in three dimensions. J. Atten. Percep. Psycho. **33**, 193–196 (1983). Springer

28. Bruce, V., Burton, M., Doyle, T., Dench, N.: Further experiments on the perception of growth in three dimensions. J. Percept. Psycho. **46**, 528–536 (1989). Springer

Coupling Dynamic Equations and Satellite Images for Modelling Ocean Surface Circulation

Dominique Béréziat[1,2](✉) and Isabelle Herlin[3,4]

[1] Sorbonne Universités, UPMC Univ Paris 06, UMR 7606, LIP6,
4 Place Jussieu, 75005 Paris, France
[2] CNRS, UMR 7606, LIP6, 75005 Paris, France
Dominique.Bereziat@inria.fr
[3] Inria, B.P. 105, 78153 Le Chesnay, France
[4] CEREA, Joint Laboratory ENPC - EDF R&D, Université Paris-Est,
Descartes Champs-sur-Marne, 77455 Marne la Vallée Cedex 2, France

Abstract. Satellite image sequences visualise the ocean surface and allow assessing its dynamics. Processing these data is then of major interest to get a better understanding of the observed processes. As demonstrated by state-of-the-art, image assimilation permits to retrieve surface motion, based on assumptions on the dynamics. In this paper, we demonstrate that a simple heuristics, such as the Lagrangian constancy of velocity, can be used and successfully replaces the complex physical properties described by the Navier-Stokes equations for assessing surface circulation from satellite images. A data assimilation method is proposed that adds an acceleration term $\mathbf{a}(t)$ to this Lagrangian constancy equation, which summarises all physical processes other than advection. A cost function is designed that quantifies discrepancy between satellite data and model values. This cost function is minimised by the BFGS solver with a dual method of data assimilation. The result is the initial motion field and the acceleration terms $\mathbf{a}(t)$ on the whole temporal interval. These values $\mathbf{a}(t)$ model the forces, other than advection, that contribute to surface circulation. Our approach was tested on synthetic data and with Sea Surface Temperature images acquired on Black Sea. Results are quantified and compared to those of state-of-the-art methods.

Keywords: Dynamic model · Optical flow · Data assimilation · Satellite image · Ocean circulation

1 Introduction

Satellite image sequences permit to visualise oceans' surface and their underlying dynamics with a high spatial resolution. Processing these images is then of major interest for a better understanding of the observed processes and the forecast of extreme events. As demonstrated by state-of-the-art, image assimilation allows to retrieve surface motion from image sequences, using heuristics on the dynamics [11,15]. Among those heuristics, the shallow water model [17] has

© Springer International Publishing Switzerland 2015
S. Battiato et al. (Eds.): VISIGRAPP 2014, CCIS 550, pp. 191–205, 2015.
DOI: 10.1007/978-3-319-25117-2_12

been proven to be suitable for representing the surface circulation of closed seas, such as Black Sea [10]. These shallow water equations have also been successfully used to estimate the upper layer circulation of Black Sea from Sea Surface Temperature (SST) images [5,6] with a data assimilation method.

In this paper, we propose to learn the surface dynamics from SST image acquisitions with a data assimilation method applied to an image model derived from the shallow water equations. In the shallow water model, surface circulation is characterised by the horizontal velocity, that is advected by itself and subject to geophysical forces such as Coriolis, Earth gravity and viscosity. The advection process is kept in the image model, but all other components are summarised in a global term, denoted $\mathbf{a}$ (letter $\mathbf{a}$ stands for acceleration) that is estimated by our approach. Adding this term to the advection is similar, from a mathematical point of view, to the weak data assimilation framework [2,12,16,18]. A data assimilation technique is then designed to compute the solution: a cost function is constructed, whose control variables are the motion field at the first acquisition date and the acceleration values $\mathbf{a}(t)$, at all dates of the acquisition interval. The minimum of the cost function is obtained thanks to optimal control techniques [9] and it is computed with the BFGS solver [20].

Section 2 provides notations that are used in the remaining of the paper and the mathematical description of the proposed approach for modelling the dynamics of the ocean's upper layer. The data assimilation method is outlined in Sect. 3. It corresponds to a weak formulation, where the non advective terms are summarised as an additional term in the evolution equation. The implementation is shortly described in Sect. 4, in order to permit that readers apply the method by themselves. Validation on synthetic data and results on SST image sequences acquired over Black Sea by NOAA-AVHRR sensors are displayed and quantified in Sect. 5.

2   Problem Statement

Image data are acquired on a bounded rectangle of $\mathbb{R}^2$, named Ω, and on a temporal interval $[0, T]$. Let define $A = \Omega \times [0, T]$ the corresponding space-time domain, on which the dynamics is modelled. A point $\mathbf{x} \in \Omega$ is defined as $\mathbf{x} = \begin{pmatrix} x & y \end{pmatrix}^T$ and the motion vector at point $\mathbf{x}$ and date $t \in [0, T]$ is written $\mathbf{w}(\mathbf{x}, t) = \begin{pmatrix} u(\mathbf{x}, t) & v(\mathbf{x}, t) \end{pmatrix}^T$. At each date t, the motion field on the domain Ω is written as $\mathbf{w}(t)$. N Sea Surface Temperature acquisitions are available at dates t_i, $i = 1 \ldots N$. They are denoted $T(t_i)$ with pixels values $T(\mathbf{x}, t_i)$.

A state vector $\mathbf{X}$ is defined on A. It includes the two components u and v of the motion vector $\mathbf{w}(\mathbf{x}, t)$ and a pseudo-temperature value $T_M(\mathbf{x}, t)$, which has properties similar to those of the Sea Surface Temperature function:

$$\mathbf{X}(\mathbf{x}, t) = \begin{pmatrix} \mathbf{w}(\mathbf{x}, t)^T & T_M(\mathbf{x}, t) \end{pmatrix}^T \tag{1}$$

The index M in T_M reminds that this is a component of the Model state vector. At the end of the data assimilation process, the discrepancy between the pseudo-temperature T_M and the satellite acquisitions T has to be small.

The heuristics on dynamics, used in the paper, are derived from the shallow water equations that express the principles of mass and momentum conservation [17]. Circulation of the upper ocean is represented by the 2D velocity $\mathbf{w} = \left(u \; v\right)^T$ and the thickness h of the mixed layer. In our model, the pseudo-temperature T_M is transported by the motion field. This provides the following set of equations:

$$\frac{\partial u}{\partial t} = -u\frac{\partial u}{\partial x} - v\frac{\partial u}{\partial y} + fv - g'\frac{\partial \eta}{\partial x} + K_{\mathbf{w}}\Delta u \tag{2}$$

$$\frac{\partial v}{\partial t} = -u\frac{\partial v}{\partial x} - v\frac{\partial v}{\partial y} - fu - g'\frac{\partial \eta}{\partial y} + K_{\mathbf{w}}\Delta v \tag{3}$$

$$\frac{\partial \eta}{\partial t} = -\frac{\partial(u\eta)}{\partial x} - \frac{\partial(v\eta)}{\partial y} - h_m\left(\frac{\partial u}{\partial x} + \frac{\partial v}{\partial y}\right) \tag{4}$$

$$\frac{\partial T_M}{\partial t} = -u\frac{\partial T_M}{\partial x} - v\frac{\partial T_M}{\partial y} \tag{5}$$

with η the thickness anomaly $\eta = h - h_m$, h_m the average value of h, f the Coriolis parameter, $K_{\mathbf{w}}$ the viscosity and $g' = g(\rho_0 - \rho_1)/\rho_0$ the reduced gravity. ρ_0 corresponds to the reference density and ρ_1 to the average density of the mixed layer.

As explained in the introduction, we propose to group all geophysical forces that do not correspond to advection in a unique term, that corresponds to the acceleration and is denoted by $\mathbf{a}$. The variable η is then considered as an hidden variable of the system. In such way, System (2, 3, 4) reduces to:

$$\frac{\partial u}{\partial t} = -u\frac{\partial u}{\partial x} - v\frac{\partial u}{\partial y} + a_u \tag{6}$$

$$\frac{\partial v}{\partial t} = -u\frac{\partial v}{\partial x} - v\frac{\partial v}{\partial y} + a_v \tag{7}$$

where $\mathbf{a} = \left(a_u \; a_v\right)^T$ expresses the discrepancy to the Lagrangian constancy of velocity:

$$\frac{d\mathbf{w}}{dt} = \frac{\partial \mathbf{w}}{\partial t} + (\mathbf{w}.\nabla)\mathbf{w} = \mathbf{a} \tag{8}$$

From Eqs. (2) and (3), we get:

$$a_u = fv - g'\frac{\partial \eta}{\partial x} + K_{\mathbf{w}}\Delta u \tag{9}$$

$$a_v = -fu - g'\frac{\partial \eta}{\partial y} + K_{\mathbf{w}}\Delta v \tag{10}$$

where η verifies Eq. (4).

Our approach estimates $\mathbf{w}(0)$ and the acceleration term $\mathbf{a}(t)$ at each date $t \in [0, T]$, thanks to the data assimilation process summarised in Sect. 3. Deriving the values $\mathbf{a}(t)$ permits describing empirically the physical processes generating the image sequence.

Equations (8) and (5) are further contracted in an evolution model $\mathbb{M}$ of the state vector $\mathbf{X}$:

$$\frac{\partial \mathbf{X}}{\partial t} + \mathbb{M}(\mathbf{X}) = \begin{pmatrix} \mathbf{a} \\ 0 \end{pmatrix} \tag{11}$$

An observation equation links the state vector to the observed Sea Surface Temperature images acquisitions T:

$$\mathbb{H}\mathbf{X} = T + \varepsilon_R \tag{12}$$

The observation operator $\mathbb{H}$ projects the state vector into the space of image observations and consequently: $\mathbb{H}\mathbf{X} = T_M$. The term $\varepsilon_R(\mathbf{x}, t)$ models the acquisition noise and the uncertainty on the state vector value. This last comes from the approximation of the model and from the discretization errors.

Some approximate knowledge of the value $\mathbf{X}(0)$ could be available and named background $\mathbf{X}_b$. However, the result of the state vector at date 0 is not exactly equal to that background value and a term ε_B is therefore introduced:

$$\mathbf{X}(\mathbf{x}, 0) = \mathbf{X}_b(\mathbf{x}) + \varepsilon_B(\mathbf{x}) \tag{13}$$

The variables ε_R and ε_B are supposed independent, unbiased, Gaussian and characterised by their respective covariance matrices R and B.

Equations (11, 12, 13) summarise the whole knowledge that is available for modelling the surface dynamics. This knowledge is processed by our approach thanks to the data assimilation algorithm that is shortly described in the next section.

3   Data Assimilation

In the data assimilation scientific community, an approach, named weak 4D-Var, has been defined in order to obtain the solution $\mathbf{X}$ that solves System (11, 12, 13). A cost function is first designed:

$$J[\varepsilon_B, \mathbf{a}] = \langle \varepsilon_B, B^{-1}\varepsilon_B \rangle + \int_t \gamma \|\nabla \mathbf{a}(t)\|^2$$
$$+ \int_t \langle \mathbb{H}\mathbf{X}(t) - T(t), R^{-1}(\mathbb{H}\mathbf{X}(t) - T(t)) \rangle \tag{14}$$

where $\langle \cdot, \cdot \rangle$ denotes the canonical inner product in an abstract Hilbert space on which the state vector is defined, with norm $\|.\|^2$ and $\|\nabla \mathbf{a}\|^2 = \langle \nabla a_u, \nabla a_u \rangle + \langle \nabla a_v, \nabla a_v \rangle$. The function J is then minimised with control on ε_B and on the values of the acceleration term $\mathbf{a}$.

The first term of J comes from Eq. (13) and expresses that the value $\mathbf{X}(0)$ at date 0 should stay close to the background value $\mathbf{X}_b$. It should be noted that the control on ε_B is equivalent to the control on initial condition as $\varepsilon_B(\mathbf{x}) = \mathbf{X}(\mathbf{x}, 0) - \mathbf{X}_b(\mathbf{x})$. The second term constrains the acceleration term $\mathbf{a}(t)$ to be spatially smooth. The last term, coming from Eq. (12), expresses that the pseudo-temperature value T_M has to be close to that of satellite acquisitions at the end of the assimilation process.

The gradient of J is derived with the calculus of variations, as given in [9]. Its two components are:

$$\frac{\partial J}{\partial \varepsilon_B}[\varepsilon_B, \mathbf{a}] = 2\left(B^{-1}\varepsilon_B + \lambda(0)\right) \tag{15}$$

$$\frac{\partial J}{\partial \mathbf{a}(t)}[\varepsilon_B, \mathbf{a}] = 2\left(-\gamma \Delta \mathbf{a}(t) + \lambda(t)\right) \tag{16}$$

with $\lambda(t)$ being the adjoint variable, that is computed backward in time with the two following equations:

$$\lambda(T) = 0 \tag{17a}$$

$$-\frac{\partial \lambda}{\partial t} + \left(\frac{\partial \mathbb{M}}{\partial \mathbf{X}}\right)^{*} \lambda = \mathbb{H}^{T} R^{-1}(\mathbb{H}\mathbf{X} - T) \tag{17b}$$

The adjoint operator $\left(\frac{\partial \mathbb{M}}{\partial \mathbf{X}}\right)^{*}$ verifies the following property. For all functions η and λ of the studied Hilbert spaces, it comes:

$$\langle Z\eta, \lambda \rangle = \langle \eta, Z^{*}\lambda \rangle. \tag{18}$$

Proof: For sake of simplicity, we suppose in this proof that Eq. (11) is written as $\frac{\partial \mathbf{X}}{\partial t} + \mathbb{M}(\mathbf{X}) = \mathbf{a}$, $\mathbf{a}$ denoting simultaneously the acceleration involved in motion evolution and $\begin{pmatrix} \mathbf{a} \\ 0 \end{pmatrix}$.

The state vector and the functional J depend on ε_B and $\mathbf{a}(t)$. Let δJ and $\delta\mathbf{X}$ be the perturbations on J and $\mathbf{X}$ obtained if ε_B and $\mathbf{a}(t)$ are respectively perturbed by $\delta\varepsilon_B$ and $\delta\mathbf{a}(t)$.

From the definition of J, we obtain:

$$\begin{aligned}
\frac{\delta J}{2} =& \langle \delta\varepsilon_B, B^{-1}\varepsilon_B \rangle + \int_t \gamma \langle \nabla \delta\mathbf{a}(t), \nabla \mathbf{a}(t) \rangle \\
&+ \int_t \langle \delta\mathbf{X}(t), \mathbb{H}^{T} R^{-1}[\mathbb{H}\mathbf{X}(t) - T(t)] \rangle
\end{aligned} \tag{19}$$

The evolution equation of $\mathbf{X}$, Eq. (11), gives:

$$\frac{\partial \delta\mathbf{X}(t)}{\partial t} + \frac{\partial \mathbb{M}}{\partial \mathbf{X}} \delta\mathbf{X}(t) = \delta\mathbf{a}(t) \tag{20}$$

and that of background, Eq. (13):

$$\delta\mathbf{X}(0) = \delta\varepsilon_B \tag{21}$$

Equation (20) gives, after multiplication by $\lambda(t)$ and integration on the space-time domain, the following equality:

$$\int_t \left\langle \frac{\partial \delta\mathbf{X}(t)}{\partial t}, \lambda(t) \right\rangle + \int_t \left\langle \frac{\partial \mathbb{M}}{\partial \mathbf{X}} \delta\mathbf{X}(t), \lambda(t) \right\rangle = \int_t \langle \delta\mathbf{a}(t), \lambda(t) \rangle \tag{22}$$

Integration by parts is applied on the first term and the definition of adjoint operator is used in the second one in order to obtain:

$$\langle \delta \mathbf{X}(T), \lambda(T) \rangle - \langle \delta \mathbf{X}(0), \lambda(0) \rangle - \int_t \left\langle \delta \mathbf{X}(t), \frac{\partial \lambda(t)}{\partial t} \right\rangle$$
$$+ \int_t \left\langle \delta \mathbf{X}(t), \frac{\partial \mathbf{M}^*}{\partial \mathbf{X}} \lambda(t) \right\rangle = \int_t \langle \delta \mathbf{a}(t), \lambda(t) \rangle \tag{23}$$

From Eq. (17a), it comes that $\langle \delta \mathbf{X}(T), \lambda(T) \rangle$ has a null value. From Eq. (21) it comes that $\langle \delta \mathbf{X}(0), \lambda(0) \rangle$ is equal to $\langle \delta \varepsilon_B, \lambda(0) \rangle$. Eq. (17b) is then used to obtain:

$$- \left\langle \delta \mathbf{X}(t), \frac{\partial \lambda(t)}{\partial t} \right\rangle + \left\langle \delta \mathbf{X}(t), \frac{\partial \mathbf{M}^*}{\partial \mathbf{X}} \lambda(t) \right\rangle = \langle \delta \mathbf{X}(t), \mathbb{H}^T R^{-1}(\mathbb{H} \mathbf{X}(t) - T(t)) \rangle \tag{24}$$

and rewrite Eq. (23) as:

$$\int_t \langle \delta \mathbf{X}(t), \mathbb{H}^T R^{-1}(\mathbb{H} \mathbf{X}(t) - T(t)) \rangle = \langle \delta \epsilon_B, \lambda(0) \rangle + \int_t \langle \delta \mathbf{a}(t), \lambda(t) \rangle \tag{25}$$

From this and Eq. (19), we derive:

$$\frac{\delta J}{2} = \langle \delta \epsilon_B, B^{-1} \epsilon_B \rangle - \int_t \gamma \langle \delta \mathbf{a}(t), \Delta \mathbf{a}(t) \rangle + \langle \delta \epsilon_B, \lambda(0) \rangle + \int_t \langle \delta \mathbf{a}(t) \lambda(t) \rangle \tag{26}$$

and obtain the gradient of J, as written in Eqs. (15, 16).

The cost function J is minimised with an iterative steepest descent method. At each iteration, the forward time integration of $\mathbf{X}$ is done, according to Eq. (11). This forward time integration provides the value of J. Then a backward time integration of λ, according to Eqs. (17a) and (17b), computes the value of ∇J. An efficient solver, described in [1, 20], is used to perform the optimisation given values of J and ∇J. To our knowledge, [7] is the first paper of the literature that describes the use of such method for estimating the initial state vector value.

4   Numerical Implementation

Time integration of Eq. (11) relies on an explicit Euler scheme. The space discretization of motion advection, described by the two following equations:

$$\frac{\partial u}{\partial t} + u \frac{\partial u}{\partial x} + v \frac{\partial u}{\partial y} = 0 \tag{27}$$

$$\frac{\partial v}{\partial t} + u \frac{\partial v}{\partial x} + v \frac{\partial v}{\partial y} = 0 \tag{28}$$

involves a source splitting method [19], as explained below.

Given an integration interval $[t_1, t_2]$, Eqs. (29) (where n stands for non linear advection) and (30) (where l stands for linear advection) are first independently integrated:

$$\frac{\partial u^n}{\partial t} + u^n \frac{\partial u^n}{\partial x} = 0 \quad t \in [t_1, t_2] \tag{29}$$

$$\frac{\partial u^l}{\partial t} + v \frac{\partial u^l}{\partial y} = 0 \quad t \in [t_1, t_2] \tag{30}$$

with $u^n(x, y, t_1) = u^l(x, y, t_1) = u(x, y, t_1)$. Then $u(x, y, t_2)$ is obtained as:

$$u(x, y, t_2) = u^l(x, y, t_2) + u^n(x, y, t_2) - u(x, y, t_1)$$

The linear advection involved in Eq. (30) is discretized by a first-order upwind scheme, as described in [4]. The nonlinear advection of Eq. (29) is first rewritten in a conservative form:

$$\frac{\partial u}{\partial t} + \frac{\partial}{\partial x}\left(\frac{1}{2}u^2\right) = 0 \tag{31}$$

and approximated by a first-order Godunov scheme [8].

The backward time integration of the adjoint variable λ involves the adjoint operator $\left(\frac{\partial M}{\partial X}\right)^*$ (see Eq. (17b)). In order to be accurate, the method requires the adjoint of the discrete model and not the discretization of the continuous adjoint. For that reason, the discrete adjoint operator $\left(\frac{\partial M}{\partial X}\right)^*$ is derived with the automatic differentiation software Tapenade described in [3].

The background used in Eq. (13) is defined as null for motion and as the first image of the studied sequence for the pseudo-temperature. As we have no information on motion, we do not want to constrain $\mathbf{w}(0)$ to stay close to that null value. For that reason, the first term of J reduces to $\langle \varepsilon_{B_T}, B_T^{-1} \varepsilon_{B_T} \rangle$. B_T and R are taken diagonal with standard deviation values corresponding to 25 % of the image brightness range. γ is given a value as small as possible: smaller values would not verify the Courant-Friedrichs-Lewy condition during the advection of the state vector.

5   Results

For sake of simplicity, we name in the following AM, or Acceleration Model, the model described in Sects. 2 and 3: it includes the advection and the acceleration term $\mathbf{a}$ summarising the Coriolis, gravity and viscosity forces. We denote ME the Motion Estimation method that is obtained if suppressing $\mathbf{a}(t)$ in Eq. (11). In that case, the second term of the cost function J, in Eq. (14), vanishes and minimisation is only controlled by ε_B.

The approach was experimented on a data base of Sea Surface Temperature sequences, acquired over Black Sea by NOAA-AVHRR sensors. The resolution of the AVHRR data is 1.1 km at nadir. Such image sequences are moreover able to display the dynamics of the submesoscale phenomena like eddies, jets, filaments and mushroom-like structures with a rather high temporal frequency. Our app-roach computes the surface motion fields and the trajectories of these structures by assimilating images. This high-level information is critical for improving the forecast of 3D ocean models such as NEMO[1] when they are tuned to resolve submesoscale processes.

---

[1] http://www.nemo-ocean.eu/.

The approach is first validated on two synthetic experiments. Then, results on four satellite image sequences are displayed and discussed in the remaining of the section.

5.1   Synthetic Experiments

Given initial conditions at time 0, displayed in Fig. 1, and given acceleration values $\mathbf{a}_{\text{ref}}(t)$, Eq. (11) is integrated in time and produces the ground-truth on motion and image fields. These initial conditions are similar to those on real data described in the next subsections.

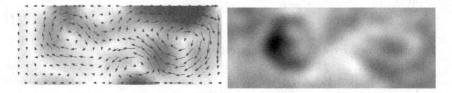

Fig. 1. Initial conditions on motion and image. The arrow representation is superposed to the coloured one (Color figure online).

Image snapshots are extracted from the simulation results and used for the assimilation process. The discrete assimilation window has temporal indexes from 0 to $N = 84$ and observation images are available at indexes k_1 to $k_5 = 2$, 22, 42, 62 and 82.

In a first experiment, the acceleration values $\mathbf{a}_{\text{ref}}(t)$ are null. Snapshots are displayed on Fig. 2.

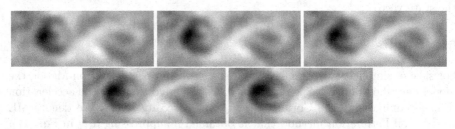

Fig. 2. Image observations.

Results are displayed on Fig. 3 for ME, AM and for the motion estimation method of Sun *et al.* [13]. The estimated motion and acceleration fields are respectively denoted $\mathbf{w}_e$ and $\mathbf{a}_e$.

Table 1 provides statistics on the discrepancy between results and ground truth. They concern the norm $\|\mathbf{w}\|$ and orientation θ of motion.

This table shows that AM estimates motion field with a quality comparable to ME: it demonstrates that the acceleration estimated by AM is almost null. Statistics on acceleration confirm this property as the average value of $\|\mathbf{a}_e\|$ is 3.6×10^{-5} on the whole temporal interval.

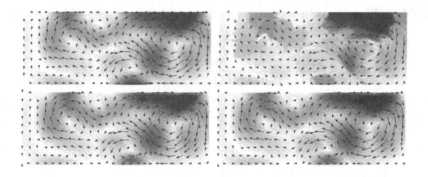

Fig. 3. First line: Groundtruth, Sun - Second line: ME, AM.

Table 1. Statistics on motion errors for Sun, ME (without **a**) and AM (with **a**).

	$|\theta_e - \theta_{\text{ref}}|$	$|\|\mathbf{w}_{\text{ref}}\| - \|\mathbf{w}_e\||/\|\mathbf{w}_{\text{ref}}\|$
Method	Mean	Mean
Sun	16.1	43 %
ME	3.2	9 %
AM	3.5	10 %

In a second experiment, the acceleration $\mathbf{a}_{\text{ref}}(t)$ is null for $t \neq 41$ and equal to the vertical constant 9×10^{-2} at $t = 41$. Results are displayed on Fig. 4. Table 2 provides statistics on the discrepancy between motion results and ground-truth. Statistics on the norm of acceleration $\mathbf{a}_e(t)$ are displayed on Fig. 5.

From Table 2, it can be seen that AM gives a better estimation of velocity than Sun. As the simulation generating image data includes an acceleration, ME fails to estimate an accurate velocity. It can be seen on Fig. 5 that our approach computes an acceleration having a maximal norm value at date 41. This means that AM correctly localises the acceleration in time. Moreover, the average

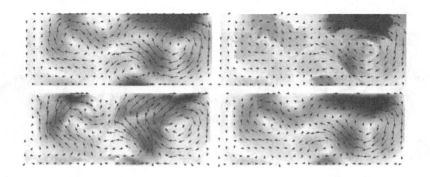

Fig. 4. First line: groundtruth, Sun - Second line: ME, AM.

Table 2. Statistics on motion errors for Sun, ME (without **a**) and AM (with **a**).

	$|\theta_e - \theta_{\mathrm{ref}}|$	$|\|\mathbf{w}_{\mathrm{ref}}\| - \|\mathbf{w}_e\||/\|\mathbf{w}_{\mathrm{ref}}\|$
Method	Mean	Mean
Sun	16.9	44 %
ME	36.3	68 %
AM	11.8	22 %

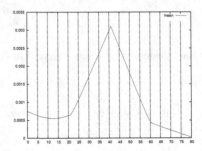

Fig. 5. Statistics on the norm of $\mathbf{a}_e$.

orientation of $\mathbf{a}_e(41)$ is $90.4°$ with a standard deviation of $11.4°$ against $90°$ for the ground truth. That further demonstrates the accuracy of the estimation.

5.2 Satellite Experiments

In a first experiment, we analyse the ability of the proposed method to correctly estimate motion, which is a natural consequence of a correct assessment of geophysical forces. For that purpose, AM results are compared with those obtained by the motion estimation method of Sun *et al.* [13]. The satellite sequence is displayed on Fig. 6. Acquisition dates are at 30 min, 6 h, 15 h, and 30 h after the beginning of the studied temporal interval.

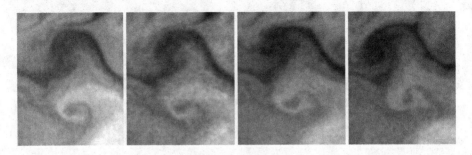

Fig. 6. Left to right: SST images acquired on October 19th 2007 over Black Sea.

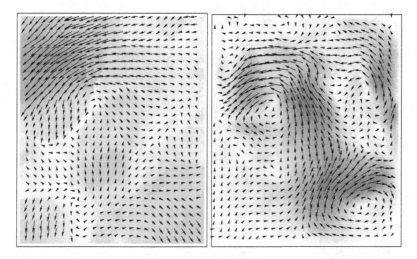

Fig. 7. Motion results computed by Sun (left) and AM (right) at first observation date.

Two gyres are clearly visible on these data. Motion results $\mathbf{w}(0)$, obtained by Sun and AM, are displayed on Fig. 7. AM successes to capture the two gyres while Sun's method fails. As $\mathbf{w}(0)$ is obtained from the analysis of the whole image sequence, its correct estimation means that the physical processes involved in $\mathbf{a}(t)$ are correctly assessed by the model. It also means that the non advective geophysical forces may be correctly described with the unique term $\mathbf{a}(t)$.

In a second experiment, the capability of AM to track features or points of interest on the whole image sequence is examined. If the method provides accurate tracking results, this means that motion is correctly estimated on the studied temporal interval and properly transports the observed structures. A sequence of four SST images acquired on October 8th 2005 is displayed on Fig. 8. They are acquired 30 min, 10 h 15 min, 12 h and 15 h 30 min after the beginning of the studied interval. Nine characteristic points are defined in white on the first observation. Points are additionally surrounded by a coloured circle that helps to discriminate them on the following observations. These points are considered as characteristic, because they sample the various types of trajectories that can be observed on the sequence. On observations 2 to 4, the position of these nine points obtained with Sun's method are displayed in red while those obtained with AM are in blue. On the fourth acquisition, in the "light pink circle" on the upper right, the point obtained with Sun's method is outside of the image domain. Looking at the trajectories, it can be observed that Sun's algorithm fails to track these characteristic points, due to a wrong estimation of motion.

Another sequence of five SST images, acquired on July 27th 2007 is displayed on Fig. 9. The sensor acquired data 30 min, 8 h 15 min, 13 h, 22 h 30 min and 24 h 30 min after the beginning of the studied interval. Seven characteristic points are defined in white on the first observation. On the next ones, positions obtained with Sun's method are displayed in red while those obtained with AM are in blue. At the second date, two points are at the same position with Sun

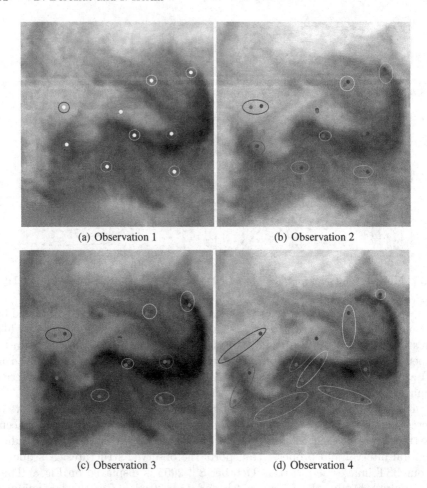

(a) Observation 1   (b) Observation 2

(c) Observation 3   (d) Observation 4

Fig. 8. Tracking of characteristic points. Results of Sun correspond to red points, those of AM to blue points (Color figure online).

and AM: only the red point is visible as the blue one is hidden behind it. On the fourth observation, one red point has disappeared from the display as it is located outside of the image domain. On the last frame, the colour of the ellipse surrounding each couple of points gives an additional information on the quality of the result: a blue ellipse means that our method gives the best result while the white one means that both methods are equivalent.

Again, Sun's motion results fail to track characteristic points on these data as physical processes are not correctly assessed by the underlying image model.

Last, AM is also compared with the optical flow estimation of Suter [14], that is dedicated to fluid flows, on a sequence of five images acquired on May 14th 2005. The acquisitions were obtained 30 min, 2 h and 45 min, 5 h and 15 min, 7 h and 15 min, and 16 h and 15 min after the beginning of the studied interval. As previously, six feature points are chosen on the first observation and displayed

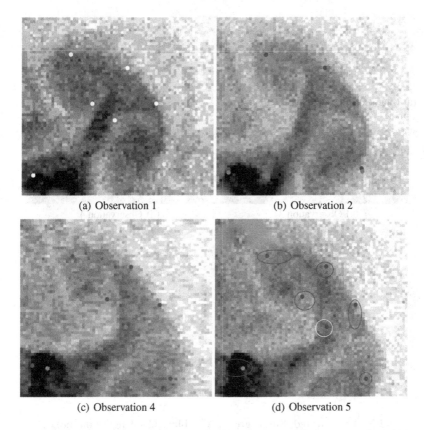

(a) Observation 1   (b) Observation 2

(c) Observation 4   (d) Observation 5

Fig. 9. Tracking of characteristic points. A blue ellipse on (d) expresses that AM is the best while the white ellipse expresses that results are equivalent (Color figure online).

on the upper image of Fig. 10. Their final position on the fifth observation is given in the lower part of the same figure.

Suter's and Sun's methods are both only constrained by grey level values and do not rely on the underlying dynamics. However, Suter's algorithm provides better result than Sun's method, because it is specifically designed for fluid flows motion. In particular, it correctly assesses rotational motion. From left to right in Fig. 10(e): AM gives the best result for the first, third, fourth and fifth points (blue ellipses). For the second and sixth (white ellipses) points, it is not possible to determine which one from Suter and AM provides the best result. The same conclusion is valid for all studied image sequences.

6   Conclusion

This paper describes how to learn the ocean surface dynamics from an image model, AM, that summarises the shallow water equations by an advection term and an acceleration term **a**. This last represents physical processes such as the

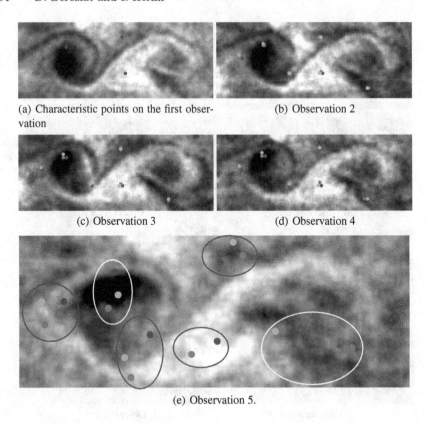

(a) Characteristic points on the first observation

(b) Observation 2

(c) Observation 3

(d) Observation 4

(e) Observation 5.

Fig. 10. Sun: red, Suter: green, AM: blue (Color figure online).

Coriolis force, the gravity force and the viscosity. A data assimilation algorithm was designed for AM that estimates the velocity field at the first acquisition date and the acceleration $\mathbf{a}(t)$ at each date of the studied interval. The function $\mathbf{a}(t)$ is of major importance for correctly assessing the hidden physical processes and accurately estimating motion on the whole image sequence.

The method has been quantified on synthetic data and illustrated on several SST sequences of Black Sea. On these last experiments, a display was given on motion fields and on the tracking of characteristic points. Moreover, the approach was compared with state-of-the-art optical flow algorithms. The conclusion is that modelling the acceleration, even as a simple unique term, improves motion estimation and allows tracking of structures.

The short-term perspectives of this research work will be to compare the acceleration term $\mathbf{a}(t)$ with forces involved in the shallow water model, in order to further validate the ability of the empirical model to assess geophysical processes.

Acknowledgements. Data have been provided by E. Plotnikov and G. Korotaev from the Marine Hydrophysical Institute of Sevastopol, Ukraine.

References

1. Byrd, R.H., Lu, P., Nocedal, J.: A limited memory algorithm for bound constrained optimization. J. Sci. Stat. Comput. **16**(5), 1190–1208 (1995)
2. Dee, D.: Bias and data assimilation. Q. J. R. Meteorol. Soc. **131**, 3323–3343 (2005)
3. Hascoët, L., Pascual, V.: The tapenade automatic differentiation tool: principles, model, and specification. ACM Trans. Math. Softw. **39**(3), 20 (2013)
4. Hundsdorfer, W., Spee, E.: An efficient horizontal advection scheme for the modeling of global transport of constituents. Mon. Weather Rev. **123**(12), 3554–3564 (1995)
5. Huot, E., Herlin, I., Mercier, N., Plotnikov, E.: Estimating apparent motion on satellite acquisitions with a physical dynamic model. In: International Conference on Image Processing, pp. 41–44, August 2010
6. Korotaev, G.K., Huot, E., Le Dimet, F.X., Herlin, I., Stanichny, S.V., Solovyev, D.M., Wu, L.: Retrieving ocean surface current by 4-D variational assimilation of sea surface temperature images. Remote Sens. Environ. **112**(4), 1464–1475 (2008) (special issue on data assimilation)
7. Le Dimet, F., Talagrand, O.: Variational algorithms for analysis and assimilation of meteorological observations: theoretical aspects. Tellus 97–110 (1986)
8. LeVeque, R.: Numerical Methods for Conservative Laws, 2nd edn. Lectures in Mathematics. ETH Zürich, Birkhäuser Verlag (1992)
9. Lions, J.L.: Optimal Control of Systems Governed by Partial Differential Equations. Springer, Berlin (1971)
10. Oguz, T., La Violette, P., Unluata, U.: The upper layer circulation of the Black Sea: its variability as inferred from hydrographic and satellite observations. J. Geophys. Res. **78**(8), 12569–12584 (1992)
11. Papadakis, N., Corpetti, T., Mémin, E.: Dynamically consistent optical flow estimation. In: Proceedings of International Conference on Computer Vision. Rio de Janeiro, Brazil (2007)
12. Sasaki, Y.: Some basic formalisms in numerical varational analysis. Mon. Weather Rev. **98**(12), 875–883 (1970)
13. Sun, D., Roth, S., Black, M.: Secrets of optical flow estimation and their principles. In: Proceedings of European Conference on Computer Vision, pp. 2432–2439 (2010)
14. Suter, D.: Motion estimation and vector splines. In: Proceedings of Conference on Computer Vision and Pattern Recognition, pp. 939–942 (1994)
15. Titaud, O., Vidard, A., Souopgui, I., Le Dimet, F.X.: Assimilation of image sequences in numerical models. Tellus A **62**, 30–47 (2010)
16. Trémolet, Y.: Accounting for an imperfect model in 4D-Var. Q. J. R. Meteorol. Soc. **132**(621), 2483–2504 (2006)
17. Vallis, G.K.: Atmospheric and Oceanic Fluid Dynamics, p. 745. Cambridge University Press, Cambridge (2006)
18. Valur Hólm, E.: Lectures notes on assimilation algorithms. Technical report, European Centre for Medium-Range Weather Forecasts Reading, UK (2008)
19. Wolke, R., Knoth, O.: Implicit-explicit Runge-Kutta methods applied to atmospheric chemistry-transport modelling. Environ. Model. Softw. **15**, 711–719 (2000)
20. Zhu, C., Byrd, R., Lu, P., Nocedal, J.: L-BFGS-B: Algorithm 778: L-BFGS-B, FORTRAN routines for large scale bound constrained optimization. ACM Trans. Math. Softw. **23**(4), 550–560 (1997)

Affine Coordinate-Based Parametrized Active Contours for 2D and 3D Image Segmentation

Qi Xue[1,2], Laura Igual[1,3]([✉]), Albert Berenguel[1,3], Marité Guerrieri[4],
and Luis Garrido[1,3]

[1] Department of Applied Mathematics and Analysis, Univerity of Barcelona,
Gran Via de les Corts Catalanes 585, 08007 Barcelona, Spain
{xueqi.bhlt,abcworld22}@gmail.com, {ligual,lluis.garrido}@ub.edu
[2] Department of Mathematics, Tongji University, Shanghai, China
[3] Computer Vision Center,
Edificio O, Campus UAB, Bellaterra (Cerdanyola), 08193 Barcelona, Spain
[4] Department of Computer Science and Applied Mathematics,
Univerity of Girona, Girona, Spain
mariteg@ima.udg.edu

Abstract. In this paper, we present a new framework for image segmentation based on parametrized active contours. The contour and the points of the image space are parametrized using a set of reduced control points that form a closed polygon in two dimensional problems and a closed surface in three dimensional problems. The active contour evolves by moving the control points. The parametrization, that uses mean value coordinates, stems from the techniques used in computer graphics to animate virtual models. The proposed framework allows to easily formulate region-based energies as the one proposed by Chan and Vese in both two and three dimensional segmentation problems. We show the usefulness of our approach with several experiments.

Keywords: Active contours · Affine coordinates · Mean value coordinates

1 Introduction

Active contours have been proved to be a powerful tool for segmentation in image processing. In active contours an evolving interface is propagated in order to recover the shape of the object of interest. In a two dimensional problem, the evolving interface is a contour whereas in a three dimensional problem the evolving interface is a surface. The interface is evolved by minimizing an energy that mathematically expresses the properties of the object to be segmented. In this energy functional, the terms corresponding to image features can be either edge-based and/or region-based. Edge-based terms measure features on the evolving interface to identify object boundaries and are usually based on a function of the image gradient [2]. They are known to be sensitive to noise and

© Springer International Publishing Switzerland 2015
S. Battiato et al. (Eds.): VISIGRAPP 2014, CCIS 550, pp. 206–222, 2015.
DOI: 10.1007/978-3-319-25117-2_13

thus energies based on such type of features usually need the evolving interface to be initialized near the solution.

Region-based terms were introduced by Chan and Vese [3]. In this case some features that are measured inside and outside of the regions allows to evolve the interface and thus drive the energy to its minimum. Chan and Vese minimize the variance of the gray-level in the interior and exterior regions. Region-based terms are known to be more robust to noise than edge-based terms and thus they usually do not need the initialization to be near the solution. Since the work of Chan and Vese, other approaches have extended the idea by measuring other types of features [19,20]. However, the latter approaches do measure the features in the whole inner and outer regions and thus may fail if these features are not spatially invariant. This problem has been tackled in [16], in which the inner and outer regions are defined by means of a band around the evolving interface. Thus, their approach is suited for segmenting objects having heterogeneous properties.

Depending on the representation of the evolving interface, active contours may be classified into parametric or geometric ones. In parametric approaches the interface in two dimensional problems may be described by means of a set of discrete points [15] or using basis functions such as B-splines [11]. In three dimensional problems, parametrizations using B-Splines [1] have also been used. In general, using basis functions such as B-splines allows to use less parameters to represent the curve rather than directly discretizing the curve or surface, have inherent regularity and hence do not require additional constraints to ensure smoothness. Parametric contours are able to deal easily with edge-based energies. However, dealing with region-based energies is more difficult, see [1,11] for instance. Moreover, parametric approaches require the user to define the number of control points that will be used to evolve the contour and it is difficult to deal with topological changes. This has favored that many works have been tackled using geometric approaches.

Geometric approaches represent the evolving interface as the zero level set of a higher dimensional function, which is usually called level set function. Therefore geometric approaches are also called level set approaches. Level set approaches are able to cope with the change of the curve topology and thus are able to segment multiple unconnected regions. This property has made level sets very popular approaches. However, level set approaches are usually computationally more complex and difficult to deal with since they increase the dimension of the problem by one. This makes level sets a difficult approaches in three-dimensional applications which are common in medical image segmentation.

In this paper, we focus on parametric representations due to its computational advantages and simplicity. A direct consequence of the explicit formulation is the loss of topological flexibility. However, this limitation is a mild constraint in many applications, where the goal may be to simply segment one connected object and thus the topological flexibility of the level sets is not needed.

We contribute with a novel framework for segmenting two and three dimensional connected objects using a new class of parametric active contours.

The parametrization is based on a class of deformable models well known in computer graphics such as the animation of characters for video games or movies. Such models are usually made up of millions of triangles. The motion of the character is controlled by a reduced number of control points: when these control points move the associated character deforms accordingly. A similar idea is applied in our paper: the evolving interface, the interior and exterior regions are parametrized by a set of control points. When these control points move the interface evolves correspondingly to the object to be segmented.

Our work stems from the ideas of free-form deformations [4,5]. Free-form deformations have been actively used for medical image registration [21]. However, to the best of our knowledge, free-form deformations have not been used for parametric active contours. In our work we use the *mean value coordinates* as the parametrization tool for the evolving interface [6]. Mean value coordinates have several advantages over free-form deformation, namely that control points only need to form a closed polygon in two dimensional problems (or surface in three dimensional problems) that may have any shape. Any point of the space, inside or outside of this polygon (or surface), can be parametrized with respect to the control points. For free-form deformations the control points need to form a regular shape and only interior points can be parametrized.

The rest of the paper is organized as follows: Sect. 2 reviews the related state-of-the-art work. Section 3 introduces the proposed segmentation method. Section 4 gives the implementation details. Section 5 shows the experimental results. Finally, Sect. 6 concludes the paper.

2   Related Work

In this Section, we review the state-of-the-art literature in level set techniques and computer graphic techniques related to our work.

2.1   Active Contours

The classic method of segmentation of Kass et al. [15] minimizes the following energy

$$E(\mathcal{C}) = \alpha \int_0^1 \|\mathcal{C}'(p)\|^2 \, dp + \beta \int_0^1 \|\mathcal{C}''(p)\|^2 \, dp - \lambda \int_0^1 \|\nabla I(\mathcal{C}(p))\| \, dp. \qquad (1)$$

where $\mathcal{C}(p) : [0,1] \to R^2$ is an Euclidean parametrization of the evolving interface and $I : R^2 \to R$ is the gray level image. As commented previously two different approaches may be used to represent $\mathcal{C}$, namely *parametric active contours* and *level sets*. The former is based on directly discretizing the curve $\mathcal{C}$ by means of a set of points and letting these points evolve independently. The level set methods are based on embedding the curve $\mathcal{C}$ in a higher dimensional function ϕ which is defined over all the image. Instead of evolving the curve $\mathcal{C}$, the function ϕ is evolved. The curve $\mathcal{C}$ corresponds to the zero level curve of ϕ.

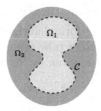

Fig. 1. For a region-based energy term the contour (dashed line) is evolved by minimizing an energy measured in the interior region, Ω_1, and the exterior region, Ω_2. For edge-based energies the contour is evolved by minimizing an energy on the evolving interface.

Rather than only using information on the interface, Chan and Vese proposed a method to evolve a curve by minimizing the variance in the interior region, Ω_1, and exterior region, Ω_2, defined by $\mathcal{C}$ [3], see Fig. 1. The energy the authors minimize is

$$E(\mathcal{C}) = \frac{1}{2} \iint_{\Omega_1} (I - \mu_1)^2 \, dx \, dy + \frac{1}{2} \iint_{\Omega_2} (I - \mu_2)^2 \, dx \, dy, \tag{2}$$

where the original terms based on the contour length and area of Ω_1 have been dropped for simplicity. In Eq. (2), the image $I : R^2 \rightarrow R$ corresponds to the observed data. The μ_1 and μ_2 refer to mean intensity values in the interior and exterior regions, respectively. In order to minimize the previous energy, authors use a level set method. They first compute the corresponding Euler-Lagrange equations and then discretize the evolution equations. The successive iterations of the evolution equations allows to evolve ϕ and thus the interface $\mathcal{C}$.

2.2   Mean Value Coordinates

An important problem in computer graphics is to define an appropriate function to linearly interpolate data that is given at a set of vertices of a closed contour or surface. Such interpolants can be used in applications such as parametrization, shading or deformation. The latter application is our main interest in this work and thus we will discuss it next. In this section we will first focus on the two dimensional case and afterwards we will discuss some details regarding the extension to the three dimensional case.

Gouraud [8] presented a method to linearly interpolate the color intensity at the interior of a triangle given the colors at the triangle vertices. Assume the triangle has vertices $\{\mathbf{v}_1, \mathbf{v}_2, \mathbf{v}_3\}$ and corresponding color values $\{f_1, f_2, f_3\}$. The color value of an interior point $\mathbf{p} = (x, y)$ of this triangle can be computed as follows

$$\hat{f}[\mathbf{p}] = \frac{\sum_j w_j f_j}{\sum_j w_j}, \tag{3}$$

where w_j is the area of the triangle given by vertices $\{\mathbf{p}, \mathbf{v}_{j-1}, \mathbf{v}_{j+1}\}$.

Many researchers have used these type of interpolants for mesh parametrization methods [6,10]) as well as free-form deformation methods [4] among others. Both applications require that a point $\mathbf{p}$ be represented as an *affine combination* of the vertices of an closed contour given by vertices $\mathbf{v}_i$:

$$\mathbf{p} = \frac{\sum_i w_i \mathbf{v}_i}{\sum_i w_i}, \tag{4}$$

and we say that the coordinate function

$$\frac{w_i}{\sum_j w_j} \tag{5}$$

is the corresponding *affine coordinate* of the point $\mathbf{p}$ with respect to the vertices $\mathbf{v}_i$.

A wealth of approaches have been published for the computation of the affine coordinates w_i. We may mention those that have interest for deformation applications, namely mean value coordinates [6], positive mean values coordinates [17], harmonic coordinates [12] or Green coordinates [18]. Among them, the mean value coordinates are easy to compute and therefore we have selected them in our work.

Mean value coordinates were initially proposed for mesh parametrization problems [6]. The author demonstrated that the interpolation generated smooth coordinates for any point $\mathbf{p}$ inside the kernel of a star-shaped polygon. Later on, in [9] it was demonstrated that mean value coordinates extended to any simple planar polygon and to any point $\mathbf{p}$ of the plane.

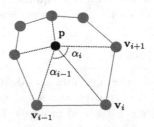

Fig. 2. For a given point $\mathbf{p}$, the mean value coordinate associated to point $\mathbf{v}_i$ is computed using $\mathbf{v}_i$, $\mathbf{v}_{i-1}$ and $\mathbf{v}_{i+1}$.

Assume the set of vertices $\mathbf{v}_j$, $j = 1 \ldots N$, of a simple closed polygon, is given. For a point $\mathbf{p} \in R^2$, its mean value coordinates $\varphi_i(\mathbf{p})$ are computed as

$$\varphi_i(\mathbf{p}) = \frac{w_i}{\sum_{j=1}^N w_j} \quad i = 1 \ldots N, \tag{6}$$

where

$$w_i = \frac{\tan(\alpha_{i-1}/2) + \tan(\alpha_i/2)}{||\mathbf{v}_i - \mathbf{p}||}, \tag{7}$$

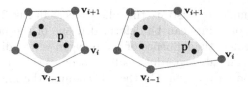

Fig. 3. Example of the deformation of a region by means of a closed polygon. From left to right: a polygon vertex $\mathbf{v}_i$ is moved producing the consequent deformation of the region after applying the interpolation function. Pixels near the moved vertex $\mathbf{v}_i$ are affected more than pixels father away due to the distance in the denominator of Eq. (7).

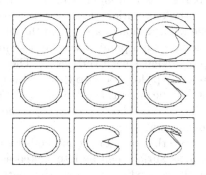

Fig. 4. The curve deformation for different polygon movements: external polygon (first and second row) and internal polygon (third row). The polygon is shown as solid line and the evolving curve as dashed line.

and $\|\mathbf{v}_i - \mathbf{p}\|$ is the distance between the vertex $\mathbf{v}_i$ and $\mathbf{p}$, and α_i is the *signed* angle of $[\mathbf{v}_i, \mathbf{p}, \mathbf{v}_{i+1}]$, see Fig. 2.

Given the affine coordinates $\varphi_i(\mathbf{p})$ of a point $\mathbf{p}$, the point $\mathbf{p}$ can be recovered with

$$\mathbf{p} = \sum_{i=1}^{N} \varphi_i(\mathbf{p})\mathbf{v}_i. \tag{8}$$

If the vertices $\mathbf{v}_i$ of the polygon move to positions $\mathbf{v}_i'$, the "deformed" point $\mathbf{p}'$ (see Fig. 3) can be recovered as

$$\mathbf{p}' = \sum_{i=1}^{N} \varphi_i(\mathbf{p})\mathbf{v}_i'. \tag{9}$$

For a given a set of points $\{\mathbf{p}\}$ the affine coordinates are computed in an independent way for each point as described in this section. If a point $\mathbf{v}_i$ of the polygon is stretched in a particular direction, all the points $\{\mathbf{p}\}$ follow the same direction with an associated weight given by $\varphi_i(\mathbf{p})$. The points $\mathbf{p}$ that are near the moved vertex have higher weights (see denominator of Eq. (7)) and thus suffer a larger "deformation" than the points which are farther away. Figures 3 and 4 show an example to illustrate this phenomena. In Fig. 3 we schematically show how a cloud of points can be deformed if one of the polygon vertices is moved. Figure 4 depicts the influence of the polygon vertex distance to a curve. In this example mean value coordinates have been used to represent the curve with respect the closed polygon. On the left column, the initial configurations are shown. The polygon, made up of 16 vertices, is shown with a solid line whereas the curve to deform with a dashed line. From top to bottom, different initializations for the polygon are shown: the first two rows are associated to polygon outside the curve whereas the last row is associated to a polygon inside the curve. The second and third column show how the curve is deformed when a control point of the polygon is moved. As can be seen, the movement of the polygon produces a smooth deformation of the curve. The closer the polygon to the curve, the higher the deformation is applied to the curve.

In the three dimensional case a point $\mathbf{p}$ will be described with respect to a closed surface made up of triangles with vertices $\mathbf{v}_i$. The algorithm to compute the mean value coordinates $\varphi_i(\mathbf{p})$ for a point $\mathbf{p} \in R^3$ is described in [13]. However, we would like to point out here that Eqs. (6) and (9) are still valid in the three dimensional case. Thus, it is relatively easy to extend any two dimensional method to the three dimensional case.

Equations (6) and (9) are the basis for the methods we have developed in our work. The closed polygon or surface can be interpreted as a cage that encloses the object to deform. In the context of our work, the control points are the cage vertices. The interior and exterior region points will be described with respect to these cage points using the affine coordinates. The cage vertices will evolve according to the minimization of an energy.

3   Method

Let us denote with $I(\mathbf{p}) : R^m \to R$ a gray-level image, where $m = 2$ (resp. $m = 3$) refers to a two-dimensional (resp. three-dimensional) data. Let $\mathbf{p}$ be a point of R^m and let us denote $\mathbf{v} = \{\mathbf{v}_1, \ldots, \mathbf{v}_N\}$ the set of cage vertices (or control points) associated to our parametrization. As commented previously, for $m = 2$ the cage is a closed polygon whereas for $m = 3$ the cage is a closed surface made up of triangles.

Let Ω_1 and Ω_2 be the initial set of pixels of the interior and exterior, respectively, of the evolving interface $\mathcal{C}$. For each point $\mathbf{p}$ of Ω_1 and Ω_2 the affine coordinates are computed. When a cage vertex $\mathbf{v}_i$ is moved, the points inside Ω_1 and Ω_2 are deformed accordingly (see Fig. 3). Thus, in our work the evolving interface does not need to be explicitly used to determine Ω_1 and Ω_2.

For simplicity we concentrate on the Chan and Vese model. However, the method can be easily extended to other types of models. The Chan and Vese model assumes that the gray-level values of pixels inside Ω_1 and Ω_2 can be modelled with different mean values. The energy functional to minimize is:

$$E(\mathbf{v}) = \frac{1}{|\Omega_1|} \sum_{\mathbf{p} \in \Omega_1} \frac{1}{2}(I(\mathbf{p}) - \mu_1)^2 + \frac{1}{|\Omega_2|} \sum_{\mathbf{p} \in \Omega_2} \frac{1}{2}(I(\mathbf{p}) - \mu_2)^2.$$

where the mean gray-level value of Ω_h with $h \in \{1, 2\}$ is defined as

$$\mu_h = \frac{1}{|\Omega_h|} \sum_{\mathbf{p} \in \Omega_h} I(\mathbf{p}), \tag{10}$$

and $| \cdot |$ denotes the cardinal of the set. The energy model is based on a direct discretization of the energy functional of Eq. (2). In order to minimize such equation, one approach is to compute its gradient and use it to drive the evolution of the interface to the minimum. This approach is called discretized optimization approach in the context of energy minimization problems [14].

Let $\mathbf{v}_j = (v_{j,x}, v_{j,y})^T$ for $m = 2$ (resp. $\mathbf{v}_j = (v_{j,x}, v_{j,y}, v_{j,z})^T$ for $m = 3$) be the coordinates of a cage vertex. The gradient of the energy with respect to $v_{j,x}$ is given by

$$\frac{\partial}{\partial v_{j,x}} \frac{1}{2} (I(\mathbf{p}) - \mu_h)^2 = (I(\mathbf{p}) - \mu_h) \left(\frac{\partial I(\mathbf{p})}{\partial x} \frac{\partial \mathbf{p}}{\partial v_{j,x}} - \frac{\partial \mu_h}{\partial v_{j,x}} \right). \tag{11}$$

A similar expression is obtained for the partial derivative with respect to $v_{j,y}$ and $v_{j,z}$. The partial derivatives with respect to $\mathbf{v}_j$ can be expressed in a compact form as follows

$$\nabla_{\mathbf{v}_j} \frac{1}{2} (I(\mathbf{p}) - \mu_h)^2 = (I(\mathbf{p}) - \mu_h)(\varphi_j(\mathbf{p})\nabla I(\mathbf{p}) - \nabla_{\mathbf{v}_j}\mu_h), \tag{12}$$

where

$$\nabla_{\mathbf{v}_j}\mu_h = \frac{1}{|\Omega_h|} \sum_{\mathbf{p} \in \Omega_h} \varphi_j(\mathbf{p})\nabla I(\mathbf{p}). \tag{13}$$

Thus, the gradient of E can be expressed as

$$\nabla_{\mathbf{v}_j} E(\mathbf{v}) = \frac{1}{|\Omega_1|} \sum_{\mathbf{p} \in \Omega_1} (I(\mathbf{p}) - \mu_1)(\varphi_j(\mathbf{p})\nabla I(\mathbf{p}) - \nabla_{\mathbf{v}_j}\mu_1)$$
$$+ \frac{1}{|\Omega_2|} \sum_{\mathbf{p} \in \Omega_2} (I(\mathbf{p}) - \mu_2)(\varphi_j(\mathbf{p})\nabla I(\mathbf{p}) - \nabla_{\mathbf{v}_j}\mu_2).$$

Note that for the computation of the gradient we have assumed that the cardinal of sets Ω_1 and Ω_2 do not depend on the cage vertex position. Indeed, as commented before, Ω_1 and Ω_2 are made up of individual pixels that deform as the vertex positions are moved. As vertices move the corresponding evolving interface may change its length (in two dimensional problems) or area (in three dimensional problems), but in our implementation no pixels are added or removed to Ω_1 and Ω_2 as the interface evolves. This approximation may be correct if the deformation applied to the cage is not grand. If the cage deforms with a big zoom, it may be necessary to recompute, at a certain iteration, the new set of pixels Ω_1 and Ω_2. This idea is similar to the one used in level sets implementation, in which the distance function is used for the evolving function ϕ. Efficient implementations re-initialize ϕ to a distance function only every certain iterations.

4   Implementation

In this Section, we describe the implementation issues associated to our proposed method.

Assume a gray-level image I and a mask Ω_1 are available. The mask Ω_1 can be for instance a binary image that is an approximation to the object we want to segment. Thus, Ω_1 will be used as initialization to our algorithm. In the case of medical images, for instance, this mask can be obtained by means of a registration of the image to be segmented with an atlas. In some cases, the mask can be manually defined by the user. In this paper we will assume that Ω_1 is a connected component. However, the method presented here is not restricted to this case. That is, Ω_1 may be composed of multiple connected components. However, note that our method is not able to deal with topological changes of Ω_1.

Several choices are available to obtain the outer pixels Ω_2. For instance, Ω_2 can be taken as the set $\Omega_2 = \Omega \setminus \Omega_1$, where Ω is the whole image support, see Fig. 1. This is the way in which Ω_2 is defined for many level set approaches and is the case in our paper. However, other approaches are possible. For instance, only a band around Ω_1 may be taken. This is useful in order to avoid taking pixels that are too far away from Ω_1, see [16]. Note that pixels of Ω_1 or Ω_2 may fall outside the polygon (2D) or surface (3D) formed by the cage vertices.

Once the set of pixels of Ω_1 and Ω_2 have been obtained, the affine coordinates of the pixels $\mathbf{p} \in \Omega_1$ and $\mathbf{p} \in \Omega_2$ have to be computed. The affine coordinates are computed using the cage vertices $\mathbf{v}$ which are assumed to be given as input. Recall that for each point $\mathbf{p} \in R^m$, $m = 2$ or $m = 3$, a set of N coordinates (i.e. floating point values) are obtained, where N is the number of cage vertices. These coordinates only have to be computed once, before the optimization algorithm is initiated.

For a given $E(\mathbf{v})$, the energy is minimized using a gradient descent. The gradient method iteratively updates $\mathbf{v}^{k+1}$ from $\mathbf{v}^k$ as follows

$$\mathbf{v}^{k+1} = \mathbf{v}^k + \alpha \mathbf{s}^k \tag{14}$$

where $\mathbf{s}^k$ is the so called search direction which is the negative gradient direction $s = -\nabla E(\mathbf{v}^k)$. The step α is computed via a back-tracking algorithm. Starting with an $\alpha = \alpha_0$, the algorithm iteratively computes $E(\mathbf{v}^k + \alpha \mathbf{s}^k)$ and reduces α until $E(\mathbf{v}^k + \alpha \mathbf{s}^k) < E(\mathbf{v}^k)$.

The computation of α_0 is a critical issue for good performance of the algorithm. In this paper α_0 is automatically computed at each iteration k so that the cage vertices move at most β pixels from its current position. That is,

$$\alpha_0 = \max_{\alpha}\{\alpha \mid \|\mathbf{v}_j^{k+1} - \mathbf{v}_j^k\| \le \beta\} \quad j = 1 \dots N \tag{15}$$

where β is a user given parameter (usually set to 5 in the experiments). The backtracking algorithm iteratively reduces the value of α until a value of α that decreases the energy value is found. The gradient descent stops when the backtracking reduces the α below a threshold associated to a cage movement, for instance 0.05 pixels.

For a given $\mathbf{v}$, the corresponding "deformed" pixel positions $\mathbf{p}$ of Ω_1 and Ω_2 are recovered using (9). The energy $E(\mathbf{v})$ and gradient $\nabla E(\mathbf{v})$ can be then computed. Note that the recovered pixel positions $\mathbf{p}$ may be non-integer positions. Hence, we apply linear interpolation to estimate $I(\mathbf{p})$ and $\nabla I(\mathbf{p})$ at such points. In addition, those pixels that fall outside of the image support are back projected to the image border (i.e. pixels at the image border are assumed to extend to infinity).

To recover the final position of the interface we parametrize the interface at the beginning (before optimization starts) with respect to the cage vertices. After convergence, the resulting interface can be easily recovered by using Eq. (9).

5   Experimental Results

In this section we present the experimental results for two and three dimensional segmentation problems. In particular, we perform different experiments to show different characteristics of the proposed segmentation method: the influence of the number of cage vertex selection, the importance of the cage shape, and the performance on some real images.

5.1   Two Dimensional Segmentation

We begin with the two dimensional segmentation, that is, the image $I : R^2 \rightarrow R$. We consider first some synthetic images, namely a square and a star, to analyze the behaviour of our approach. For these experiments Ω_1 is taken to be the whole interior of the evolving interface and Ω_2 the whole exterior. We would like to point out that in order to generate these images the cage and the evolving contour has been painted rounding the coordinates to nearest integers. Thus, within the resulting images shown in this section there may seem to be noisy boundaries.

Figure 5 is devoted to show the influence of the number of cage points to the segmentation result. The initializations are shown on the left column and the resulting segmentations are shown on the right. The cages are drawn with light gray whereas the evolving interface with dark gray. Experiments show how the interface deforms when using 4, 8 and 16 cage points. As can be seen, increasing the number of cage points allows better adaptation to the shape of the object of interest. The number of cage points influences the regularization of the evolving contour.

Figure 6 shows the robustness of the method to simple Gaussian noise. As expected, the model proposed in Eq. (10) is robust to Gaussian noise.

Figure 7 shows the effect of the cage distance to the evolving interface. For both experiments the initial evolving interfaces are the same, but the cages are created at a different distance from the evolving interface: the distance on the top is smaller than the bottom. The images on the left show initializations, whereas the images on the right show the results. It can be observed that the distance between the evolving interface and the cage plays an important role in

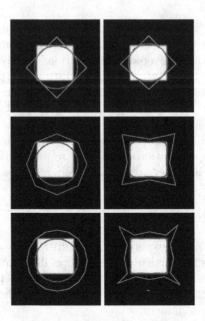

Fig. 5. Results for a synthetic image using different number of cage points. Top, 4 cage points. Middle, 8 cage points. Bottom, 16 cage points. On the left column the initial images are shown, while the right correspond to the results.

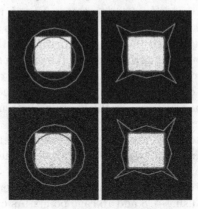

Fig. 6. Results for a synthetic image and 16 cage points using different levels of Gaussian noise. Top, noise variance is $\sigma^2 = 0.3$. Bottom, $\sigma^2 = 0.5$, where the image gray-level is assumed to be normalized to the range $[0, 1]$.

the deformations that can be applied to the evolving interface. In addition, note that the result is inherently regular (smooth ends instead of sharp ends) due to the nature of the parametrization. Therefore no regularization terms are needed in the energy function.

We now show results on real images. We have compared our method with a level-set based Chan and Vese implementation available in [7]. The code implements the following energy

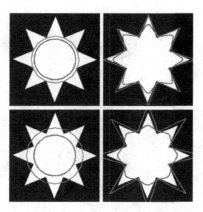

Fig. 7. Results for a synthetic image with cages at different distances from the evolving contour. Images on the left are the initial images and on the right are corresponding results.

Fig. 8. Example of level-set Chan and Vese segmentation. On the left, original image is shown. Initialization, a checker board shape for the embedded function, is shown in the middle. On the right, resulting segmentation shows contour in white color.

$$E(\mathcal{C}) = \mu \, \mathrm{Length}(\mathcal{C}) + \nu \, \mathrm{Area}(\mathcal{C}) \tag{16}$$

$$+\lambda_1 \frac{1}{2} \iint_{\Omega_1} (I - \mu_1)^2 \, dx \, dy + \lambda_2 \frac{1}{2} \iint_{\Omega_2} (I - \mu_2)^2 \, dx \, dy. \tag{17}$$

The first term of this energy controls the regularity of the evolving contour $\mathcal{C}$ by penalizing the length. The second term penalizes the enclosed area of $\mathcal{C}$ to control its side. The third and fourth terms are the terms we use in our energy and penalize the discrepancy between the piecewise constant models μ_1 and μ_2 and the input image.

The previous energy is implemented by means of a level-set, please see details in [7]. As an example, we show in Fig. 8 a segmentation result. The initialization of the level-set is performed by using a checkerboard shape and is shown in the middle Figure. The resulting segmentation is shown in white on the right Figure and has been obtained with the parameters $\mu = 0.25$, $\nu = 0.0$, $\lambda_1 = 1.0$ and $\lambda_2 = 1.0$. As can be seen, level-set based methods are able to deal with topology changes and thus multiple connected components may be obtained in the segmentation. Observe that here two contours have been obtained as result: the contour around the elephant and the one (small circle) over the head of the elephant.

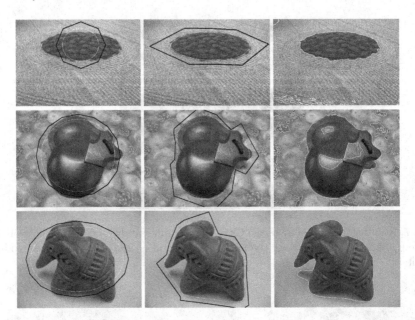

Fig. 9. Results for some real images. On the left column, the initialization is shown: the cage is drawn in black color whereas the evolving contour in white color. From top to bottom, the number of cage vertices used is 8, 16 and 16. In the middle, the resulting segmentation obtained with our method is shown: the resulting cage and segmentation are again shown in black and white, respectively. On the right, the resulting segmentation using a level-set based Chan and Vese segmentation is shown.

The results for the same image and two other real images are shown in Fig. 9. On the left, the initialization is shown: the cage is drawn in black color and the initial evolving contour in white. In the middle, the resulting segmentation obtained with our method is shown. On the right, the segmentation obtained with the Chan and Vese level-set method is shown. The parameters that have been used for the latter are $\mu = 0.25$, $\nu = 0.0$, $\lambda_1 = 1.0$, $\lambda_2 = 1.0$ for the second and third row, whereas $\mu = 0.1$, $\nu = 0.0$, $\lambda_1 = 1.0$, $\lambda_2 = 1.0$ has been used for the first row.

As can be seen, our method is able to obtain a regular boundary without using any specific energy term. Regularization is obtained thanks to the used parametrization. In the level-set method, regularization has to be explicitly introduced in the energy term.

On the other hand, the level-set method is able to deal with topology changes (see the rather high number of regions obtained in the images on the right column), whereas our method does not. As commented before, this limitation is a mild constraint in many applications, where the goal may be to simply segment one connected object and thus the topological flexibility of the level sets is not needed.

Fig. 10. Results for a synthetic 3D image. First column, original binary image. Second column, the initial surface. The cage has the same number of triangles but has a slightly higher radius. Third column, resulting surface after evolution. Fourth and fifth columns, the same experiment is repeated but with a higher number of triangles for both the initial surface and the cage.

5.2   Three Dimensional Segmentation

We now show some examples for the three dimensional segmentation problem. As commented before, in this case the cage is a surface made up of triangles.

Figure 10 shows an example for a synthetic image. The original image is shown on the first column of the Figure and represents a binary image with a cross. The original surface is a ball which constructed using the marching cube algorithm, is drawn on the second column. The cage is created from the previous surface just by inflating it a bit (0.5 pixels). The surface is then evolved using the proposed energy and the resulting segmentation is drawn on the third column. At the fourth and fifth columns the same experiment is shown but in this case the cage (as well as the mask) are modelled with a higher number of vertices. As can be seen, the number of vertices plays an important role in order to adapt the evolving surface to the object of interest.

Finally, we show the result with real images, see Fig. 11. We have recorded a video in which a ball moves along a line. The initial and end position of the ball is shown on the middle and right columns. The individual images of the video are then stacked as a 3D image, and an initial mask is created by defining a 3D ball whose center is approximately at the middle point joining the initial and end position and whose radius (in 3D) is the radius of the moving ball. On the first column of Fig. 11, it can be seen the result after convergence has reached. As can be observed, our method is able to properly extract the "shape" of the 3D object formed by the moving ball.

Fig. 11. Result for a real 3D image. First column, segmentation result for the moving ball. Middle and right columns, two of the images of the video. Video images a stacked to form a 3D image.

5.3 Computation Time

Currently, the code for 2D has been implemented in C language whereas the 3D only has been implemented in MATLAB. Regarding the 2D segmentation, the total amount of time spend by the algorithm is less than one minute. This includes the computational time of the mean value coordinates for all pixels of the image and the gradient descent to evolve the contour. With respect the 3D segmentation, the bottleneck is currently at the computation of the mean value coordinates for all the voxels of the volume, which may take a long time. Gradient descent takes then about a minute to converge.

Our method can be easily parallelized. Indeed, two issues use most of the computation resources in our approach, namely the interpolation of the image at non-integer values (see Eq. (9)) and the computation of the mean value coordinates. These two issue will receive our focus of attention in future work.

6 Conclusions

In this paper, we have presented a new parametrized active contour approach for two and three dimensional segmentation problem. The interface is evolved by moving a set of cage points. In the two dimensional problem the cage is a closed polygon whereas in the three dimensional problem the cage is a closed surface (made up of triangles). Mean value coordinates are used to parametrize the points of the space, inside or outside the cage. Other parametrization options exist, such as harmonic coordinates or Green coordinates, but mean value coordinates are simple to compute compared to other methods. Note that the parametrization has an intuitive interpretation. By moving a cage point, the associated points are moved correspondingly. This allows to introduce into the segmentation process the user interactivity: the user may manually move the control points to the correct position so that the system automatically learns from them.

In this paper, we have applied our framework for the implementation of a region-based energy, namely the classical Chan and Vese one; however, the framework is also suitable for the implementation of edge-based energies. In addition, we have shown that, within our framework, the regularization of the evolving interface can be controlled via the cage itself: the larger the distance of the cage to the evolving contour, the higher the contour regularization. Thus, there is really no need to include regularization terms within the energy.

Moreover, we think our method can be easily embedded in a shape-constrained approach, that is, an approach in which the movement of the cage is constrained so as to ensure certain shapes for the evolving contour. Our future work is to apply our method for 3D medical image segmentation problems. The method may be also extended to deal with multiple evolving interfaces: in this case one may impose restrictions on the shape of each of the evolving interfaces as well as interdependent restrictions on the location of each interface. We will study multi-resolution techniques iterating with different cages defined in each iteration by an increasing number of cage points. Finally, we also want to parallelize the method to improve its speed.

Acknowledgments. Q. Xue would like to acknowledge support from Erasmus Mundus BioHealth Computing, L. Igual and L. Garrido from MICINN projects, reference TIN2012-38187- C03-01, MTM2012-30772 and TIN2013-43478-P, and from Catalan Government award 2014-SGR-1219.

References

1. Barbosa, D., Dietenbeck, T., Schaerer, J., D'hooge, J., Friboulet, D., Bernard, O.: B-spline explicit active surfaces: an efficient framework for real-time 3D region-based segmentation. IEEE Trans. Image Process. **21**(1), 241–251 (2012)
2. Caselles, V., Kimmel, R., Sapiro, G.: Geodesic active contours. Int. J. Comput. Vis. **22**, 61–79 (1997)
3. Chan, T., Vese, L.: Active contours without edges. IEEE Trans. Image Process. **10**(2), 266–277 (2001)
4. Coquillart, S.: Extended free-form deformation: a sculpturing tool for 3d geometric modeling. SIGGRAPH **24**(4), 187–196 (1990)
5. Faloutsos, P., van de Panne, M., Terzopoulos, D.: Dynamic free-form deformations for animation synthesis. IEEE Trans. Visual. Comput. Graph. **3**(3), 201–214 (1997)
6. Floater, M.S.: Mean value coordinates. Comput. Aided Geom. Des. **20**(1), 19–27 (2003)
7. Getreuer, P.: Chan-vese segmentation. Image Processing On Line (2012)
8. Gouraud, H.: Continuous shading of curved surfaces. IEEE Trans. Comput. **C–20**(6), 623–629 (1971)
9. Hormann, K., Floater, M.: Mean value coordinates for arbitrary planar polygons. ACM Trans. Graph. **25**(4), 1424–1441 (2006)
10. Hormann, K., Greiner, G.: Continuous shading of curved surfaces. In: Curves and Surfaces Proceedings, pp. 152–163, Saint Malo, France (2000)
11. Jacob, M., Blu, T., Unser, M.: Efficient energies and algorithms for parametric snakes. IEEE Trans. Image Process. **13**(9), 1231–1244 (2004)
12. Joschi, P., Meyer, M., DeRose, T., Green, B., Sanocki, T.: Harmonic coordinates for character articulation. In: SIGGRAPH (2007)
13. Ju, T., Schaefer, S., Warren, J.: Mean value coordinates for closed triangular meshes. Proc. ACM SIGGRAPH **24**, 561–566 (2005)
14. Kalmoun, M., Garrido, K., Caselles, V.: Line search multilevel optimization as computational methods for dense optical flow. SIAM J. Imaging Sci. **4**(2), 695–722 (2011)
15. Kass, M., Witkin, A., Terzopoulos, D.: Snakes: active contour models. Int. J. Comput. Vis. **1**(4), 321–331 (1988)
16. Lankton, S., Tannenbaum, A.: Localizing region-based active contours. IEEE Trans. Image Process. **17**(11), 2029–2039 (2008)
17. Lipman, Y., Kopf, J., Cohen-Or, D., Levin, D.: GPU-assisted positive mean value coordinates for mesh deformations. In: Eurographics Symposium on Geometry Processing (2007)
18. Lipman, Y., Levin, D., Cohen-Or, D.: Green coordinates. In: SIGGRAPH (2008)
19. Michailovich, O., Rathi, Y., Tannenbaum, A.: Image segmentation using active contours driven by the Bhattacharyya gradient flow. IEEE Trans. Image Process. **16**(11), 2787–2801 (2007)

20. Rousson, M., Deriche, R.: A variational framework for active and adaptive segmentation of vector valued images. In: Proceedings of the Workshop on Motion and Video Computing, pp. 56–61. IEEE Computer Society (2002)
21. Rueckert, D., Sonoda, L., Hayes, C., Hill, D., Leach, M., Hawkes, D.: Nonrigid registration using free-form deformations: application to breast MR images. IEEE Trans. Med. Imaging **18**(8), 712–721 (1999)

Extended Bayesian Helmholtz Stereopsis for Enhanced Geometric Reconstruction of Complex Objects

Nadejda Roubtsova[(✉)] and Jean-Yves Guillemaut

Centre for Vision, Speech and Signal Processing,
University of Surrey, Guildford, UK
{n.roubtsova,j.guillemaut}@surrey.ac.uk

Abstract. Helmholtz stereopsis is an advanced 3D reconstruction technique for objects with arbitrary reflectance properties that uniquely characterises surface points by both depth and normal. Traditionally, in Helmholtz stereopsis consistency of depth and normal estimates is assumed rather than explicitly enforced. Further, conventional Helmholtz stereopsis performs *maximum likelihood* depth estimation without neighbourhood consideration. In this paper, we demonstrate that reconstruction accuracy of Helmholtz stereopsis can be greatly enhanced by formulating depth estimation as a Bayesian *maximum a posteriori probability* problem. In re-formulating the problem we introduce neighbourhood support by formulating and comparing three priors: a depth-based, a normal-based and a novel depth-normal consistency enforcing one. Relative performance evaluation of the three priors against standard maximum likelihood Helmholtz stereopsis is performed on both real and synthetic data to facilitate both qualitative and quantitative assessment of reconstruction accuracy. Observed superior performance of our depth-normal consistency prior indicates a previously unexplored advantage in joint optimisation of depth and normal estimates. Further, we highlight several known artefacts of Helmholtz stereopsis due to sensor saturations, normal corruption by 2D texture and by intensity sampling at grazing angles and enrich the initially proposed pipeline of Bayesian Helmholtz stereopsis with simple yet effective extensions to tackle the artefacts.

Keywords: 3D reconstruction · Helmholtz stereopsis · Complex reflectance

1 Introduction

As evidenced by the formidable volume of past and active research, reconstruction of 3D geometry is both challenging and much desirable for practical applications. A tremendous progress has been made in the field with sub-millimetre accurate geometries being obtained when capture conditions and surface properties are tailored for reconstruction. All prior algorithms rely on multiple images

© Springer International Publishing Switzerland 2015
S. Battiato et al. (Eds.): VISIGRAPP 2014, CCIS 550, pp. 223–238, 2015.
DOI: 10.1007/978-3-319-25117-2_14

to resolve inherent depth ambiguity, while differing in acquisition and view correlation employed to formulate ambiguity resolving constraints. Also variable is the degree of neighbour support used when characterising a surface point by its depth, normal or both.

The oldest of 3D reconstruction techniques is conventional stereopsis thoroughly surveyed in [14,15] for single- and multi-view respectively. Conventional stereopsis computes disparity through feature-point intensity matching between acquired views in the presence of sufficient texture. In this approach, surface points are characterised by depth, which is reciprocal to disparity. Conventional intensity-based stereo strongly relies on the inter-viewpoint constancy of feature appearance (i.e. Lambertian reflectance) failing when the assumption is violated. Alternative SIFT features are robust to intensity variations, although they only facilitate sparse representation.

Another reconstruction approach called photometric stereopsis [20] permits an arbitrary reflectance model as long as it is *a priori* known. Photometric constraints are linked to the response of a point to varying illumination at a *constant* viewpoint. The sought surface orientation is the one best reconciling intensity predictions and measurements. Essentially, the reflectance issue of conventional stereo is not solved by photometric stereo but rather the burden of it is shifted to the calibration phase of which a surface-orientation-dependent reflectance model is required. Reflectance modelling is a tedious task, often impossible to the desired accuracy for real objects. With an accurate model, photometric stereo directly outputs highly descriptive surface normals. Individual normals however need to be integrated into a surface often resulting in drift (global shape distortion) due to accumulation of numerical integration errors. Unlike conventional stereo, photometric stereo with its single viewpoint avoids the task of feature-point matching.

There are few techniques bypassing the need for reflectance modelling. One old example is the shape-from-silhouette algorithm [1] which computes a rough 3D outline of the object, its visual hull [10], by intersecting visual cones of multiple views. Visual hull is used for initialisation by many advanced reconstruction algorithms. A more recent highly promising photometric technique of Helmholtz stereopsis (HS) [23] addresses the fundamental problem of reflectance modelling by enforcing consistency of *reflectance-model-independent* Helmholtz reciprocity observation: i.e. swapping the sensor and the light source does not alter reflectance response. Besides its independence of the reflectance model, HS has the unique feature of point characterisation by both depth and normal. However, in its conventional formulation, HS is sequential with depth estimates uniquely determining the normals. Such unidirectional indexation of normals by depth estimates means that the typically noisy HS depth maps result in normal inaccuracies and hence in local and global reconstruction errors (although a certain degree of robustness to normal errors has been observed). Conceptually, the depth-normal dependency need not be unidirectional: depth and normal estimation can be unified in a single framework enforcing consistency and comparable accuracy levels of both estimates.

Conventional HS essentially performs *maximum likelihood estimation* (MLE). Even in the absence of noise, inherent point depth ambiguities exist, for instance, due to coincidental symmetries in the acquisition set-up configuration relative to the sampled surface. Since local evidence is ambiguous, neighbourhood support is clearly needed and warranted as there is always a degree of local smoothness in real objects. Yet, to our knowledge, MAP optimisation in the context of HS has not been previously attempted. In this paper, we propose a novel MAP formulation embedding HS into a Bayesian framework with a prior that for the first time explicitly enforces consistency between depth and normals. We show that, with the consistency prior, MRF optimisation of MAP HS indeed results in superior reconstruction accuracies. Unlike alternative depth-based or normal-based priors, the depth-normal consistency prior capitalises on the unique ability of HS to provide both depths and normals and, combining the two, produces the most correct geometries coherent in both depth and integrated normal representation. These conclusions are based on the quantitative and qualitative evaluation comparing conventional ML HS to MAP optimisation with 1. classical depth-based 2. normal-based and 3. novel depth-normal consistency priors. The evaluation also revealed a common set of artefacts caused by missing or corrupted intensity information. So, we address these artefacts caused by sensor saturation, textured surfaces (i.e. printed patterns) and inaccurate intensity sampling at grazing angles by means of a pair of artefact handling extensions inserted into our Bayesian pipeline as either pre- or post processing to the optimisation phase.

2   Related Work

Helmholtz reciprocity states that a light ray and its reverse will undergo the same processes of reflection, refraction and absorption [6]. Let $\hat{v}_1$ be the unit vector directed from the surface point to the camera and $\hat{v}_2$ the corresponding vector from the surface point to the light source. The implication of Helmholtz reciprocity, first observed by Zickler et al. [23] in the context of multi-view reconstruction, is that interchanging the light source and camera in the set-up, thereby swapping the vector definitions, has no effect on the point's reflective behaviour. Mathematically, Bidirectional Reflectance Distribution Function (BRDF) f_r is reciprocal: $f_r(\hat{v}_2, \hat{v}_1) = f_r(\hat{v}_1, \hat{v}_2)$. The following standard image formation equations for reciprocal images I_1 and I_2 respectively:

$$i_1 = f_r(\hat{v}_2, \hat{v}_1)\frac{\hat{n} \cdot \hat{v}_2}{r_2}, \quad i_2 = f_r(\hat{v}_1, \hat{v}_2)\frac{\hat{n} \cdot \hat{v}_1}{r_1} \tag{1}$$

express a surface point's image intensities i_1 and i_2 as a function of BRDF, surface normal $\hat{n}$, the two reciprocal unit vectors and the radiation fall-off factor r. Reciprocity of BRDF in conjunction with Eq. (1) result in the following constraint $\hat{w}$ notably without any dependency on the BRDF:

$$\left(i_1\frac{\hat{v}_1}{r_1} - i_2\frac{\hat{v}_2}{r_2}\right) \cdot \hat{n} = \hat{w} \cdot \hat{n} = 0. \tag{2}$$

With a single $\hat{w}$ per reciprocal pair, 3 or more reciprocal pairs result in constraint matrix W to which singular value decomposition (SVD) can be applied:

$SVD(W) = U\Sigma V^*$ where U, V are unitary and Σ is a rectangular diagonal matrix. The last column of V gives the normal at the sampled point. The last diagonal value of Σ, the SVD residual σ_3, tends to 0 when there is a mutual constraint consistency. For outlier elimination, Zickler et al. also involve σ_2 in their consistency measure: the SVD coefficient $\frac{\sigma_2}{\sigma_3}$ tends to infinity for consistent W (those of true surface points).

The requirement of multiple reciprocal pairs and the need for careful offline calibration led to discussions on acquisition impracticality of HS. In response, Zickler in [22] devises an auto-calibration algorithm using specular highlights and intensity patches, two inherent easily identifiable regions of interest. Unlike conventional stereo, intensity matching in HS is not conditional on the validity of the Lambertian assumption making intensity patches stable calibration markers.

The inherent appearance predictability of an intensity patch of one reciprocal image based on the other is also employed in [17] to formulate prediction error for registration in full 3D HS. Work on full 3D HS is scarce. The work of Delaunoy et al. [2] and the more recent publication of Weinmann et al. [19] are two notable examples. Both are variational approaches requiring computationally intensive optimisation over the entire surface with long execution times. The method of Weinmann et al. is more cumbersome due to fusion with structured light at acquisition. The impracticalities are however outweighed by the impressive degree of demonstrated reconstruction detail.

To address the issue of multiple constraint requirement, Zickler et al. in [24] propose binocular HS i.e. reconstruction from a single reciprocal pair. The method is a differential approach where the (single) constraint is formulated as a PDE of depth over the surface coordinates. The PDE requires initialisation and results in a family of solutions. The ambiguity is resolved through multi-pass optimisation. Although interesting as an exercise on acquisition simplification, additional computational complexity is perhaps not outweighed by the advantages.

Binocular HS and full 3D HS methods exploit the advantages of optimisation over a set of surface points. In the original HS, the depth label at each point is computed independently of its neighbours (without a Bayesian prior). Logically, there is a strong correlation between neighbouring surface points. By formulating a prior, the problem is turned into a MAP one, solvable by numerous mature MRF optimisation techniques [16]. The value of MRF optimisation for other 3D reconstruction approaches has been established. In conventional stereo, the top-performers are global optimisation algorithms solved through MRF-based graph-cuts and belief propagation [14]. [21] is an interesting work where MRF optimisation is performed in the context of photometric stereo with a normal-based (rather than depth-based) prior achieving remarkable robustness in the face of noisy input, complex geometries, shadows, transparencies etc.

Conventional HS will similarly benefit from MRF-based optimisation, since geometries obtained by ML HS lack in smoothness and fine structural detail due to inherent depth ambiguities, intensity measurement noise, sensor saturation and calibration/discretisation errors. In order to tackle the issues of conventional HS, we make the following contributions. Firstly, we devise a MAP framework allowing to apply MRF-based optimisation in the context of HS. Secondly, we introduce a novel smoothness prior enforcing coherence of depth and normal estimates of HS and show superiority of the prior over the purely depth-based or normal-based ones. Finally, we incorporate simple but effective extensions into the proposed pipeline for artefact elimination in the regions of locally missing or corrupted information because of sensor saturation, discretisation and calibration errors on textured surfaces as well as due to inaccurate intensity sampling at grazing angles. This extends our previous work in this area [13] which focused on the general MAP formulation and suffered from the aforementioned artefacts.

3   Methodology

In reconstruction by HS, we begin with N reciprocal image pairs $(N \geq 3)$ and a discrete volume V of $N_X \times N_Y \times N_Z$ voxels $v(x, y, z)$ containing the object. Each $v(x, y, z)$ is sampled by projection onto the reciprocal images to acquire a set of N intensity 2-tuples $\{(i_1, i_2)_1, \ldots, (i_1, i_2)_N\}$ and formulate N constraints $\hat{w}$ as in Eq. (2). Only those $v(x, y, z) \in V$ containing surface points will have N mutually consistent constraints. In standard HS, voxel sets $\mathcal{R}_{p(x,y)} = \{v(x, y, z) : x = x^*, y = y^*\}$ with constant 2D coordinates x^* and y^* are sampled exhaustively over V. Each $\mathcal{R}_{p(x,y)}$ defines the depth search space for random variable $p(x, y)$ with the optimal depth value $z_{p(x,y)} = z^*_{p(x,y)}$ corresponding to the surface point $P(x, y, z^*_{p(x,y)})$. In our work, the object's visual hull (VH) restricts the search space to $\mathcal{R}^{VH}_{p(x,y)} = \{v(x, y, z) : x = x^*, y = y^*, v(x^*, y^*, z_{p(x,y)}) \in VH\}$ (implicit in Eqs. (4) and (5)) limiting sampling in both 2D ($\mathcal{R}_{p(x,y)} \cap VH = \emptyset$ for some $p(x, y)$) and depth.

We postulate that depth estimation of HS is in fact a *labelling* problem where each random variable $p(x, y)$ is assigned depth label $z^*_{p(x,y)}$. The optimal solution to such a labelling problem is known [11] to be one maximising the *a posteriori* probability (MAP) i.e. the likelihood of a parameter assuming a certain value given the observation. The problem is typically translated into the equivalent one of posterior energy minimisation where the total energy is a sum of likelihood (data) and prior energy terms. Likelihood is related to the noise model of the observation representing its quality, viewed independently of the other observations. The prior term (typically local smoothness) is the knowledge of the problem encapsulating interaction between observations. In our work, the interaction is contained within the Markovian neighbourhood of $p(x, y)$ restricted by VH:

$$\mathcal{N}(p(x, y)) = \{p(x + k, y + (1 - k)) : k \in \{0, 1\}, \mathcal{R}^{VH}_{p(x+k, y+(1-k))} \neq \emptyset\} \qquad (3)$$

In this paper, we formulate depth estimation of HS as a MAP problem in Eq. (4). For each random variable $p(x, y)$ and given $p(x', y') \in \mathcal{N}(p(x, y))$, we define normalised data and smoothness costs, $E_d(x, y, z_{p(x,y)})$ and $E_s(x, y, z_{p(x,y)}, x', y', z_{p(x',y')})$ respectively, weighted by the normalised parameter α. The total optimisation is over all random variables within VH. The solution to the labelling problem is the label configuration $f^* = \{z^*_{p(x,y)} \in [Z_1, \ldots, Z_{N_Z}] : x \in [X_1, \ldots, X_{N_X}], y \in [Y_1, \ldots, Y_{N_Y}]\}$ selected from a set of S such configurations.

$$f_{MAP}^* = \arg\min_{f \in S} \sum_{p(x,y)} ((1-\alpha)E_d(x,y,z_{p(x,y)}) \tag{4}$$

$$+ \sum_{p(x',y') \in \mathcal{N}(p(x,y))} \alpha E_s(x,y,z_{p(x,y)}, x',y',z_{p(x',y')}))$$

In contrast to our approach, conventional HS solves a simpler *maximum likelihood* (ML) optimisation problem, without the smoothness prior, resulting a sub-optimal solution because each random variable is optimised independently:

$$f_{ML}^* = \arg\min_{f \in S} \sum_{p(x,y)} E_d(x,y,z_{p(x,y)}) \tag{5}$$

Sub-optimality leads to noisy depth maps and hence lacking surface smoothness and structural detail. The global shape may be reasonable, but the reconstruction finesse of conventional HS is limited because noisy depth labels index approximate normals. Through Bayesian formulation in Eq. (4), we endeavour to obtain cleaner depth maps improving accuracy by more accurate normal indexing. A Bayesian framework is clearly more suitable because of the strong statistical dependency between neighbouring depth estimates. The following two sections define data energy and the three investigated smoothness priors. Section 3.3 subsequently describes the methodology of the proposed artefact handling extensions to the framework.

3.1   Data Term

Depth hypothesis z of each $p(x,y)$ has got a likelihood $E_d(x,y,z)$ defined through the SVD coefficient $\frac{\sigma_2(v(x,y,z))}{\sigma_3(v(x,y,z))}$ associated with the corresponding voxel. The coefficient tends to infinity as z approaches the correct depth z^*. Since MRF optimisation is formulated as a minimisation, $E_d(x,y,z)$ is a decaying function of $\frac{\sigma_2(v(x,y,z))}{\sigma_3(v(x,y,z))}$. Throughout the paper, we adhere to the exponential formulation with the decay factor $\mu = 0.2 \times log(2)$:

$$E_d(x,y,z) = e^{-\mu \times \frac{\sigma_2(v(x,y,z))}{\sigma_3(v(x,y,z))}} \tag{6}$$

As it indicates likelihood, the function is bounded in the range $[0, 1]$.

3.2   Smoothness Term

Unlike conventional *data-based* HS, we introduce priors. We devise three prior types reflecting the unique ability of HS to generate both depth and normal estimates.

1. Depth-based Prior. The prior is known from conventional stereo. In this work, we define depth-based smoothness cost $E_{s\_d}(x,y,z,x',y',z')$ of voxel $v(x,y,z)$ relative to $v(x',y',z')$ by the discontinuity-preserving truncated squared difference of the neighbouring depth hypotheses normalised by the total number of labels (N_Z):

$$E_{s\_d}(x, y, z, x', y', z') = \lambda \times \min(S_{max}, (z - z')^2) \qquad (7)$$

where $S_{max} = (0.5 \times N_Z)^2$ is the truncation value and $\lambda = (N_Z)^{-2}$ is the normalising constant. With penalties for much different labels at neighbouring surface points, the prior encourages piece-wise constant depth and biases towards a fronto-parallel representation. Discontinuities are preserved by truncation moderating depth fluctuation penalties.

2. Normal-based Prior. Surface characterisation through normals is typical of photometric techniques. A suitable normal-based prior would enforce locally constant normals encouraging locally flat, though not necessarily fronto-parallel, surfaces. Hence, this prior is less restrictive of reconstructed surfaces than the depth-based one. The complication in formulating normal-based priors arises because 1. normals are continuous 3D quantities that cannot be labels in our discrete framework and 2. normal correlations are irregular expressions not optimisable by graph cuts [9]. In this work, discrete depths are still the labels, but rather than depth information, we use the corresponding normal similarity to assess label compatibility. Sequential tree re-weighted message passing (TRW-S) [8,18] MRF optimisation is used consistently throughout the paper because it does not require regularity of prior.

Given photometric normals $\hat{n}(x, y, z) = (n_x, n_y, n_z)^\top$ and $\hat{n}(x', y', z') = (n'_x, n'_y, n'_z)^\top$ of voxels $v(x, y, z)$ and $v(x', y', z')$ respectively, we formulate the smoothness-based constraint:

$$E_{s\_n}(x, y, z, x', y', z') = \pi^{-1} \arccos(\,\hat{n}(x, y, z) \cdot \hat{n}(x', y', z')\,) \qquad (8)$$

The cost function of Eq. (8) is the normalised correlation angle between normals.

3. Depth-normal Consistency Prior. The depth-based prior seeks to clean up depth maps by enforcing their smoothness, while the normal-based approach promotes gradual spatial evolution of the normal field. Both approaches are one-sided: the depth is optimised indexing the normals or vice versa. Depth and normal estimation processes are however not independent and must be consistent with each other. We formulate a prior explicitly enforcing consistency between depths and normals, for the first time performing joint depth map and normal field optimisation.

Each depth transition between $v(x, y, z)$ and $v(x', y', z')$ uniquely defines local *geometric* normal $\hat{n}_g(x, y, z, x', y', z')$, always contained in the transition plane (xz or yz). If the depth transition is correct, the geometric normal correlates well with the projections of the photometric normals of $v(x, y, z)$ and $v(x', y', z')$ (their normal estimates), respectively $\hat{n}_{ph\_prj}(x, y, z)$ and $\hat{n}_{ph\_prj}(x', y', z')$, onto the corresponding depth transition plane. Mathematically, the correlation degree can be expressed by the correlation angle. For example, for $v(x, y, z)$ we have:

$$\phi_{ph-g} = \pi^{-1} \arccos(\hat{n}_{ph\_prj}(x, y, z) \cdot \hat{n}_g(x, y, z, x', y', z')) \qquad (9)$$

Since the range of the arc-cosine function is $[0, \pi]$, the orientation of the geometric normal must be forced to be consistent with the photometric normals

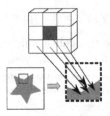

Fig. 1. The idea of patch-based averaging.

(i.e. out of the surface, $z > 0$). Hence, the depth-normal consistency prior $E_{s\_dn}(x, y, z, x', y', z')$ is formulated as follows:

$$E_{s\_dn}(x, y, z, x', y', z') = \frac{1}{2}(\phi_{ph-g} + \phi'_{ph-g}). \tag{10}$$

3.3 Artefact Handling Extensions

HS is sensitive to calibration and discretisation errors. As described in [5], these result in a surface point being projected onto somewhat mismatched pixel locations in the reciprocal images substituting inconsistent intensities i_1 and i_2 into Eq. (2). In a region with 2D (imprinted) texture, a projection mismatch occurs when, within the same reciprocal pair, a point is projected into different intensity fields (see Fig. 1) resulting in reconstruction of 2D texture as geometrical detail (stripes and flowers from Fig. 2 in Fig. 5(b), (c)).

To counteract the problem we use patch-based averaging - a simplified version of the method of Guillemaut *et al.* [5]. As illustrated in Fig. 1, the method involves replacing the intensities of individually projected $v(x, y, z)$ in Eq. (2) by the intensity average of all projections within a fronto-parallel rectangular $p_x \times p_y$ patch: $\sum\limits_{v(x',y',z') \in P(v(x,y,z))} \left(i_{v(x',y',z')} \right)$ where

$$P(v(x, y, z)) = \{v(x', y', z') : x' \in [x - p_x, x + p_x], y' \in [y - p_y, y + p_y], z' = z\}.$$

In addition to removing texture from geometry as reported previously [5], in this work the extension is introduced to exert a mitigating effect on the sensor saturation artefacts since instead of the indiscriminative sensor upper limit, the intensities in Eq. (2) acquire more meaningful values through regional support of patch-based averaging.

For points sampled at grazing incidence angles, normals cannot be resolved accurately. Due to the noisy normals, the rims of 2.5D reconstructions appear "tattered". It is possible to repair the rims by replacing the noisy photometric normals by geometric normals obtained from silhouette cues. It has been observed that the binary (foreground/background) mask of the visual hull projection onto the image plane of the virtual camera gives a neat outline of the reconstructed view. The projection is convolved with the vertical and horizontal Scharr gradient operators resulting in directional gradient images $G_x(x, y)$

Fig. 2. Real data: object appearance.

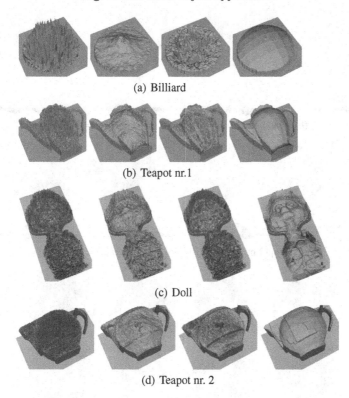

(a) Billiard

(b) Teapot nr. 1

(c) Doll

(d) Teapot nr. 2

Fig. 3. Depth maps corresponding to reconstructions in Fig. 5.

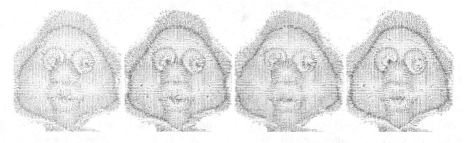

Fig. 4. Normal fields corresponding to doll reconstructions in Fig. 5.

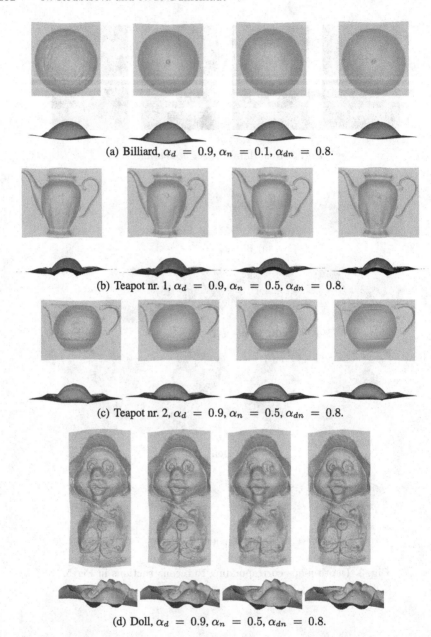

(a) Billiard, $\alpha_d = 0.9$, $\alpha_n = 0.1$, $\alpha_{dn} = 0.8$.

(b) Teapot nr. 1, $\alpha_d = 0.9$, $\alpha_n = 0.5$, $\alpha_{dn} = 0.8$.

(c) Teapot nr. 2, $\alpha_d = 0.9$, $\alpha_n = 0.5$, $\alpha_{dn} = 0.8$.

(d) Doll, $\alpha_d = 0.9$, $\alpha_n = 0.5$, $\alpha_{dn} = 0.8$.

Fig. 5. Final reconstructions (frontal and side views) by integration using the FFT-based Frankot-Chellappa (FC) algorithm. In each sequence of 4 images (left to right) standard (ML) HS is compared against proposed MAP HS formulation using MRF optimisation with depth-based (d), normal-based (n) and depth-normal consistency (dn) priors (data-smoothness weighting α as is indicated in each case). Sampling resolution $\Delta x \times \Delta y \times \Delta z$ and sampled volumes $|V| = N_X \times N_Y \times N_Z$ are as follows. Doll: $1.0\,mm \times 1.0\,mm \times 0.5\,mm$, $|V| = 160 \times 82 \times 60$; teapot no. 1: $1.0\,mm \times 1.0\,mm \times 0.25\,mm$, $|V| = 150 \times 200 \times 320$; teapot no. 2: $1.0\,mm \times 1.0\,mm \times 0.25\,mm$, $|V| = 120 \times 190 \times 480$; billiard : $1.0\,mm \times 1.0\,mm \times 0.25\,mm$, $|V| = 60 \times 60 \times 100$.

and $G_y(x,y)$. Subsequently, for all outer $p(x,y)$, the normal components n_x and n_y are replaced by $G_x(x,y)$ and $G_y(x,y)$ respectively. Random variable $p(x,y)$ with location (x,y) is considered outer if, for some window $w_x \times w_y$, there is a neighbouring location $\{(x',y') : x' \in [x - w_x, x + w_x], y' \in [y - w_y, y + w_y]\}$ in the background.

4   Evaluation

We perform evaluation of the proposed method on both real and synthetic data to enable qualitative and quantitative analysis. Throughout, our method with its 3 prior options is compared against the standard HS approach without MRF optimisation. For real data, the quality of results is assessed visually as there is no ground truth. Synthetic input imagery, on the other hand, is generated from an a priori known mesh and hence permits quantitative assessment of both local and global shape deviations of the reconstructions. Evaluation of our artefact handling mechanisms concludes the section.

4.1   Real Data

Real data is composed of 4 sets (Fig. 2) from [4], each posing different challenges. The billiard ball and the teapots are specular smooth objects. The teapots are more complex with wider specularities and 2D texture (e.g. stripes, flowers). The terracotta doll is Lambertian but has many fine geometric details (e.g. dimples, clothing).

Figures 3 and 5 respectively contrast estimated depth maps and the corresponding integrated surfaces of the proposed MAP HS formulation against standard (ML) HS. MAP optimisation priors are compared by qualitatively accessing both local and global accuracy of the generated depth maps and surfaces. The relative weight α is tuned for each prior independently but the optimal setting per prior tends to be consistent across all datasets. Surface integration is performed from the normal fields in the frequency domain using the Frankot-Chellappa (FC) algorithm [3].

The results in Fig. 5 show that, relative to ML HS, MAP optimisation clearly produces smoother reconstructions (the surface of the billiard ball and the teapots in Fig. 5(a)–(c)) with finer structural detail (the doll's dimples/eyebrows in Fig. 5(d); the corresponding normal fields in Fig. 4 are rectified to reveal structural detail). On the relative performance of the priors, the key observation is that only the depth-normal consistency prior generates geometrically correct depth maps. Global accuracy of the depth-based prior ranges from poor (ball, Fig. 3(a)) to reasonable (teapots, Fig. 3(b), (d)) and high (doll, Fig. 3(c)). The depth maps of the depth-based prior are however universally noisy. Corresponding normal-based prior results are consistently heavily distorted. Correct reconstruction is only possible from a geometrically (globally and locally) accurate depth map. While this is evident for the doll dataset, in other cases distorted depth maps of the normal-based approach may, in the frontal view, seem to produce decent reconstructions (Fig. 5(b), (c)). The deceiving appearance results from optimisation accidentally finding well-correlating normals at wrong depths. These integrate into locally smooth surfaces, yet distort the global shape for all four datasets (side views in Fig. 5(a)–(d)). As the depth map accuracy assessment suggests, the best global shape reconstruction belongs to the depth-normal consistency prior optimisation.

Presented reconstructions show two notable artefacts: erroneous printed pattern reconstruction as geometric detail (e.g. stripes and flowers on the teapots) and inaccurate normal estimation at the rims resulting in their "tattered" appearance. In Sect. 4.3 we show improved reconstructions by means of proposed artefact handling extensions from Sect. 3.3.

4.2   Synthetic Data

Our synthetic object of choice is a sphere (radius $= 200\,mm$). The mesh of the sphere is rendered in POV-Ray [12] with reflectance consisting of diffuse and specular components and under controlled camera/light source configurations to produce 8 noise-free reciprocal pairs. For robustness assessment, two additional image sets are produced by corrupting the original set by Gaussian zero-mean *intensity* noise with the variance levels of respectively 0.001 and 0.01. Reconstruction quality is quantified by the geometric error: the reconstructed mesh is sampled for each point seeking the closest distance on the target ground truth mesh. Figure 6 shows the RMS error as a function of noise variance comparing standard ML HS and our MRF (MAP) formulation with 3 different priors (sampled volume $|V| = 82 \times 82 \times 251$ at resolution $5\,mm \times 5\,mm \times 1\,mm$). The input intensity range is $[0, 65535]$. Hence, the normalised noise variance of 0.001 and 0.01 translates into the absolute intensity standard deviations of 2072 and 6554 respectively. Surface integration is performed using unscreened Poisson surface reconstruction [7] because the method outputs surfaces characterised in the absolute world coordinate system facilitating their easy comparison unlike FC surfaces in their individual relative reference frames needing alignment.

In Fig. 6 MRF optimisation with the depth-normal consistency prior clearly outperforms the other approaches by an order of magnitude margin achieving reconstruction accuracies of roughly 0.5 mm and 5 mm for noise-free and noise-corrupted sets respectively. The accuracy of the other methods is the order of a few mm and a few cm for the two respective cases. For the noise-free case, standard HS comes closest to the depth-normal prior with the 2 mm error. The second observation to be made is that the normal-based prior is utterly ineffective with the accuracy always below standard HS, consistent global shape distortion and the greatest performance deterioration with increased noise levels owing to the high susceptibility of normals as continuous 3D quantities to input noise via SVD. Even for the noise-free case, the error with the normal-based prior is high because there is no theoretical guarantee that the best correlating normals correspond to the correct depth positions. The depth-based prior appears to facilitate an improvement compared to standard HS for noise-corrupted data.

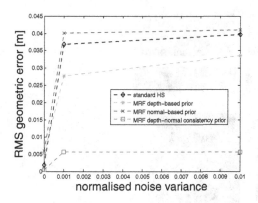

Fig. 6. Quantitative evaluation on synthetic data. RMS geometric error as a function of normalised intensity noise variance for our MRF (MAP) HS formulation with different priors and standard (ML) HS.

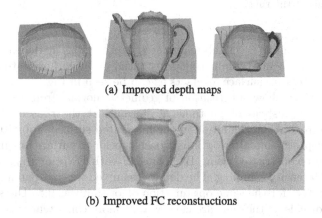

(a) Improved depth maps

(b) Improved FC reconstructions

Fig. 7. Mitigation of saturation and mis-calibration artefacts through patch-based averaging. Settings: MAP HS with depth-normal consistency prior, patch-based averaging and $\alpha_{dn} = 0.5$.

On visual inspection however, the depth-based reconstruction is seen to retain the global shape distortion of the standard HS result, albeit mitigated as its lower RMS indicates. In the noisy case, only our depth-normal consistency prior produces a global shape visually acceptable as a sphere.

4.3 Extension Evaluation

Along with sharpening true geometries, optimisation is seen to strengthen the effect of calibration errors: e.g. printed stripes and flowers on the teapots (Fig. 2) appear embedded into geometry (Fig. 5(b), (c)). Performing patch-based intensity averaging during input data sampling to bring inconsistent measurements into closer correspondence counteracts the problem. In Fig. 7(a) we show how

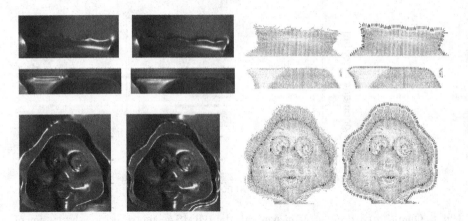

Fig. 8. Rim repair by substitution of geometric normals from silhouette cues. Integrated surfaces and normal fields with respectively photometric (left) and geometric (right) normals at the rims.

patch-based averaging effectively eliminates depth errors in the saturated and pattern regions, hence eliminating/substantially mitigating saturation and embedded 2D texture surface artefacts as can be seen in Fig. 7(b).

Figure 8 shows how substitution of geometric normal from silhouette cues (marked in red) instead of photometric normals at the rims of 2.5D reconstruction improves their definition in the integrated surface. Substitution takes place as post-processing after optimisation but prior to surface integration. The teapots were reconstructed using Bayesian HS with DNprior ($\alpha_{dn} = 0.5$) and patch-based intensity sampling. The doll was sampled without patch-based averaging and its α_{dn} for the same optimisation algorithm was 0.8. The substituted geometric normals at the rims are clearly a lot more consistent with the object outlines than the original photometric normals. The resultant 2.5D reconstructions have much better defined rims than the original reconstructions.

5   Conclusions and Future Work

We have proposed a novel MRF optimisable MAP formulation of HS, instead of its standard ML form. To this end, we have formulated and compared a depth-based, a normal-based and a specially tailored depth-normal consistency prior. We conclude that correctly utilising the given of piece-wise surface smoothness in the MAP formulation greatly improves both local and global reconstruction accuracy relative to ML results. Both quantitative and qualitative results indicate our depth-normal consistency prior to be the correct formulation of the smoothness term, which by enforcing consistency between depth and normal information produces the best results in terms of both local smoothness and global object shape. The results generated with the prior are uniquely consistent in both depth and integrated normal domain with the normals being indexed from a geometrically correct depth map. The computational overhead for the prior is dependent on the size of the sampled voxel volume and for the presented real data ranges from 2 min (billiard ball) to 4 hours (teapot no. 2). In future work we can reduce the run-times substantially by embedding MRF optimisation into a coarse-to-fine framework using octrees and/or by parallelising and porting prior cost pre-computation (the bottleneck of the pipeline) onto the GPU.

In addition, we have introduced extensions to deal with artefacts having the generic common cause of locally missing or corrupted information. Specifically, we have addressed the issues with sensor saturation, manifestation of calibration and discretisation errors in textured surface reconstruction and inaccurate rim reconstruction due to sampling at grazing angles by introducing patch-based averaging and geometric normal substitution mechanisms. The extensions have been shown effective, however several directions for further exploration are possible. Firstly, at the current resolution, with a single pass (as opposed to the stated as future work coarse-to-fine) implementation, the applicability of patch-based averaging is limited to smooth objects as it may oversmooth structural detail (e.g. of the doll). Also as future work, in a manner similar to occlusion handling in conventional stereo, one can explore sensor saturation mechanisms integrated directly into the MRF optimisation and involving a label for missing information. Lastly, performing geometric normal substitution prior to MRF optimisation instead of as post-processing may result in the positive effect on the global shape as well as the demonstrated local accuracy improvement.

References

1. Baumgard, B.: Geometric Modeling for Computer Vision. Ph.D. thesis, University of Stanford (1974)
2. Delaunoy, A., Prados, E., Belhumeur, P.N.: Towards Full 3D Helmholtz Stereovision Algorithms. In: Kimmel, R., Klette, R., Sugimoto, A. (eds.) ACCV 2010, Part I. LNCS, vol. 6492, pp. 39–52. Springer, Heidelberg (2011)
3. Frankot, R., Chellappa, R.: A method for enforcing integrability in shape from shading algorithms. PAMI 10(4), 439–451 (1988)
4. Guillemaut, J.Y., Drbohlav, O., Illingworth, J., Šára, R.: A maximum likelihood surface normal estimation algorithm for Helmholtz stereopsis. VISAPP 2, 352–359 (2008)
5. Guillemaut, J.Y., Drbohlav, O., Šára, R., Illingworth, J.: Helmholtz stereopsis on rough and strongly textured surfaces. In: 3DPVT, pp. 10–17 (2004)
6. Helmholtz, H.: Treatise on Physiological Optics, vol. 1. Dover, New York (1925)
7. Kazhdan, M., Bolitho, M., Hoppe, H.: Poisson surface reconstruction. In: SGP, pp. 61–70 (2006)
8. Kolmogorov, V.: Convergent tree-reweighted message passing for energy minimization. PAMI 28(10), 1568–1583 (2006)
9. Kolmogorov, V., Zabih, R.: What energy functions can be minimized via graph cuts? PAMI 26, 65–81 (2004)
10. Laurentini, A.: The visual hull concept for silhouette-based image understanding. PAMI 16(2), 150–162 (1994)
11. Li, S.: Markov random field models in computer vision. In: ECCV, vol. B, pp. 361–370 (1994)

12. POV-Ray: POV-Ray - The Persistence of Vision Raytracer. http://www.povray. org/ (2013)
13. Roubtsova, N., Guillemaut, J.Y.: A Bayesian framework for enhanced geometric reconstruction of complex objects by Helmholtz Stereopsis. In: VISAPP (2014)
14. Scharstein, D., Szeliski, R.: A taxonomy and evaluation of dense two-frame stereo correspondence algorithms. IJCV 47(1–3), 7–42 (2002)
15. Seitz, S., Curless, B., Diebel, J., Scharstein, D., Szeliski, R.: A comparison and evaluation of multi-view stereo reconstruction algorithms. CVPR 1, 519–528 (2006)
16. Szeliski, R., Zabih, R., Scharstein, D., Veksler, O., Kolmogorov, V., Agarwala, A., Tappen, M., Rother, C.: A comparative study of energy minimization methods for markov random fields with smoothness-based priors. PAMI 30(6), 1068–1080 (2008)
17. Tu, P., Mendonça, P.R., Ross, J., Miller, J.: Surface registration with a Helmholtz reciprocity image pair. In: Proceedings of IEEE Workshop on Color and Photometric Methods in Computer Vision (2003)
18. Wainwright, M.J., Jaakkola, T.S., Willsky, A.S.: Map estimation via agreement on trees: message-passing and linear-programming approaches. IEEE Trans. Inf. Theory 51(11), 3697–3717 (2005)
19. Weinmann, M., Ruiters, R., Osep, A., Schwartz, C., Klein, R.: Fusing structured light consistency and helmholtz normals for 3D reconstruction. In: BMVC, pp. 108.1–108.12. BMVA Press (2012)
20. Woodham, R.J.: Shape from shading, chap. Photometric Method for Determining Surface Orientation from Multiple Images, pp. 513–531. MIT Press, Cambridge, MA, USA (1989)
21. Wu, T.P., Tang, K.L., Tang, C.K., Wong, T.T.: Dense photometric stereo: a markov random field approach. PAMI 28(11), 1830–1846 (2006)
22. Zickler, T.: Reciprocal image features for uncalibrated Helmholtz stereopsis. In: CVPR, pp. 1801–1808 (2006)
23. Zickler, T., Belhumeur, P.N., Kriegman, D.J.: Helmholtz stereopsis: exploiting reciprocity for surface reconstruction. IJCV 49(2–3), 215–227 (2002)
24. Zickler, T.E., Ho, J., Kriegman, D.J., Ponce, J., Belhumeur, P.N.: Binocular helmholtz stereopsis. ICCV 2, 1411–1417 (2003)

Efficient Region-based Classification
for Whole Slide Images

Grégory Apou[1](✉), Benoît Naegel[1], Germain Forestier[2], Friedrich Feuerhake[3],
and Cédric Wemmert[1]

[1] Icube, University of Strasbourg, 300 bvd Sébastien Brant, 67412 Illkirch, France
gapou@unistra.fr
[2] MIPS, University of Haute Alsace, 12 rue des Frères Lumiere,
68093 Mulhouse, France
[3] Institute for Pathology, Hannover Medical School, Carl-Neuberg-Straße 1,
30625 Hannover, Germany

Abstract. For the past decade, new hardware able to generate very high
spatial resolution digital images called Whole Slide Images (WSIs) have
been challenging traditional microscopy. But the potential for automation is hindered by the large size of the files, possibly tens of billions of
pixels. We propose a fast segmentation method coupled with an intuitive
multiclass supervised classification that captures expert knowledge presented as morphological annotations to establish a cartography of a WSI
and highlight biological regions of interest. While our primary focus has
been the development of a proof of concept for the analysis of breast cancer WSIs acquired after chromogenic immunohistochemistry, this method
could also be applied to more general texture-based problems.

Keywords: Whole slide images · Biomedical image processing ·
Segmentation · Classification

1 Introduction

In recent years, the advent of digital microscopy deeply modified the way certain diagnostic tasks are performed. While the initial diagnostic assessment
and the interpretation histopathological staining results remain a domain of
highly qualified experts, digitization paved the way to semi-automated image
analysis solutions for biomarker quantification and accuracy control. With the
expected increase of the number and quality of slide scanning devices, pathologists are facing the challenge to integrate complex sets of relevant information,
partially based on conventional morphology, and partially on molecular genetics
and computer-assisted readout of single immunohistochemistry (IHC) parameters [1].

Despite their potential for automation to help reduce bias [2], the integration
of WSIs in routine diagnostic workflows in the clinical setting is not straightforward [3]. Indeed, these images can contain hundreds of millions or even billions

© Springer International Publishing Switzerland 2015
S. Battiato et al. (Eds.): VISIGRAPP 2014, CCIS 550, pp. 239–256, 2015.
DOI: 10.1007/978-3-319-25117-2_15

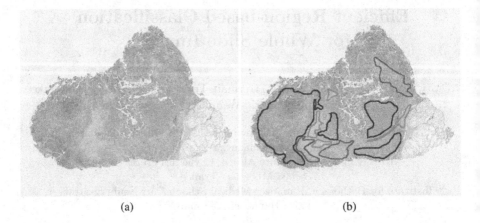

(a)   (b)

Fig. 1. Example of a whole slide image: (a) raw; (b) manually annotated by an expert: the regions are outlined, each color represents a class; excluded regions (dark green) are given as examples, which is why the image is not fully annotated (Color figure online).

of pixels, causing practical difficulties for the storage, transmission, visualization and processing by conventional algorithms in a reasonable time. Figure 1 presents an example of a "small" WSI of 18000 by 15000 pixels. Moreover, this new technology is still perceived as ineffective by pathologists who are more familiar with the use of classical light microscopy.

This paper presents a new approach, based on an object-oriented analysis (segmentation, classification) to establish an automatic cartography of WSI. The main objective is to propose a decision support tool to help the pathologist to interpret the information contained by the WSI.

Compared to former work on WSI analysis, our contributions are: (i) an efficient computational framework enabling the processing of WSI in reasonable time, (ii) an efficient texture descriptor based on quantized color histograms and (iii) a multiclass supervised classification based on expert annotations allowing a complete cartography of the WSI.

The paper is organized in 4 sections. First, existing approaches to analyze WSI are presented (Sect. 2), followed by the different steps of the method (Sect. 3). Then, experiments on WSI of breast cancer samples are described to evaluate the benefits of this approach (Sect. 4). Finally, we conclude and present some perspectives (Sect. 5).

2   Related Work

As optical microscopy image analysis is a specific field of image analysis, a great variety of general techniques to extract or identify regions already exists. The main distinctive characteristic of the whole slide images (WSI) is their very large size, which makes impossible the application of number of conventional processing, despite their potential interest.

Signolle and Plancoulaine [4] use a multi-resolution approach based on wavelet theory to identify the different biological components in the image, according to their texture. The main limitation of this approach is its speed: about 1 hour to analyze a sub-image of size 2048×2048 pixels, and several hundred hours for a complete image (60000×40000 pixels).

To overcome this drawback, several methods have been developed to avoid the need for analyzing entire images at full resolution. Thus, Huang et al. [5] noted that, to determine the histopathological grade of invasive ductal breast cancer using a medical scale called *Nottingham Grading System* [2,6], it is important to detect areas of "nuclear pleomorphism" (i.e. areas presenting variability in the size and shapes of cells or their nuclei), but such detection is not possible at low resolutions. So, they propose a hybrid method based on two steps: (i) the identification of regions of interest at a low resolution, (ii) multi-scale algorithm to detect nuclear pleomorphism at a high resolution in the regions of interest identified previously. In addition, through the use of GPU technology, it is possible to analyze a WSI in about 10 min, which is comparable to the time for a human pathologist.

Indeed, the same technology is used by Ruiz [7] to analyze an entire image (50000×50000) in a few dozen seconds by splitting the image into independent blocks. To manage even larger images (dozen of gigapixel) and perform more complex analyzes, Sertel [8] uses a classifier that starts on low-resolution data, and only uses higher resolutions if the current one does not provide a satisfactory classification. In the same way, Roullier [9] proposed a multi-resolution segmentation method based on a model of the pathologist activity, starting from the coarsest to the finest resolution: each region of interest determined at one resolution is partitioned into 2 at the higher resolution, through a clustering performed in the color space. This unsupervised classification can be performed in about 30 min (without parallelism) on an image of size 45000×30000 pixels.

More recently, Homeyer [10] used supervised classification on tile-based multi-scale texture descriptors to detect necrosis in gigapixel WSI in less than a minute. While tessellating an image with square tiles is simple and effective, the resulting contours can exhibit a lack of smoothness. In order to address that, our method combines supervised classification with a fast segmentation method close to the superpixel lattices described by Moore [11].

The characteristics of our method compared to others are summarized in Table 1.

3   Method

To achieve a fast and efficient classification of whole slide images, we propose a methodology enabling to partition the initial image in relevant regions. This approach is based on a superpixel segmentation algorithm and on a supervised classification using a textural characterization of each region.

Table 1. Comparison of our method with some existing ones. H&E means Hematoxylin and Eosin, a widely used staining.

Method	Pixels	Coloration	Classes	Performance	Parallelism
[7]	10^9	H&E	2 (supervised)	145 s (GeForce 7950 GX2)	GPU
[4]	10^9	H, DAB	5 (supervised)	>100 h (Xeon 3 GHz)	Unknown
[8]	10^{10}	H&E	2 (supervised)	8 min (Opteron 2.4 GHz)	Cluster of 8 nodes
[5]	10^9	H&E	4 (supervised)	10 min (GeForce 9400M)	GPU
[9]	10^9	H&E	5 (unsupervised)	30 min (Core 2.4 GHz)	Parallelizable
[10]	10^9	H&E	3 (supervised)	<1 min (Core 2 Quad 2.66 GHz)	Unknown
Proposed method	10^8	H, DAB, PRD	6 (supervised)	10 min (Opteron 2 GHz)	Parallelizable

3.1   Overview

The proposed method relies on two successive steps: (i) image segmentation into segments or "patches"; (ii) supervised classification of these segments. The main challenge of the segmentation step is to provide a relevant partitioning of the image in an efficient way due to the large size of whole slide images. To cope with this problem, we propose to partition the image by using a set of horizontal and vertical optimal paths following image high gradient values. We make the assumption that the distribution of biological objects that a pathologist would use to produce a decision can be described by a textural representation of a region. Thus, the classification step is based on a textural approach where each region is labelled according to its texture description.

A training set of texture descriptors is computed from a set of manually annotated images enabling a supervised classification based on a k-nearest neighbor strategy. The method overview is illustrated in Figs. 2 and 3.

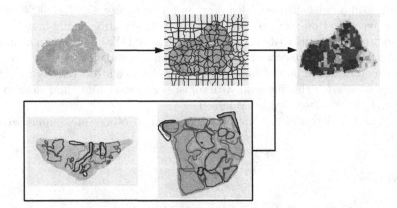

Fig. 2. Method overview. Left: image under analysis, image partitioning and regions classification. Right: manually annotated images allowing the construction of the training set.

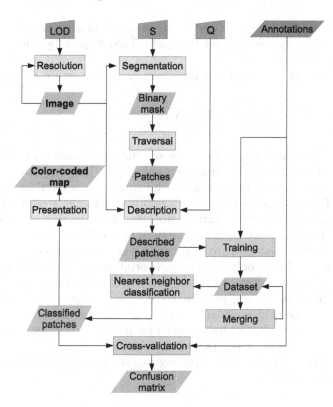

Fig. 3. Data structures and processes. Red boxes represent parameters, green box represents expert annotations (ground truth). Blue boxes are image-related operations, yellow boxes are data mining-related. Gray boxes are data structures input, exchanged, modified and output by the processes (Color figure online).

3.2 Segmentation

Let $f : E \rightarrow V$ be a 2D discrete color image defined over a domain $E \subseteq \mathbb{Z}^2$ with $V = [0, 255]^3$. Let f_i denotes the scalar image resulting of the projection of f on its ith band. We suppose that E is endowed with an adjacency relation. A path is a sequence of points $(p_1, p_2, \ldots, p_n)$ such that, for all $i \in [1, \ldots, n-1]$, p_i and p_{i+1} are adjacent points. Let W, H be respectively the width and the height of f. The segmentation method is based on two successive steps.

First, the image f is partitioned into W/S vertical and H/S horizontal strips, with $S > 1$ an integer controlling the width of a strip.

Second, a path of optimal cost is computed from one extremity of the strip to the other in each image strip. The cost function is related to the local variations, hence favoring the optimal path to follow high variations of the image. More precisely, the local variation of f in the neighborhood of p is computed as:

$$g(p) = \max_{q \in N(p)} d(f(p), f(q)), \tag{1}$$

where d is a color distance, and $N(p)$ the set of points adjacent to p. In our experiments we used $d(a, b) = \max_i |a_i - b_i|$ (L_∞ norm) and $N(p) = \{q \mid \|p - q\|_\infty \leq 1\}$ (8-adjacency).

The global cost associated to a path $(p_i)_{i \in [1...n]}$ of length n is defined as:

$$G = \sum_{i=1}^{n} g(p_i) \tag{2}$$

From an algorithmic point of view, an optimal path maximizing this summation can be retrieved using dynamic programming [12] in linear time with respect to the number of points in the strip, hence requiring to scan all image pixels at least once. To speed up the process, a suboptimal solution is computed instead by using a greedy algorithm: starting from an arbitrary seed at an extremity of the strip, the successive points of the path are added by choosing, in a local neighborhood, the point q where $g(q)$ is maximal. By doing so, the values of g are computed on the fly, only for the pixels neighboring the resulting path.

Some notable properties of this algorithm are: (i) its speed due to the fact that not all pixels need to be processed, (ii) a low memory usage even for large images because the only required structures are the current strip and a binary mask to store the result, (iii) its potential for parallelization, because all strips in a given direction can be processed independently.

Figure 4 illustrates the steps of the segmentation method and Fig. 5 gives an example of the end result.

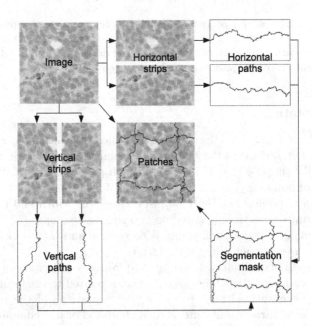

Fig. 4. Path computation in horizontal and vertical strips, leading to an image partition.

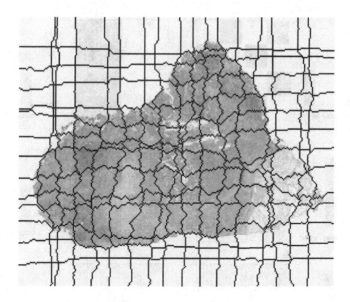

Fig. 5. Example of image partitioning based on our algorithm.

3.3 Training

To create a training base, the reference images are segmented into patches. Using the expert annotations, each patch is associated with a class or label. Then, by computing a texture descriptor for each patch, we can create an association between a texture and a group of labels. A texture can have several labels if it is present in regions of different classes. As a result, the training base can be modeled as a function $B : T \to G$ where T is the set of texture descriptors and $G = \mathcal{P}(L)$ is the power set of all labels L.

When $|B(t)| \neq 1$, the texture t is ambiguous. Section 4.2 describes how to measure this phenomenon and thus quantify the validity of the model. In order to perform the classification, all $B(t)$ must be singletons. To that end, B is updated so that ambiguous textures are classified as excluded elements.

3.4 Classification

Some authors use distributions of descriptors to describe textures [13]. The chosen descriptors can be arbitrarily complex, and, as a starting point, we decided to use simple color histograms that are functions $H : V \to [0, 1]$ that associate each pixel color to its frequency in a given patch. The ability of histogram to discriminate between textures is illustrated in Fig. 6. Given the fact that all images were obtained using the same process and equipment with the same settings, no image preprocessing was deemed necessary.

In order to perform a supervised multi-class classification, we opted for a one nearest neighbor classification because of its simplicity (no assumption needs to be made on the distribution of the textures in the descriptor space) and the sufficient amount of training samples available (which turned out to be a bit excessive

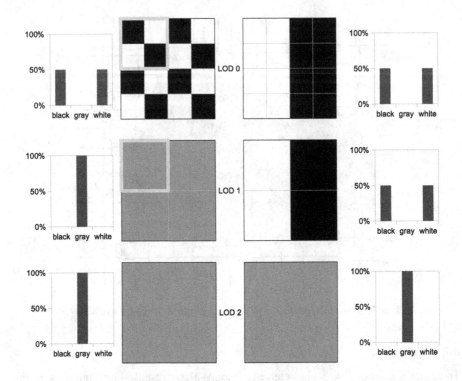

Fig. 6. The most common argument against the use of histograms for texture characterization can be visualized in the top row: the two presented images cannot be distinguished by their histograms. But by performing a subsampling operation to reduce the resolution (example outlined in red), local constraints are introduced and the middle row shows that the two images can now be distinguished by their histograms. This ability disappears in this example after further subsampling, suggesting the possible existence of an optimal level of resolution to characterize a given set of textures with their histograms (Color figure online).

with sometimes up to several millions of elements). For this kind of classification, we measure the distance between histograms using the euclidean metric:

$$d\left(h_1, h_2\right) = \sqrt{\sum_v \left(h_2\left(v\right) - h_1\left(v\right)\right)^2} \tag{3}$$

This choice of metric is arbitrary and will serve as a reference for future work.

4   Experiments and Results

4.1   Data

Unlike most of the work published on the subject, our images are obtained using a double-staining process [14]. More precisely, we used formalin-fixed paraffin-embedded breast cancer samples obtained from Indivumed®, Hamburg,

Germany. Manual immunohistochemistry staining was performed for CD8 or CD3 and Perforin, and antibody binding was visualized using 3,3-diaminoben-zidine tetrahydrochloride (DAB, Dako, Hamburg, Germany) and Permanent Red (PRD, Zytomed, Berlin, Germany). Cell nuclei were counterstained with hema-toxylin before mounting. As a result, cancerous (in this type of tumor predom-inantly large) and noncancerous (predominantly small) cell nuclei appear blue (hematoxylin). The chromogenic labeling of the lymphocyte lineage markers CD3 and CD8 results in brown (DAB) staining of cell membranes and/or cytoplasm; the antibody labeling for perforin is visualized in red (PRD) as granular cyto-plasmic color dots. Due to the small size of cells and sectioning effects, the blue nuclei may sometimes be covered by brown or red color.

For our experiments, 7 whole slide images ranging from $1 \cdot 10^8$ to $5 \cdot 10^8$ pixels have been annotated by a pathologist using 6 simplified labels (Fig. 7): invasive tumor (defined as predominantly solid formations), invasive tumor (simplisticly defined as less coherent tumor cell groups diffusely infiltrating pre-existing tis-sues), intersecting stromal bands (defined as the non-malignant mesenchymal tissue component regardless whether pre-existing, or induced by tumor growth), DCIS (Ductal Carcinoma In Situ; in a simplified manner this class encompasses real ductal invasive carcinoma in situ, and invasive tumor accidentally grow-ing within ductal structures), non-neoplastic glands and ducts, and edges and artifacts to be excluded.

The annotations do not explicitly provide quantifiable cell characteristics that could be used to design a medically relevant cell-based region identification. Instead, they take the form of outlines that may or may not match visual features (Fig. 1). Even though some of these seem obvious, like the background, it is not easy for an untrained eye to establish a set of intuitive rules that would explain the expert's opinion, even after trying some simple visual filters (quantization, thresholding). Moreover, the classes are not uniformly represented in our data (Fig. 8): while excluded elements are described by only a handful of annotations, they actually account for the majority of the area of the images, especially because of the background; on the other hand, ductal structures constitute a minority and are sometimes completely missing.

Nonetheless, the delimited regions appear to exhibit a texture-related behav-ior, and we can use that to decide on a model: a delimited region is made of a set of patches that can be identified by their texture, and delimited regions of the same class share the same set of textures. Thus, by partitioning the image into patches and labeling each patch based on its texture, we can draw a color-coded map like in Fig. 9.

4.2   Model Evaluation

The method is evaluated with a leave-one-out cross-validation involving all the annotated images: for each image (in our set of 7), a training base is created with the other 6. All the values given in the rest of this article are obtained by averaging the values from 7 experiments.

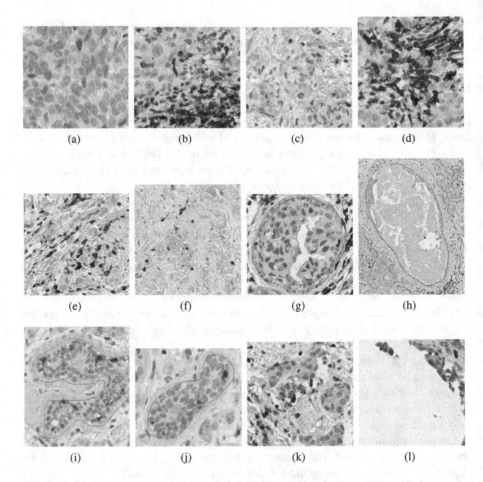

Fig. 7. (a,b) Annotation: Invasive tumor (solid formations). Description: High concentration of cancerous cells.

Note: All classes can contain foreign objects, such as the brown lymphocytic infiltrate that can be seen in (b), and sometimes the same objects (foreign or not) can be seen in several classes (compare with c,d). (c,d) Annotation: Invasive tumor (diffusely infiltrating pre-existing tissues).Description: Cancerous cells disseminated in noncancerous tissue.

(e, f) Annotation: Intersecting stromal bands. Description: Connective tissue.

(g, h) Annotation: DCIS and invasion inside ductal structures. Description: A Ductal Carcinoma In Situ refers to cancer cells within the milk ducts of the breast; the simplified definition used here encompasses "real" ductal invasive carcinoma in situ, in the sense of a ductal proliferation respecting the anatomical structure of ducts, and invasive tumor components growing withing pre-existing ducts; some examples may include central necroses as shown in (h).

(i, j) Annotation: Nonneoplastic glands and duct. Description: Noncancerous structures.

(k, l) Annotation: Edges and artefacts to be excluded. Description: Nonbiological features (background, smudges, bubbles, blurry regions, technician's hair, ...) and damaged biological features (borders, defective coloration, missing parts, ...).

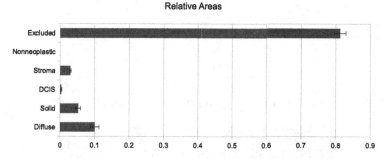

Relative Areas

Fig. 8. Relative areas of all the classes: average for each training set. Standard deviation is given as horizontal bars.

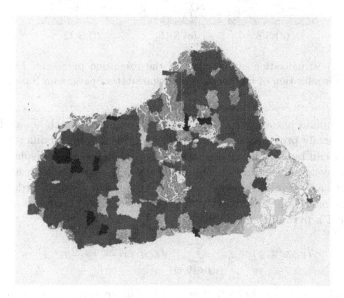

Fig. 9. Classification map obtained by the presented method; the colors match those used by the expert, except for the excluded regions which are left untouched (Color figure online).

The quality of the model can be measured by computing the certainty of the training base for each label l:

$$C(l) = \frac{|B^{-1}(\{l\})|}{|\{t \in T : l \in B(t)\}|} \qquad (4)$$

When the certainty is 100 %, it means that the only group containing the label is a singleton, and so the textures can be used to uniquely characterize the corresponding class. On the other hand, a certainty of 0 % means that the textures are too ambiguous for a one-to-one mapping.

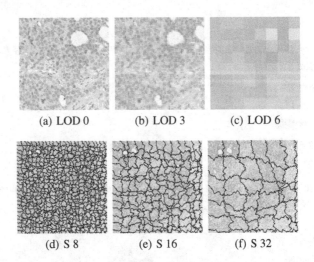

(a) LOD 0   (b) LOD 3   (c) LOD 6

(d) S 8   (e) S 16   (f) S 32

Fig. 10. (a-c) Visualization of the effect of the resolution parameter LOD on pixel data. (d-f) Visualization of the effect of the segmentation parameter S on pixel data.

Since a human pathologist uses a multi-resolution approach [9], a whole slide image is typically provided as a set of images corresponding to different magnifications that can be used by visualization software to speed up display. But they restrict the systematic study of the impact of the resolution level, and can also cause additional degradation due to lossy compression. So, in order to determine the information available at each level of detail (LOD), we compute for each image I a pyramid defined by:

$$I_{LOD}(x,y) = \frac{1}{4} \sum_{(i,j)\in\{0,1\}^2} I_{LOD-1}(2x+i, 2y+j) \tag{5}$$

The original image is at LOD 0 (Fig. 10). It can be observed that high resolution is correlated with high data set certainty for the chosen texture descriptor (Fig. 11).

Ideally, the segmentation algorithm should create patches of the right size, so that each patch would contain just enough information to identify a class-characteristic texture. Instead, we will assume the existence of a common texture scale that applies to all classes: the segmentation parameter S (Fig. 10). At high resolution, a texture described by its color histogram can help identify a class with very little doubt (Fig. 11). But at lower resolutions, larger values of S increase the ambiguity of the texture description, because the patches become large enough to contain multiple textures from adjacent regions of different classes.

Finally, despite the use of sparse structures, a one nearest neighbor classification using color histograms requires large amounts of memory and processing time. A simple yet effective technique to mitigate this issue is to use a quantization scheme where the values used as histogram keys are $2^Q \lfloor \frac{v}{2^Q} \rfloor$ instead of v,

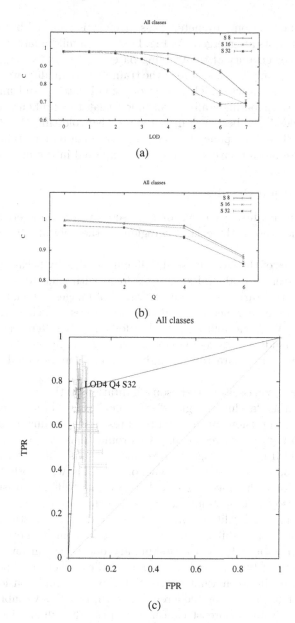

Fig. 11. (a) Visualization of the effect of the resolution parameter LOD on the overall certainty of the training base for different values of S and Q = 4. Standard deviation is given as vertical bars.

(b) Visualization of the effect of the quantization parameter Q on the overall certainty of the training base for different values of S and LOD = 4. Standard deviation is given as vertical bars (barely visible because they are small).

(c) ROC points (light gray) for each parameterization are obtained by merging experimental data points for all the classes. For clarity, a convex curve (red) has been synthesized from these points with one point that stands out; such a synthetic curve makes sense because a classifier can be built for any interpolated point [15]. Standard deviation is given as horizontal and vertical bars.

and Q is the quantization parameter. Less memory is required because textures with close descriptors are merged. As Fig. 11 shows, a mild quantization ($Q \leq 4$) barely affects the certainty of the training base.

By considering only this measure of the training base quality, we would expect to get the best results with high resolution, small patches and minimal quantification. But we experimentally determined that the configuration (LOD 4, Q 4, S 32) yielded the best overall outcome when plotting the data in ROC space (Fig. 11). The discrepancy between high training base certainty and lesser classification results can have several causes, explored in the next section.

4.3   Classification Results

By observing the confusion matrix for a chosen set of parameters (Table 2), we can see that the class of the excluded regions is the only one to be adequately detected.

The structures of the DCIS class are difficult to identify because of the ambiguity between seemingly ductal structures filled with cancerous cells, but also due to their relative rarity in the current series of images. The latter holds also true for the class of non-neoplastic ductal structures, which were rare in the current series of WSIs as well. As expected for a highly differentiated epithelial structure, the glands and ducts of the pre-existing mammary gland tissue were more prone to textural characterization, but more examples would be needed for confirmation.

The remaining 3 classes illustrate some limitations of the model. The "stroma" class is detected as "excluded regions", "stroma" and "diffuse invasion". As it turns out, "diffuse invasion" means that textures corresponding to cancerous cells are mixed with textures corresponding to stroma. This mixing creates conflicts which are resolved by assigning the "excluded regions" class to the ambiguous textures (Sect. 3.3). The same phenomenon explains why both "solid formations" (regions of high cancerous cell density) and "diffuse invasion" (regions of low cancerous cell density) are detected as a mixture of "excluded regions", "solid formations" and "diffuse invasion". Regarding the distinction between "solid formations" and "diffuse invasion", it is important to note that the current annotation is by definition a preliminary one based on few samples that needs to be iteratively improved in further studies. There is a poorly defined range of variation between what descriptive reports would consider "diffusely infiltrating" versus "solid", or "coherent" growth, and this variability will even increase with inclusion of breast cancer subtypes other than "unspecific type" (synonymous with the former designation "ductal invasive"). One possible workaround for this problem could be to merge the two classes into one "tumor are" class, which could result in a much higher detection rate of up to 90 %. Another way to address the challenge of tumor heterogeneity is to investigate a broader range of samples and work on a closer approximation towards widely accepted annotations concordant between experienced pathologists.

But the major underlying problem is that the model is flawed: a texture unit defined by one patch is not enough to identify a class. The certainty of

Table 2. Confusion matrix for a particular parameterization (LOD 4, Q 4, S 32). Results are given as mean and standard deviation computed from 7 values.

	DCIS	Excluded regions	Stroma	Solid formations	Nonneoplastic objects	Diffuse invasion
DCIS	**57 % ± 49**	29 % ± 35	2 % ± 5	4 % ± 4	0.1 % ± 0.2	8 % ± 10
Excluded regions	0.3 % ± 0.4	**85 % ± 3**	3 % ± 1	3 % ± 2	0.2 % ± 0.2	9 % ± 3
Stroma	1 % ± 3	29 % ± 10	**31 % ± 10**	4 % ± 3	0.1 % ± 0.2	25 % ± 10
Solid formations	0.9 % ± 2	19 % ± 14	0.5 % ± 0.6	**44 % ± 32**	0.0 % ± 0.1	35 % ± 24
Nonneoplastic objects	0 % ± 0	11 % ± 20	0.6 % ± 1	6 % ± 14	**71 % ± 45**	12 % ± 18
Diffuse invasion	0.2 % ± 0.2	37 % ± 12	3 % ± 3	11 % ± 9	2 % ± 3	**47 % ± 15**

Table 3. Sensitivity and specificity for a particular parameterization (LOD 4, Q 4, S 32). Results are given as mean and standard deviation computed from 7 values.

Class	Sensitivity	Specificity
DCIS	57 % ± 49	99.6 % ± 0.6
Excluded regions	85 % ± 3	67 % ± 13
Stroma	31 % ± 10	97 % ± 2
Solid formations	44 % ± 32	96 % ± 3
Nonneoplastic objects	71 % ± 45	99.8 % ± 0.2
Diffuse invasion	47 % ± 15	89 % ± 4

the training bases is high because, unexpectedly, simple color histogram are not only strong enough to describe such texture units, but they also capture small variations that can almost identify the patches themselves, hence the large size of the training bases (see next section). We conjecture that if we were to ignore these variations, we would obtain a small set of textures (maybe a few dozens) that could be used to identify regions like a pathologist does. With that in mind, the confusion matrix could suggest that regions corresponding to stroma, diffuse invasion and solid formations are made up of the same set of textures (as described by color histograms of patches), but they differ by the proportions of these textures; that phenomenon could be quantified by measuring their local density distributions.

Another clue in support of that conjecture is the specificity data (Table 3). While the presence of a given texture is not always enough to identify only one given class, its presence might still be a necessity and thus its absence can reliably be used to exclude some classes, especially for stroma (sparsely populated regions with a "sinuous" appearance) and solid formations (dense clumps of large cancerous cell nuclei). The specificity for the excluded regions is lower because of its role as "default class" to resolve ambiguities in our current method, as

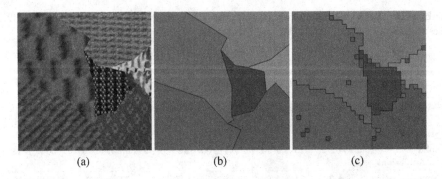

(a)   (b)   (c)

Fig. 12. Partial application of the method to an image of a texture segmentation benchmark available online http://mosaic.utia.cas.cz: (a) original image; (b) ground truth; (c) actual result.

explained previously. At this time, we believe that we don't have enough data on DCIS and nonneoplastic objects to draw a definite conclusion on these classes.

Additional experiments were performed on small computer generated texture mosaics available online as part of the Prague Texture Segmentation Datagenerator and Benchmark [16]. On the normal color set comprising 89 textures in 20 images, the parameterization (LOD 0, Q 5) with square patches of size 16×16 yielded a false positive rate of $0.2\,\%\pm0.2$ and a true positive rate of $79.8\,\%\pm15.6$. At the time of testing, the benchmark website attributed to the results a "correct segmentation" score of 55.44, corresponding to the 16th place. A detailed summary indicated that the quality of the classification depends on the nature of the texture, with textile (Fig. 12) and wood being better characterized than glass and plants. It is interesting to note that despite $(0.2\,\%, 79.8\,\%)$ being a seemingly good ROC point, the segmentation score is only 55.44, due to the distribution of the textures in the images.

4.4   Performance

The experiments were run on an AMD Opteron 2 GHz with 32 Gb of memory. The segmentation step takes at most a few minutes even on large images (less than 2 min for a gigapixel image with our current sequential Java implementation). But, as shown on Fig. 13, the main bottleneck of the method is the size of the training bases, mostly because of the time needed to search a nearest neighbor. So far, this has prevented us from testing our current algorithm with high resolutions but we have verified that capping the training base size to 10000 elements (which is still large) can bring down the computing time to less than 2 h for the highest resolution. That being said, our current results suggest that lower resolutions may already have enough information to completely analyze the image.

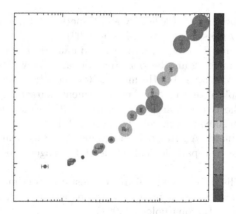

Fig. 13. Mean sequential calculation time by image according to the size of the training base. Each point corresponds to a setting (LOD, Q) for a segmentation of size 32 pixels. The colored discs symbolize the image size, related to the parameter LOD. Some experiments at high resolution were not performed because of excessive time and memory requirements. Standard deviation is given as horizontal and vertical bars (Color figure online).

5   Conclusion and Future Work

The advent of whole slides images is a great opportunity to provide new diagnostic tools and to help pathologists in their clinical analyses. However, it also comes with great challenges, mainly due to the large size of the images and the complexity of their content. To achieve a fast and efficient classification of the images, we proposed in this paper a methodology enabling to partition the initial image in relevant regions. This approach is based on a fast segmentation algorithm and on a supervised classification using a textural characterization of each region. We carried out experiments on 7 annotated images and obtained promising results.

In the future, we will extend the level of detail for annotations and increase the range of tumor variability in order to identify biologically relevant structures more precisely. We are confident that accurate automated detection of clinically relevant regions of interest in cancer-related WSIs for subsequent in-depth analysis is a key contribution to the development of novel tools for biomarker discovery and validation. Increasing speed and accuracy of digital pathology workflows will support the implementations of automated analysis modules into the diagnostic work-up and thereby help to improve cancer therapy directed against targets that are detectable in tissue biopsies.

References

1. Gurcan, M.N., Boucheron, L.E., Can, A., Madabhushi, A., Rajpoot, N.M., Yener, B.: Histopathological image analysis: a review. In: IEEE Reviews in Biomedical Engineering (2009)

2. Tavassoli, F.A., Devilee, P.: Pathology and Genetics of Tumours of the Breast and Female Genital Organs. IARCPress, Lyon (2003)
3. Ghaznavi, F., Evans, A., Madabhushi, A., Feldman, M.: Digital imaging in pathology: Whole-slide imaging and beyond. In: Annual Review of Pathology (2013)
4. Signolle, N., Plancoulaine, B., Herlin, P., Revenu, M.: Texture-based multiscale segmentation: application to stromal compartment characterization on ovarian carcinoma virtual slides. In: Elmoataz, A., Lezoray, O., Nouboud, F., Mammass, D. (eds.) ICISP 2008 2008. LNCS, vol. 5099, pp. 173–182. Springer, Heidelberg (2008)
5. Huang, C.H., Veillard, A., Roux, L., Loménie, N., Racoceanu, D.: Time-efficient sparse analysis of histopathological whole slide images. Comput. Med. Imaging Graph. (2010)
6. Elston, C.W., Ellis, I.O.: Pathological prognostic factors in breast cancer. I. the value of histological grade in breast cancer: experience from a large study with long-term follow-up. Histopathology (1991)
7. Ruiz, A., Sertel, O., Ujaldon, M., Catalyurek, U., Saltz, J., Gurcan, M.: Pathological image analysis using the gpu: stroma classification for neuroblastoma. In: 2007 IEEE International Conference on Bioinformatics and Biomedicine (BIBM 2007) (2007)
8. Sertel, O., Kong, J., Shimada, H., Catalyurek, U.: Computer-aided prognosis of neuroblastoma on whole-slide images: classification of stromal development. Pattern Recogn. (2009)
9. Roullier, V., Lézoray, O., Ta, V.T., Elmoataz, A.: Multi-resolution graph-based analysis of histopathological whole slide images: application to mitotic cell extraction and visualization. Comput. Med. Imaging Graph. (2011)
10. Homeyer, A., Schenk, A., Arlt, J., Dahmen, U., Dirsch, O., Hahn, H.K.: Practical quantification of necrosis in histological whole-slide images. Comput. Med. Imaging Graph. (2013)
11. Moore, A.P., Prince, S.J.D., Warrell, J., Mohammed, U., Jones, G.: Superpixel lattices. In: IEEE Conference on Computer Vision and Pattern Recognition (2008)
12. Montanari, U.: On the optimal detection of curves in noisy pictures. Commun. ACM (1971)
13. Ojala, T., Pietikäinen, M., Harwood, D.: A comparative study of texture measures with classification based on featured distributions. Pattern Recogn. (1996)
14. Wemmert, C., Krüger, J., Forestier, G., Sternberger, L., Feuerhake, F., Gançarski, P.: Stain unmixing in brightfield multiplexed immunohistochemistry. In: IEEE International Conference on Image Processing (2013)
15. Fawcett, T.: An introduction to roc analysis. Pattern Recogn. Lett. (2006)
16. Haindl, M., Mikeš, S.: Texture segmentation benchmark. In: Proceedings of the 19th International Conference on Pattern Recognition (2008)

Unsupervised Visual Hull Reconstruction of a Dense Dataset

Maxim Mikhnevich$^{(\boxtimes)}$ and Denis Laurendeau

Computer Vision and Systems Laboratory, Laval University, Quebec, QC, Canada
{maxim.mikhnevich.1,denis.laurendeau}@ulaval.ca
http://vision.gel.ulaval.ca/en/

Abstract. In this paper a method for the reconstruction of an objects Visual Hull (VH) is presented. An image sequence of a moving object under different lighting condition is captured and analyzed. In this analysis, information from multiple domains (space, time and lighting) is merged based on a MRF framework. The advantage of the proposed method is that it allows to obtain an approximation of an object 3D model without any assumption on object appearance or geometry. Real-data experiments show that the proposed approach allows for robust VH reconstruction of a variety of challenging objects such as a transparent wine glass or a light bulb.

Keywords: Shape from silhouette · SFS · Multi-view image segmentation · Multi-lighting · Visual hull · VH · Graph cuts

1 Introduction

Shape from silhouette (SFS) is a classic computer vision technique for 3D object reconstruction. Exploring this technique in unknown environments when no prior information is available on the object's geometry or surface properties is still a difficult problem. The advantages of using the silhouette for reconstructing the shape of an object is that it requires neither constant object appearance nor the presence of textured regions.In the current work we exploit this property in order to reconstruct the VH of a wide set of objects.

More precisely, the aim of our work is to extract the silhouette of an object from a set of views without prior knowledge of the scene content or the object properties such as appearance and geometry and use these silhouettes to build a VH. This task faces several challenges. Firstly, the object interaction with light includes many effects such as shadows, self-shadows, color bleeding, light inter-reflection, transparency and subsurface scattering. These phenomena have an impact on the appearance of the object in the image and make the separation of foreground from background a complex task. Secondly, the camera can be positioned at any viewpoint on the hemisphere above the object, which leads to the impossibility to model the background at the pixel level (as done previously in static [1] or active [2] cases) before positioning the camera even if

© Springer International Publishing Switzerland 2015
S. Battiato et al. (Eds.): VISIGRAPP 2014, CCIS 550, pp. 257–272, 2015.
DOI: 10.1007/978-3-319-25117-2_16

the viewpoints are calibrated. Finally, the scene being captured under unknown lighting conditions adds extra complexity to the silhouette extraction problem. To cope with these phenomena we propose a fundamental approach where the object moves in an unknown but static environment while the camera remains fixed for a given viewpoint. The only assumption that is made about the scene is one on the background being static while the object moves. In comparison to other approaches which consider that the scene's background is known beforehand or which assume object photometric consistency, the proposed approach does not make any assumption about the object and therefore allows the handling of a wide variety of objects with surface reflectance properties ranging from textureless to completely transparent.

The experiment is performed as follows: the object is placed on a turntable which is then rotated in order to capture the object from different viewpoints. The images captured with this procedure are processed in a time sequential manner. Assuming a constant background, the time sequence is analyzed and the *background likelihood* is estimated. Then, the *object likelihood* is iteratively updated in order to estimate object boundaries precisely. Finally, several time frames are processed simultaneously to enforce boundary consistency between frames. All the computations are based on a Markov Random Field (MRF) framework and the optimization is performed through graph cuts. The silhouettes obtained for all viewpoints are used to build the VH of the object.

The paper is organized as follows: in Sect. 2 an overview of the related work is given. Section 3 introduces research hypotheses and the notation used in the paper. In Sects. 4–6, the details of the estimation of background and object likelihoods as well as the segmentation framework are presented. Section 7 presents the experiments and discusses the results. The last section provides some conclusions on the proposed approach and identifies directions for future work.

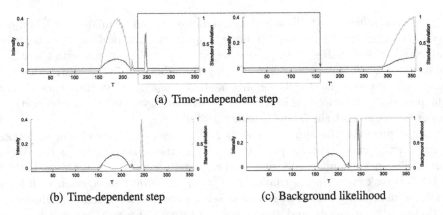

(a) Time-independent step

(b) Time-dependent step   (c) Background likelihood

Fig. 1. Background likelihood estimation. (a) - Time-independent step. First the intensity profile is sorted, then $S(x_i)$ is estimated, and finally estimated values are reordered. (b) - Time-dependent step. $S(x_i)$ is estimated on the original intensity profile. (c) - Background likelihood computed as a combination of time-independent and time-dependent steps using Eq. 6.

2 Related Work

SFS was first introduced by Baumgart [3], this concept suggests to fuse silhouettes of an object in 3D to obtain the VH. Since the object's silhouette is the key element for VH construction, the following review concentrates on silhouette extraction approaches.

The obvious and easy way to implement techniques for silhouette extraction is chroma keying [4]. This approach is based on the knowledge of the scene background. An object is imaged against a uniform or known background, then the silhouette is extracted by thresholding the background color or by background subtraction. Due to implementation simplicity, this technique was used in many SFS works [2,5]. Even though this method provides fairly good results, there are some drawbacks. Firstly, it implies preliminary scene background manipulations for each camera viewpoint, which limits possible camera positions on a hemisphere since the background has to be visible from all viewpoints. Secondly, the case when part of the object has the same color as the background may lead to incorrect segmentation.

Chroma keying was extended in other works where instead of a static background, an active background system was used [2,6]. As an active background, a controlled display was installed around an object. A scene was captured with and without an object with different background patterns for a fixed viewpoint. Even though such an approach allows the extraction of the alpha matte of the silhouette of an object made from material with complex optical properties such as glass, the hardware requirement seriously complicates the acquisition process and limits the method's application area. The major drawback is the inability to move the camera with respect to the background screens, since images with and without an object have to be aligned at the pixel level.

Another group of algorithms with explicit background modeling is based on background subtraction. A good review can be found in [7–9]. A background subtraction technique is based on the construction of a background model of a scene at first, followed by the classification of pixels that do not fit this model as foreground pixels. The major drawback of these methods is the requirement of an explicit estimation of the background. This requirement imposes that an update of the background model needs to be done every time the position of the camera is changed which can be difficult for non uniform backgrounds.

A more universal way to segment images is to rely on user initialization [10]. Here, user input is used to obtain initial information about object and background properties. This information is used to construct a graph and the object segmentation problem is considered as a graph energy minimization. A graph cuts is applied to find the global minimum. In the approach presented in this paper, an energy minimization via graph cuts is also performed to obtain optimal segmentation. However, our goal is to find the initial information required to construct an MRF automatically.

Single image segmentation by graph cut was further extended to automatic silhouette extraction in multiple views in [11,12]. Although these methods may work well, the usage of explicit object and background color modeling limits the type of objects that can be reconstructed. Another drawback related to color modeling is when the same color belongs to the object and background model. In this case, the result may lead to over- or under-estimation of the silhouette. In our work, we avoid explicit color modeling of an object and background in order to overcome these limitations.

3   Hypothesis and Notation

The proposed method is based on the estimation of the background likelihood assuming that the background is unknown but constant and the iterative update of an object likelihood. It is assumed that the camera viewpoint is fixed, and the object changes its position N_T times. In case of multiple lighting conditions, an object is captured N_L times for each frame, one time per source. Note that the proposed method is independent from background modeling, therefore the acquisition process can be repeated multiple times for different camera viewpoints.

The captured image set is organized into a 4D volume. The dimensions of this volume are: U, V, T, and L. U and V are spatial dimensions, T parameterizes object displacement and L represents lighting condition. Thus $I(u, v, t, l)$ is the intensity of a pixel (u, v) at time t under lighting condition l. For notational convenience we define a few shortcuts. $I_L \subset I$ consists of all the images captured under different lighting conditions for a given object position. $I_T \subset I$ is comprised of all the images captured from all the object positions but under fixed lighting. $I_{t,l}$ represents a single image with an object at position t under light source l.

4   Background Likelihood Estimation

In order to estimate background likelihood an "object" and "background" must be defined. A pixel can be called a background pixel if its intensity remains stable for a number of observations among all observations while the object is in motion. This definition follows from the constant background assumption. A pixel whose intensity deviates with respect to its neighbors in time is more likely to represent an object pixel. The definition of an object pixel follows from the fact that during an object motion the orientation of the surface normal of any point on an object changes with respect to the light source or a camera view or both, which is in fact the pixel intensity.

We consider a set of sequential frames as a 3D array and process all subsets of pixels along the time axis. A single subset of pixels form an intensity profile which is defined as:

$$I_T(u, v) = X = \{x_1, x_2, \ldots, x_i, \ldots x_{N_T}\} \tag{1}$$

where x_i is the intensity value of a pixel at time i. This profile is depicted by a blue curve in Fig. 1.

The core idea of measuring background likelihood is an estimation of the time stability $S(x_i)$ in the intensity profile X. It is measured by estimating the minimum standard deviation around each point. The smaller the deviation, the more stable the point is. Thus, a point with low $S(x_i)$ is most likely to belong to the background. In order to estimate the minimum deviation for a given point $x_i \in X$ a window of size w is slid around it and each time the standard deviation is measured. Among measured values, the minimum has to be found. Formally, the measurement of $S(x_i)$ is defined as follows:

$$S(x_i) = \min_{j \in [i-w+1, i]} \sigma_w(x_j), \tag{2}$$

where $\sigma_w(x_j)$ is the standard deviation calculated on the subset $\{x_j, x_{j+1}, \ldots, x_{j+w-1}\}$. $S(x_i)$ describes the constancy of a point x_i in a region with size w.

Since many factors (such as the object's unique geometry, shadows or light inter-reflection in a scene) can affect the intensity of a given pixel, the simple estimation of the stability for each point using Eq. 2 is not robust enough. Therefore, the estimation of the background likelihood is performed in two steps: "time-dependent" and "time-independent". The necessity of the time-independent step is dictated by the possibility that an object may contain gaps between its parts. In such a case the points inside the intensity profile are mixed between object and background. When the pixels's intensity is analyzed independently of its time order, then one can avoid mixing background and object intensities, as shown in Fig. 1(a). The idea of the time-dependent step is to evaluate the property of a point in its original time sequence. It is possible that at some positions, an object point may have the same color intensity as the background. Thus, considering this pixel in its original time sequence order allows a correct estimation of the point deviation as opposed to the time-independent step, see Fig. 1(b). The combination of these two steps leads to a reliable estimation of the background likelihood.

The whole algorithm for background likelihood estimation can be summarized as follows:

1. Sort all the points from the intensity profile:

$$X' = sort(X). \tag{3}$$

2. Time-independent step, see Fig. 1(a):

$$S'_g(x'_i) = \min_{j \in [i-w_g+1, i]} \sigma_{w_g}(x'_j). \tag{4}$$

3. Based on the correspondence between X' and X, reorder S'_g in order to obtain S_g
4. Time-dependent step, see Fig. 1(b):

$$S_l(x_i) = \min_{j \in [i-w_l+1, i]} \sigma_{w_l}(x_j). \tag{5}$$

5. Compute the background likelihood for each point in $x_i \in X$ as follows (Fig. 1(c)):

$$P_B(x_i) = \frac{1}{\exp\left(S_g(x_i) + S_l(x_i)\right)}. \tag{6}$$

Equation 6 is such that it tends to 0 when $S_l + S_g \to \infty$, indicating that the point is inside a varying region and most likely belongs to an object. It tends to 1 when $S_l + S_g \to 0$, meaning that the point is inside a stable region and most likely belongs to the background.

4.1 Space-Time-Light Volume Fusion

The estimation of background likelihood for space-time volume was described above. If the scene is illuminated uniformly by ambient lighting or only a single light source is used during the acquisition process, then it is enough to use Eq. 6 to compute the final background likelihood. However, if several directional light sources are exploited, then a fusion process should be applied in order to incorporate information from different light sources. The difficulty of the fusion is caused by contradictory estimations of background likelihoods from different light sources. For example with one light source, some parts of an object can be in the shadow which results in a high value for background likelihood, due to

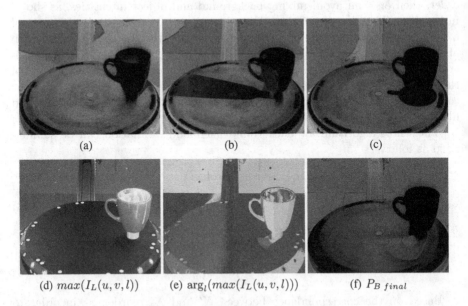

(a)   (b)   (c)

(d) $max(I_L(u, v, l))$ (e) $\arg_l(max(I_L(u, v, l)))$ (f) $P_{B\ final}$

Fig. 2. Computation of $P_{B\ final}$. (a)-(c) - Background likelihoods estimated for different light sources. (d) - Maximum image obtained from 20 images captured under different lighting conditions. (e) - The color represents the index of the light sources that corresponds to maximum intensity. (f) - Estimated background likelihood based on the index from (b) (Color figure online).

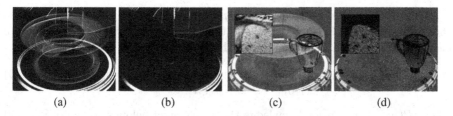

(a)   (b)   (c)   (d)

Fig. 3. Boundary terms comparison. (a) - γ for diagonal pixel neighbors using raw images: a clear object trace can be seen. (b) - γ for diagonal pixel neighbors excluding object pixels: object influence on gamma disappears. (c) - $B_{p,q}$ with γ from (a), an object trace that is present in γ also affects the boundary term. Some object background boundaries are weakly separated due to that trace. (d) - $B_{p,q}$ with γ from (b), object trace does not appear and a clearer separation between the object and the background for some parts (compare to (c)) is obtained.

low intensity deviation for such a region. Under another lighting conditions the same part of an object can be well illuminated and thus have a lower background likelihood.

In order to choose an appropriate light source we use a simple but effective rule (a similar approach was used in [13] for normal initialization). For a given view and pixel we consider all the images under different lighting conditions, and for each pixel, we find the one that corresponds to a maximum intensity. These lighting condition are used to select the background likelihood for a given pixel(see Fig. 2):

$$P_{B\ final} = P_B(u, v, max_{ind}). \tag{7}$$

where $max_{ind} = \arg_l \max(I_L(u, v, l))$.

4.2   Background to Object Likelihood

Since an estimated background likelihood through Eq. 7 is just an approximation, the object likelihood cannot rigorously be estimated as $1 - P_B$. Thus we follow the definition of an object pixel (stated in Sect. 4) defined as a high deviation of the intensity profile. The higher the deviation, the closer P_B is to 0. Therefore all the pixels whose background likelihood are close enough to 0 (less than a threshold R) are assigned a value f_1 in order to indicate that there is a possibility for an object. The other pixels are assigned the value $f_2 \approx \frac{f_1}{10}$, which indicates that these points are less likely to represent an object. The object likelihood is estimated as follows:

$$P_O = \begin{cases} f_1 : P_{B\ final} < R \\ f_2 : \quad otherwise. \end{cases} \tag{8}$$

5   Segmentation as an Optimization Process

In the previous section the estimation of prior background and object likelihoods was described. Now the whole segmentation process can be defined. The goal of segmentation is to assign to each pixel p in image $I_{t,l}$ a label m_p which can be the object or the background. Segmentation is performed by minimization of an energy function E through graph cuts [10]. Formally,

$$E(M) = \lambda \sum_{p \in I_{t,l}} P(m_p) + \sum_{p,q \in N} B(p,q)[m_p \neq m_q], \qquad (9)$$

where $P(m_p)$ is the prior knowledge that each pixel belongs to the object and background; $B(p,q)$ is a boundary term that defines the connection strength between neighboring pixels; M is the set of all labels, each element $m_p, m_q \in M$ can be either background or object with values $\{0,1\}$; λ controls the importance of prior knowledge versus the boundary term ($\lambda \in [0, \infty]$); N is the neighborhood pixel connectivity (in our experiment we use 8-neighbor connectivity).

The boundary term $B(p,q)$ characterizes the relationship between neighboring pixels. If the difference in intensity is small then it is likely that these pixels belong to the same object, therefore they have to be strongly connected. In the case of a large difference in intensity, it is likely that there is an edge and therefore it is probable that these pixels belong to different objects. In such a case $B(p,q)$ should be close to 0 in order to encourage a minimization algorithm to saturate an edge between these points.

Since the object is captured under different lighting conditions, extra information is considered. For example the same point may be in the shadow in one image and may be bright under another light source illumination. This extra data can be used to improve the accuracy of $B(p,q)$. For this purpose, we use the boundary term from [14] and modify it in order to incorporate images captured under different light sources (see Fig. 3(c)):

$$B(p,q) = \sum_{j}^{N_L} \exp\left(-\frac{||I_{p,i,j} - I_{q,i,j}||^2}{2\gamma_{p,q,j}N_L}\right) \cdot \frac{1}{D(p,q)}, \qquad (10)$$

where $I_{p,i,j} = I(u_p, v_p, t_i, l_j)$ and $I_{q,i,j} = I(u_q, v_q, t_i, l_j)$ are intensities for pixel p and q at time t under lighting l_i, $D(p,q)$ is the Euclidean distance between two pixel sites, and $|| \cdot ||$ is L2-norm. γ is constructed as an expected value over time for each connected pair of pixels. In this way γ is adapted for each viewpoint and lighting condition (see Fig. 3(a)):

$$\gamma_{p,q,j} = \sum_{i}^{N_T} \frac{||I_{p,i,j} - I_{q,i,j}||^2}{N_T}. \qquad (11)$$

The prior knowledge term $P(m_p)$ in Eq. 9 defines a preference for pixel p to be object and background:

$$P(m_p) = \begin{cases} P_B & m_p = 0 \ (background), \ Eq. 7 \\ P_O & m_p = 1 \ (object), \ Eq. 8. \end{cases} \qquad (12)$$

Finally, the energy in Eq. 9 is optimized through graph cuts and the result of this optimization is a silhouette of an object for each view which is then integrated into the VH.

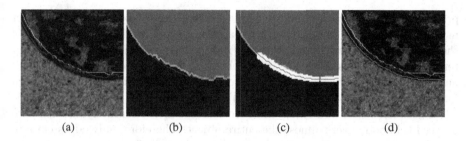

(a)   (b)   (c)   (d)

Fig. 4. Iterations for updating object likelihood. (a) - Boundary term with initial object boundary (white line) and a boundary of the computed silhouette (red line). (b) - Updated object likelihood based on Eq. 14 and the boundary of the new silhouette. (c) - Updated object likelihood based on Eq. 15 and the boundary of the new silhouette. (d) - Boundary term with final silhouette boundary (green line), intermediate boundaries (red lines) and initial object boundary (white line) (Color figure online).

6 Visual Hull Refinement

6.1 Boundary Term Refinement

Having a good approximation of an object shape and its location in each frame allows us to estimate the boundary term more precisely. One of the main parameters of the boundary term is γ. It acts as a threshold: if the difference in intensity between two neighbors is less than γ then the connection between these pixels is strongly penalized. Therefore a clever selection of γ is very crucial for weak edges (when the difference between neighbors is quite small). Thus, it is important to estimate γ as precisely as possible to obtain pure connectivity of background neighboring pixels. Therefore the following procedure was adopted: the VH is projected onto each frame and pixels that belong to the silhouette are excluded from the calculation of γ. This exclusion does not eliminate all the shading effects such as shadows, inter-reflections and color bleeding but their effect is almost negligible and is even reduced by signal averaging over time.

The result is that γ is computed almost only between non object pixels, which is in some way similar to computing γ on the background image (without an object):

$$\gamma_{p,q,j} = \sum_{i}^{N_T} \frac{||I_{p,i,j} - I_{q,i,j}||^2}{N_T}, p,q \notin Pr_i^{-1}(H), \tag{13}$$

where $Pr_i^{-1}(H)$ is a silhouette of the projected VH H on frame i. The result of such an update is shown in Fig. 3(b): the influence of an object's motion on γ almost disappears with the result that a pure background connectivity information between neighboring pixels is estimated.

Substituting γ computed with Eq. 13 in the boundary term in Eq. 10 produces more accurate results, see Fig. 3(d). In Fig. 3(c) the boundary term computed with the initial γ formulation is shown. As it can be seen, some boundary parts between the cup handle and the background are weakly separated due to the presence of the object's motion trace in γ, see Fig. 3(a). However when the object motion is eliminated from γ (Fig. 3(b)) a clearer separation is obtained. One of the issues with the new formulation of γ in Eq. 13 is that the resulting boundary term becomes more sensitive to image changes. It can be seen that much more neighboring weights inside an object receive low penalty compared to the initial formulation of γ (see Fig. 3(c) and (d)). Nevertheless, it is not critical since edges between object and background are detected more accurately and non-zero object likelihood covers almost the entire object. Therefore, only edges close to the object boundary play an important role when maxflow is computed. Note that by computing an adaptive γ for each neighboring pixel connection, most of the background edges are eliminated. In our scene a non uniform background with many edges can be observed, nevertheless almost all the background edges do not appear in boundary term (see Fig. 3(d)). The formulation of this term is one of the contributions of this work.

6.2   Iterative Refinement of Object Likelihood

One source of inaccuracy is the strong edge on the object near the boundary. It is possible that the boundary of the computed silhouette can pass through such strong internal object edges. Therefore we try to find such places and push the boundary out in order to bypass these internal edges. For that reason we apply the following strategy: first we try to push the boundary of the obtained silhouette. If some parts of the boundary move, then we adjust these parts by searching for another strong edge nearby.

As an initial step, all the points that belong to the silhouette are assigned weight w, points located no further than T_1 to the closest point of the silhouette are assigned weight $2 * w$

(a)   (b)

Fig. 5. Acquisition system. (a) - Image of the acquisition system. (b) - CAD model of the system. It consists of a turntable, a lighting system and a camera system. Red arrows show the direction of possible rotation of each element of the setup (Color figure online).

$$P_{O_1}(x_i) = \begin{cases} w & : x_i \in S \\ 2 * w & : dist(x_i, S) < T_1, x_i \notin S \\ 0 & : otherwise \end{cases} \tag{14}$$

Such an update of the object likelihood allows the potential identification of internal object edges that were accepted as the object boundary during the initial calculation of maxflow.

In the second step all the points of a computed silhouette that coincide with the zero region of P_{O_1} or with its boundary form a set C. This set represents points that are close to or belong to an internal object edge. We want to push the silhouette boundary that is inside C to overcome internal edges and move it toward the real object boundary. Therefore the object likelihood is updated as follows:

$$P_{O_2}(x_i) = \begin{cases} 3 * w & : dist(x_i, C) < T_2, \\ P_{O_1}(x_i) & : otherwise \end{cases} \tag{15}$$

We continue to update the object likelihood using Eq. 15 and maxflow calculation until set $C \neq 0$ or until the maximum number of iterations is reached.

All these steps are illustrated in Fig. 4. In Fig. 4(a) the boundary term with the initial object border (white line) and the resulting silhouette border (red line) are depicted. As it can be seen, the boundary of the silhouette goes through the edge inside the object. A new object likelihood is constructed based on Eq. 14, see Fig. 4(b) and the boundary of the resulting silhouette is depicted by the red line. It can be seen that an internal object edge was crossed. Since the resulting silhouette is not totally inside the non-zero region of P_{O_1}, set C is not empty. Therefore, the object likelihood is updated again based on Eq. 15 (see Fig. 4(c)). Finally, the resulting boundary (red line) is completely inside the non-zero region of P_{O_2} and therefore, C is empty. The final silhouette boundary (thick red line) with the boundary term is depicted in Fig. 4(d). The part of the initial boundary that was inside an object was pushed towards the object boundary and the rest of the boundary that was close to the true object-background edge was just slightly adjusted.

Note that when two object parts are separated by the background and the distance between the closest object points is less or equal to T_2, such regions are joined together in the resulting silhouette. This problem is addressed by the final step of the algorithm.

6.3 Visual Hull Completion

Finally, in order to enforce silhouette boundary smoothness and coherency between frames, a graph cuts on a set of sequential frames is performed. Several consecutive frames are considered together and treated as a 3D array. A new graph is constructed in a way similar to what was done previously for each individual frame except for two differences.

As a first difference, the object likelihood is taken from the last step of the iterative algorithm described in Sect. 6.2. All the values that belong to the

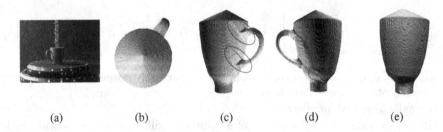

(a)   (b)   (c)   (d)   (e)

Fig. 6. VH of a cup. (a) - An image of cup. (b)-(e) - The VH of the cup from different viewpoints. In (b) a small region near the cup handle goes beyond since the cup handle hides this part from direct observation in several views. In (c) - a small bump can be observed due to target merging with the cup silhouette in some views.

silhouette of the projected VH are taken from P_{O_2}, the rest are set to zero.

$$P_{O_3}(x_i) = \begin{cases} 0 & : x_i \notin Pr_i^{-1}(H), \\ P_{O_2}(x_i) & : otherwise. \end{cases} \tag{16}$$

In using this construction of the object likelihood one can overcome the problem of merging nearby object areas mentioned in Sect. 6.2. Since points located outside of the projected VH are set to 0, a strong object enforcement is eliminated for inter-object areas while the rest of the object likelihood remains the same.

A second difference is that sequential frames have to be connected together by inter-frame arcs in the graph. Based on the object motion between two frames, we can identify which graph nodes must be connected between frames. Using the VH and calibration information for each frame allows the most common object motion directions to be found between two frames. VH voxels are first projected in each frame and then all the projected voxels falling into the boundary of the silhouette at least in one frame are used to form a set of directions:

$$D = Pr_i^{-1}(H(v_i)) - Pr_{i+1}^{-1}(H(v_i)), \forall v_i \in H. \tag{17}$$

The set of directions D may contain a large number of different directions. Therefore, only the most common directions are selected (typically between 8 and 15) to connect nodes between frames. The weight for each inter-frame arc is computed using Eq. 10. The background likelihood term and the boundary term are constructed the same way as explained previously.

7   Experimental Results

The experiments for validating the approach are performed with a roboticized system of our design, which allows the position of a turntable, the camera position on a hemisphere above the turntable and the lighting condition to be controlled by a computer, the setup is shown in Fig. 5. The background behind an

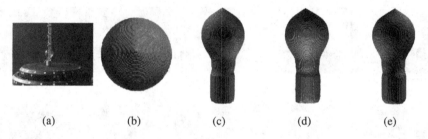

(a)   (b)   (c)   (d)   (e)

Fig. 7. VH of a light bulb. (a) - An image of a light bulb. (b)-(e) - The VH of a light bulb from different viewpoints.

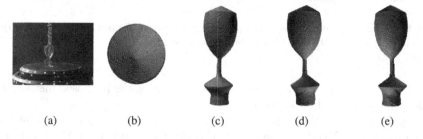

(a)   (b)   (c)   (d)   (e)

Fig. 8. VH of a wine glass. (a) - An image of a wine glass. (b)-(e) - The VH of a wine glass from different viewpoints.

object is not uniform, it consists of: a wall, different parts of the setup and a turntable with white calibration discs. The camera viewpoint is not constant and can be easily changed which leads to a complete change of the observed background. The proposed approach was tested on several objects with complex surface properties. In a typical experiment, an object is rotated 360 times by 1° increments and a grayscale image is captured under 30 different lighting conditions. In cases when the object surface shows specular properties it may reflect light to an area near its base and thereby violate the constant background assumption near this location. By resting the object on a small pedestal on a turntable, this effect is reduced significantly and therefore can be neglected.

Figure 6 shows the VH of a cup. The cup has a smooth conical shape, is made from ceramic and its surface is covered by uniform glossy paint which causes specular reflections and non constant appearance to be observed during image acquisition. Another complication is the difficulty of finding distinctive features on the object for multi-view matching. Such object properties highly complicate the reconstruction of the geometry for feature-based methods. A few traces (enclosed in red ellipses) near the cup handle can be seen (Fig. 6(c)). They appear due to the fact that this area is hidden by the cup handle from direct camera observation in several consecutive views. Also a small bump can be observed near the bottom of the cup base (Fig. 6(d)). It is caused by some circular targets on the turntable treated as part of the silhouette since they match

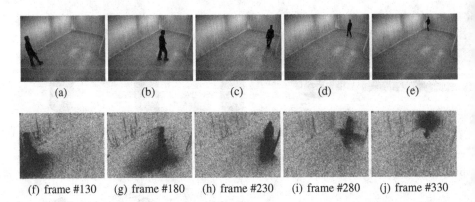

(a)   (b)   (c)   (d)   (e)

(f) frame #130   (g) frame #180   (h) frame #230   (i) frame #280   (j) frame #330

Fig. 9. Walking human dataset. (a)-(e) - Original image frames and detected object silhouette. (f)-(j) - Estimated background likelihood.

the definition of an object. Despite these small errors, the proposed method was able to reconstruct the cup correctly.

Finally the algorithm was tested with fully transparent objects: a light bulb and a wine glass. Due to transparency, it is practically useless to try to estimate the distribution of object colors or to search for distinctive object features, as only the properties of the scene located behind the object will be observed. Another complication with a transparent object is that during its motion, a different background is observed, which makes it difficult to estimate a consistent feature and color model between several views. As it can be seen in Figs. 7 and 8, the body of the light bulb and the wine glass are transparent and the background is visible through them. Since our approach is not based on object color features modeling, it is possible to obtain a reliable reconstruction of the geometry of both the bulb and the wine glass.

Although this work is presented in the context of the exploitation of the special setup in Fig. 5, it is possible to extend the method to the more common case in which a standard camera is installed inside a building, see Fig. 9. The scene was captured by a color camera under a fixed view and single lighting condition. Each color image was decomposed into 3 separate grayscale images and each image channel was treated as a different lighting condition. Since an image sequence was captured under a single light source, a human shadow was also detected as an object. Despite the limited number of lighting conditions, the proposed method was able to separate the human from the background in each frame, without any modification or parameter tuning with respect to the experiments reported in the paper.

8   Conclusions

In this paper an approach for the reconstruction of the Visual Hull of an object with complex photometric properties was described. The proposed approach is based on two principles: modeling scene background based on signal stability which is independent of camera viewpoint and then iterative updating the object likelihood to refine the estimated silhouette boundary accurately. The advantage of the proposed approach is that instead of attempting to model the object color space or matching object features, the evolution of pixel intensity over time is analyzed. Such an analysis avoids the use of standard objects property, such as color and edges and allows the VH to be reconstructed for a wide range of objects with different shapes and reflective properties without any prior knowledge. We show that the proposed method is capable of dealing with objects with complex reflectance properties such as textureless objects or completely transparent ones. The requirement for handling a wide variety objects with completely different photometric properties is that a dense set of images is required for the construction of the Visual Hull. As a future work, we plan to use photometric information for estimating object reflectance properties and fuse this information with the VH to obtain complete object description.

References

1. Snow, D., Viola, P., Zabih, R.: Exact voxel occupancy with graph cuts. In: IEEE Conference on Computer Vision and Pattern Recognition (CVPR) vol. 1, p. 1345 (2000)
2. Matusik, W., Pfister, H., Ngan, A., Beardsley, P., Ziegler, R., McMillan, L.: Image-based 3d photography using opacity hulls. ACM Trans. Graph. **21**, 427–437 (2002)
3. Baumgart, B.G.: Geometric modeling for computer vision. Ph.D. thesis, Stanford, CA, USA (1974)
4. Smith, A.R., Blinn, J.F.: Blue screen matting. In: ACM International Conference on Computer Graphics and Interactive Techniques (SIGGRAPH), pp. 259–268 (1996)
5. Jagers, M., Birkbeck, N., Cobzas, D.: A three-tier hierarchical model for capturing and rendering of 3d geometry and appearance from 2d images. In: International Symposium on 3-D Data Processing, Visualization, and Transmission (3DPVT) (2008)
6. Zongker, D.E., Werner, D.M., Curless, B., Salesin, D.H.: Environment matting and compositing. In: ACM International Conference on Computer Graphics and Interactive Techniques (SIGGRAPH), pp. 205–214 (1999)
7. Piccardi, M.: Background subtraction techniques: a review. In: International Conference on Systems, Man and Cybernetics (SMC), pp. 3099–3104 (2004)
8. Radke, R.J., Andra, S., Al-Kofahi, O., Roysam, B.: Image change detection algorithms: a systematic survey. IEEE Trans. Image Process. **14**, 294–307 (2005)
9. Parks, D.H., Fels, S.S.: Evaluation of background subtraction algorithms with post-processing. In: International Conference on Advanced Video and Signal Based Surveillance, pp. 192–199 (2008)
10. Boykov, Y., Jolly, M.P.: Interactive graph cuts for optimal boundary and region segmentation of objects in n-d images. In: Eighth IEEE International Conference on Computer Vision (ICCV), vol. 1, pp. 105–112 (2001)
11. Campbell, N., Vogiatzis, G., Hernndez, C., Cipolla, R.: Automatic 3d object segmentation in multiple views using volumetric graph-cuts. In: British Machine Vision Conference, vol. 1, pp. 530–539 (2007)

12. Lee, W., Woo, W., Boyer, E.: Identifying foreground from multiple images. In: Yagi, Y., Kang, S.B., Kweon, I.S., Zha, H. (eds.) ACCV 2007, Part II. LNCS, vol. 4844, pp. 580–589. Springer, Heidelberg (2007)
13. Wu, C., Liu, Y., Ji, X., Dai, Q.: Multi-view reconstruction under varying illumination conditions. In: Proceedings of the IEEE International Conference on Multimedia and Expo, pp. 930–933 (2009)
14. Rother, C., Kolmogorov, V., Blake, A.: "Grabcut": interactive foreground extraction using iterated graph cuts. ACM Trans. Graph. **23**, 309–314 (2004)

Rock Fragment Boundary Detection Using Compressed Random Features

Geoff Bull$^{(\boxtimes)}$, Junbin Gao, and Michael Antolovich

School of Computing and Mathematics, Charles Sturt University, Bathurst, Australia
{gbull,jbgao,mantolovich}@csu.edu.au
http://www.csu.edu.au/faculty/business/comp-math

Abstract. Sections of the mining industry depend on regular analysis of rock fragmentation to detect trends that may affect safety or production. The limitations inherent in 2D imaging analysis mean that human input is typically needed for delineating individual rock fragments. Although recent advances in 3D image processing have diminished the need for human input, it is often infeasible for many mines to upgrade their existing 2D imaging systems to 3D. Hence there is still a need to improve delineation in 2D images. This paper proposes a method for delineating rock fragments by classifying compressed Haar-like features extracted from small image patches. The optimum size of the image patches and the number of compressed features are determined empirically. Experimental results show the proposed method gives superior results to the commonly used watershed algorithm, and compressing features improves computational efficiency such that a machine learning approach is practical.

Keywords: Compressed sensing · Random projections · Sparse representation · Image patches · Feature extraction · Image segmentation · Classification

1 Introduction

Monitoring rock fragmentation is a very important process in the mining industry. Knowledge of fragmentation can improve the economics of operating a mine through optimizing the operation of crushing equipment, and can be used to improve estimates of the volume of ore remaining in a mine. Most importantly, changes in fragmentation can alert the operators of a mine to potentially fatal conditions developing in the mine.

A standard technique to perform fragmentation analysis has been to sieve particles through progressively finer sieves and then weigh the contents of each sieve. However, increasingly image processing and statistical techniques are being applied to this problem to reduce costs and to make the collection of more data in a timely manner possible.

Most algorithms for performing fragmentation analysis on rock images rely on finding the boundaries that delineate individual rock fragments. Unfortunately simple edge detectors perform poorly at this task and more sophisticated

S. Battiato et al. (Eds.): VISIGRAPP 2014, CCIS 550, pp. 273–286, 2015.
DOI: 10.1007/978-3-319-25117-2_17

approaches are necessary. There are a number of commercial software packages that have been widely used for analyzing fragmentation of rock particles on conveyor belts. One of these is based on a combination of edge detection techniques and watershed segmentation [16]. Unfortunately when used for less constrained images, e.g. of broken rock in an underground draw-point, many systems often require significant manual editing to correct false delineation of fragment boundaries, often taking more than 30 min per image [12]. An underground mine with 300 draw-points would require a total processing time of less than 5 min per image if 1 image per draw-point was to be analyzed per 24 hour period. Recent research [20,23] has focused on using 3D imaging, with promising results, to overcome the limitations of 2D imaging used in traditional analysis systems. However, upgrading the imaging systems already deployed in existing underground mines would be expensive and disrupt operations, and so there remains a need to improve the results of fragmentation analysis using 2D images.

The watershed algorithm [4] treats an image like a topographic surface, and simulates flooding with water from local minima. The lines where waters from different sources meet are called watershed lines, and water is not permitted to cross these. The watershed lines form the boundaries between image segments. A common problem with the watershed algorithm is that it is susceptible to noise and tends to over-segment images. One solution to this uses markers [19], but this solution needs the markers to be manually specified. The mean shift algorithm [10] has been proposed a suitable method for generating markers [2], however that approach is computationally expensive.

Segmentation of grey scale rock images is often very challenging due to poor lighting and shadowing, and color and texture variation, overlapped rocks, fine material and determination of scale [22]. Fragmentation analysis is difficult with 2D images [22] and typically a human operator often must be involved in the analysis. It has been argued that 2D images contain insufficient information to differentiate between overlapping and non-overlapping rocks [23]. However, the fact that a human operator is able to manually edit the results [21] suggests that improved algorithms may still yield better results with 2D images. Watershed segmentation has been extended by incorporating 3D surface data [24] and this approach overcomes some of the issues observed with 2D images. Despite ongoing improvements to this technique [22,23], the fact remains that it is not always practical to collect the necessary 3D data.

A fundamental task for image processing and machine learning is the selection of appropriate features that generalize well and have a low computational overhead. Recent advances have seen Compressed Sensing (CS) [8,14] used to learn features for image analysis and computer vision. For sparse signals, CS allows the sampling rate to be reduced well below the usual Nyquist rate while still allowing almost perfect reconstruction. Storage requirements and computational overhead are reduced accordingly. For CS, signals are "measured" by compressing them as they are acquired.

Classification in compressed space can achieve accuracies close to those achieved by classification in the original signal space [7]. A recent example of

CS in machine learning that confirms this is the use of compressed sensing features to assist in data dimensionality reduction [15]. Randomly projecting data onto lower dimension subspaces has been found [5] to be as effective as conventional dimensionality reduction methods such as principal components analysis (PCA). Moreover, while random projections are significantly less computationally expensive than PCA, they also do not introduce significant distortions to the data. An issue with PCA is that if the data contains outliers the projected subspace can lie an arbitrarily large distance from the true subspace [26].

CS can be used to design projections that increase the level of compression leading to individual features that are more informative than components of the original signal. For example, random feature selection has been used to get more accurate texture classification than with features that had been specifically designed for the task [18].

A recently demonstrated algorithm for tracking objects in video achieves real-time performance by using compressed features [27]. These features have been shown to be also useful for image segmentation [6]. Although the generalized Haar-like feature used has a very high dimensionality, the computational burden is actually very low because features are randomly projected into a low dimensional subspace. The feature is derived from the generalized Haar-like wavelet of [?] which is in turn derived from the Haar-like wavelet popularized by [25]. These features are very useful because they are very sparse [27], enabling dimensionality reduction by random projection, and they are very efficient to compute via the method of integral images [25].

In this paper, the use of machine learning together with compressed random features to solve the problem of finding fragment boundaries is proposed. Manually delineated images of rocks are used to train a support vector machine (SVM) [11], and the model produced is used to predict boundary regions in test images. A compressed Haar-like feature vector is used and compared to using simple brightness patches as features. The minimum length of feature vector needed for good classification is empirically determined, and the optimum size of image patches to use for feature extraction is also found. The use of these techniques is justified by comparing the results with those achieved with the watershed algorithm.

The remainder of this paper is structured as follows: The proposed classification algorithm and the compressed Haar-like feature are described in Sect. 2. The experimental investigation is then presented, along with a discussion of the results, in Sect. 3. The results are summarized and conclusions drawn in Sect. 4.

2   The Proposed Approach

2.1   Compressed Features

In this section the construction of random compressed features is described.

Compressed sensing (CS) [8,14] is based on the idea that a signal can be reconstructed from a very limited number of measurements if the signal has a sparse representation in some basis. When such a sparse representation exists, the signal is said to be compressible. The Restricted Isometry Property (RIP) and the related Johnson–Lindenstrauss (JL) lemma [3] are two well-known theorems of CS. RIP determines the conditions under which a compressed signal can be efficiently reconstructed. From JL it can be shown that if a signal is sampled using a properly designed measurement matrix the distances between signals after compression will be very close to the distances between the same signals before compression [3]. The proposed method relies on this observation.

The properties of compressed sensing only hold for signals that are sparse (i.e. most values being zero). However, many images are not sparse in their original domain and they must be represented in some basis where they are sparse. Consider a signal consisting of a $n \times n$ square patch surrounding a pixel in an input image, that can be represented as an n^2-dimensional vector $\mathbf{x} \in \mathbb{R}^{n^2}$. If the patch $\mathbf{x}$ is compressible, it can be transformed to a sparse vector $\mathbf{f} = \mathbf{\Psi x} \in \mathbb{R}^D$ by a sparsifying matrix $\mathbf{\Psi}$. The sparse vector will have most of the coefficients close to zero. Usually $D \gg n^2$, in our case $D = n^4$, and it would be very inefficient to perform computations directly on the sparsified vector. To make the calculations tractable, a dimensionality reducing transform $\mathbf{\Phi}$, is applied to reduce the signal down to a k-dimensional vector

$$y = \mathbf{\Phi f} = \mathbf{\Phi \Psi x} \in \mathbb{R}^k .\qquad(1)$$

$\mathbf{y}$ is called a Compressed Sensing Feature and $\mathbf{f}$ the Sparse Coding Feature [15]. $\mathbf{\Phi}$ is known as the measurement matrix.

For the measurement matrix to preserve projected distances between feature vectors, the Johnson–Lindenstrauss lemma must be satisfied and Achlioptas [1] found that is the case if the entries in $\mathbf{\Phi}$ are

$$\phi_{ij} = \sqrt{s} \begin{cases} 1 & \text{with probability} \frac{1}{2s} \\ 0 & \text{with probability} 1 - \frac{1}{2s} \\ -1 & \text{with probability} \frac{1}{2s}, \end{cases}\qquad(2)$$

for values of $s = 2$ and 3. In fact, according to Li et al. [17], $s \gg 3$ will still satisfy JL with a limit of $s = D/\log D$ for approximately normal data. A value of $s = D/4$ has been demonstrated to give good results [27].

To generate the sparsifying matrix $\mathbf{\Psi}$, the same method as Zhang et al. [27] is used. The $n \times n$ square around each pixel is convolved with all possible box filters and the responses of the filters are concatenated into a single feature vector $\mathbf{f} \in \mathbb{R}^{n^4}$. When this vector is multiplied by the sparse measurement matrix $\mathbf{\Phi}$, most of the entries in $\mathbf{f}$ are discarded, and this would be very wasteful if the calculations were performed explicitly. To avoid this, only the non-zero entries of the matrix $\mathbf{\Phi\Psi} \in \mathbb{R}^{k \times n^2}$ are stored, and only the locations, sizes and weights of the rectangles for the box filters that ultimately contribute to a compressed featured are calculated and stored. This approach for calculating a representation of $\mathbf{\Phi\Psi}$ is shown Algorithm 1. Zhang et al. [27] observed that the box filter outputs randomly multiplied by ± 1 results in features very similar to the Haar-like features [25], and are very efficient to calculate by the method of integral images [25].

For an image with N pixels, the $n \times n$ square patch around pixel i is the signal x_i. If all x_i patches in the image are gathered together, a matrix $\mathbf{X} = [\mathbf{x}_1, \ldots, \mathbf{x}_N] \in \mathbb{R}^{n^2 \times N}$ is formed. Then the compressed features for the entire image are $\mathbf{Y} = [\mathbf{y}_1, \ldots, \mathbf{y}_N] \in \mathbb{R}^{k \times N}$, where y_i is given by Eq. 1. Y is calculated according the method in Algorithm 2. The number of compressed features k and the size of the patches n^2 are determined empirically, as discussed in Sect. 3.

2.2 Classification

In this section, the approach for classifying features extracted from rock images is described.

The objective is to identify boundaries between rock fragments in a grey-scale image. A supervised classification method is used to label small image patches as either "boundary" or "non-boundary" according to the following steps. Different images of rocks were selected for training, validation of parameters and final testing. The pixels in these images were manually designated as being on a boundary between rocks, or as not being on a boundary. An area around each boundary was masked so it would not be used for training and classification, as there is some uncertainty about the exact location of the boundary and what classification should be assigned to pixels close to the boundary.

There are many more non-boundary pixels than boundary pixels in an image. Ten percent of boundary pixels in each image were randomly selected for processing, and a similar number of non-boundary pixels were also selected for the training set. This equalization is to avoid the following scenario. If the dataset is very unbalanced, with say 90 % of pixels in one class and 10 % in the other, 90 % accuracy can be achieved simply by classifying *all* pixels as being in the dominant class. It is also possible to handle this situation by weighting training samples; however sampling the data also reduces the processing time so is the preferred method when sufficient data is available.

At each selected pixel location both square patches of brightness features, and compressed Haar-like features, were extracted. A support vector machine (SVM) was trained using a radial basis function (RBF) kernel with the extracted features on the training set. An SVM with RBF has two parameters: C and γ. A grid search was performed to find the best combination of parameters, with the validation dataset being used to measure which was best. A separate validation image was used rather than cross-validation on the training data because the training data, being spatially dependent, are not independent. Cross-validation on the training data would give a biased model that may not generalize well to other data. The methods for training an SVM model and for performing a test using the trained model are shown in Fig. 1.

3 Experimental Results

The aims of the experiments were to establish the overall accuracy of identifying the boundaries of the rock fragments; to compare the accuracy of classification using compressed features with the accuracy using raw intensity patches; to find the optimum sized image patch; to determine the number of compressed features needed to get good accuracy; to understand the variability of results due to random generation of features; and compare the time taken for classification using compressed features with the time taken using raw intensity patches.

3.1 Training, Validation and Testing Data

Unfortunately, to the best of our knowledge, there is not a generally available dataset for this application. For training data, a grey-scale image of some rocks was selected as shown in Fig. 2(a). To create training labels, the edges of rocks were manually delineated using a four pixel wide line, as shown in Fig. 2(d). An

---

Algorithm 1. Calculation of $\mathbf{\Phi\Psi}$ compression matrix representation.

Input:
> patch size n
> number of features k
> minimum number of box filters nf_{min}
> maximum number of box filters nf_{max}

Output: $\mathbf{\Phi\Psi}$ representation

> **for all** feature $j \in \{1 \ldots k\}$ **do**
> > Generate random number of box filters nf
> > $nf \in \{nf_{min} \ldots nf_{max}\}$
> > $weight = 1/\sqrt{nf}$
> > **for all** box filter $bf \in \{1 \ldots nf\}$ **do**
> > > filter location $= random$ within patch
> > > filter size $= random$
> > > assign weight with random sign to filter
> > **end for**
> **end for**
> **return** set of box filters representing $\mathbf{\Phi\Psi}$

Algorithm 2. Calculation of $\mathbf{Y} = \mathbf{\Phi\Psi X}$ for an image.

Input:
> Image
> $\mathbf{\Phi\Psi}$ representation

Output: $\mathbf{Y}$ compressed features for image

> Calculate integral image to be used for filter response calculations.
> **for all** pixel patch x_i **do**
> > **for all** feature $j \in \{1 \ldots k\}$ **do**
> > > $y_{ji} =$ weighted filter response for feature j
> > **end for**
> **end for**
> **return** $\mathbf{Y}$ compressed feature vectors

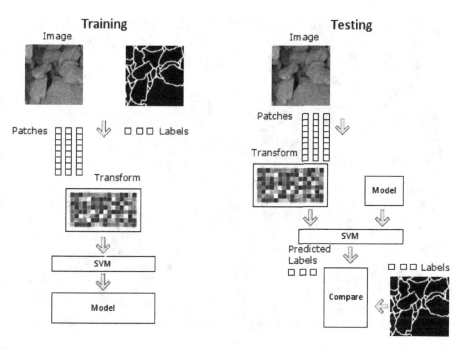

Fig. 1. SVM training consists of creating a model using a training image and associated pixel labels. Features for training are created by taking small image patches and applying the transforms in Eq. (1). When the raw intensity of the image is used as features, the transform is omitted. SVM validation and testing involves taking a test image and extracting features according to Eq. (1). The previously trained SVM model is used to predict labels for the pixels in the test image and the these labels are compared against a ground truth reference.

$n \times n$ square patch was taken around each pixel and the n^2 pixel intensities in that square used as a feature. Pixels within 11 pixels of designated edges were excluded from the training dataset, because of the uncertainty as to whether they should be classified as edge or non-edge. The mask used for this purpose was generated by a morphological dilation of the edges in Fig. 2(d). Pixels from edges were labelled $+1$ and pixels from non-edge regions were labelled -1. Each image in Fig. 2 is 4 Mega-pixels.

Validation and testing data were generated using the same method as for training data, and this is also shown in Fig. 2.

3.2 SVM Training and Parameter Selection

For the experiments, the LIBSVM [9] support vector machine was used with a radial basis function (RBF) kernel. For this, two parameters are needed: C and γ. To choose the best values for the parameters, a grid search was performed using the validation image. The parameters that achieved the best accuracy for classifying the validation image were selected for use in testing.

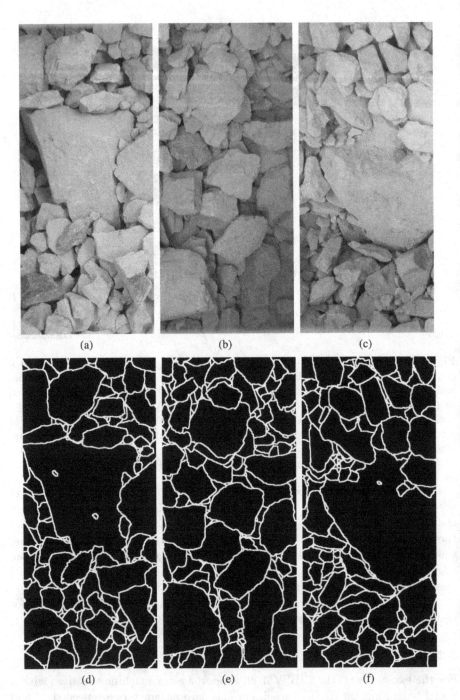

Fig. 2. Data used for the experiments. The image data used for training, validation and testing is shown in (a), (b) and (c) respectively. The pixel labels used are shown below each image, in (d), (e), and (f).

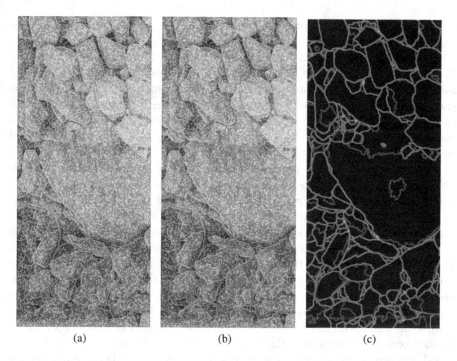

(a)   (b)   (c)

Fig. 3. (a) Result for classifying sampled 15 × 15 patches from the test image using compressed Haar-like features (20D), (correctly classified samples are marked with a green square, and incorrect samples with a red square); (b) result for classifying raw intensity patches (225D); and (c) comparison of watershed boundaries (red) with human drawn boundaries (green) in test image, yellow indicates the two coincide, and blue dots are watershed markers (Color figure online).

To balance the training set, all pixels on edges were included, pixels not on edges were randomly sampled so that the training set includes features for edge and non-edge pixels in approximately equal proportions. Then to reduce the size of the training set, 10 % of the data were randomly selected to be actually used for training.

3.3 Results and Discussion

To establish the overall accuracy of identifying the boundaries of the rock fragments and to compare the accuracy of classification using compressed features with the accuracy using raw intensity patches two SVMs were trained and validated using the appropriate images from Fig. 2, using 15 × 15 image patches. The first SVM used compressed Haar-like features of dimension 20, while the second used the raw intensity patch (dimension 225). Pixels from the test image in Fig. 2 were sampled and classified by both SVMs. The resulting classifications are displayed in Fig. 3. In each classified image, correctly classified pixels are displayed in green and incorrect pixels are red. The pixels that were not sampled are shown in their original color.

The test accuracy for the compressed features in Fig. 3(a) is 86.7 % and for the raw intensity features in Fig. 3(b) is 82.5 %. The result images for compressed features and for the raw intensity patch classifications have a different appearance partly because the two types of features having different precision/recall characteristics. With the raw intensity patches, more rock boundaries are recalled correctly but this is offset by poorer precision causing solid rock to misclassified as boundaries.

The performance of the proposed method was compared with the watershed algorithm on the test image. A satisfactory algorithm for generating watershed markers was not available, so markers were placed at the center of each human identified fragment. This means that the watershed results were unrealistically good. The boundaries found by the watershed algorithm compared to human drawn boundaries are shown in Fig. 3(c). If a watershed boundary was within 11 pixels of a "true" edge, the edge was deemed as having been recalled. Using this method, recall was calculated as 79.9 %. However there were many watershed lines that did not correspond to edges, resulting in a precision (the % of predicted edges that are actually edges) being only 67.8 %. For classification of compressed features in Fig. 3(a), the recall is 80.7 % and the precision is 89.8 %. As a proxy for the accuracy of watershed, the F score is 82.5 %. The F score for classification is 85 %, which is very close to the measured accuracy. The two approaches are comparable (assuming proper markers can be generated for watershed), except that the watershed tends to generate more edges that are well away from the true edges.

The accuracy of classification using compressed features are compared with the accuracy using raw intensity patches are compared in Table 1. Similar accuracies were achieved, identical for a 15×15 patch. This is expected as the accuracy of classification of compressed signals should be close to that achieved for uncompressed signals [7].

To determine the optimum size of the image patches, compressed features of length 50 were generated for patch sizes from 5×5 through to 23×23. The validation accuracy is plotted in Fig. 4, together with the test accuracy. The validation accuracy was used to select the optimum size image patch: 15×15. The choice of image patch size is confirmed using the test accuracy. The 13×13 patch actually had a slightly better test accuracy than the chosen patch size of 15×15.

To determine how many features were needed in the compressed feature vector, tests were performed with many different patch sizes and feature vector lengths. The testing and validation accuracies for 7×7, 15×15 and 21×21 are plotted against the feature vector length in Fig. 5. It was found for all patch sizes that accuracy was very poor with only a few features, where under-fitting would be occurring. The accuracy increased rapidly with the number of features. However, increasing feature lengths above 20 did not result in significant further increases in accuracy, regardless of the patch size. The accuracy for the 7×7

Table 1. Validation accuracy for three sizes of image patches using 20 compressed features per patch, compared to using raw image intensity features. The classification accuracies for the compressed and raw patches are comparable, as expected.

	CS (20)	Raw
7×7 Accuracy	80.4 %	83.5 %
15×15 Accuracy	85.3 %	85.3 %
21×21 Accuracy	85.3 %	83.8 %

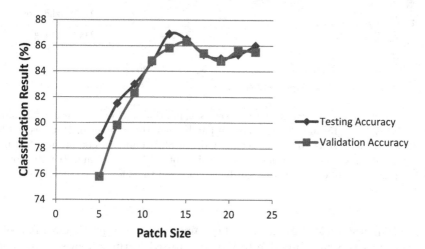

Fig. 4. Validation and test accuracy for patch sizes from 5×5 to 23×23, with the intensity patch compressed to 50 features. The optimum size patch is selected by finding the patch size with the highest validation accuracy, which is 15×15.

patch, which of course contains 49 pixels, dropped off sharply, and rose again, when the compressed feature length increased above 50. A possible cause of this is overfitting.

Since the features are randomly generated, and the dimensionality reduced by using random projections, a test was performed to determine whether results varied greatly due to the random variations inherent in the method. Tests with varying random number generator seeds were run for images with compressed features of length 20 generated from 15×15 patches. The accuracy for the test images, across five tests, varied between 85.6 % and 86.7 %. This small variation is expected with a properly constructed measurement matrix and sparse input signal.

A benefit of the compressed feature is that there is a much smaller computational burden. For a 15×15 patch with 20 compressed features the grid search and evaluation needed to train the SVM model (requiring training 90 models and performing 270 classifications) could be completed in approximately 6.74 hours. For the same patch, using the raw image intensity the grid search took in 74.5

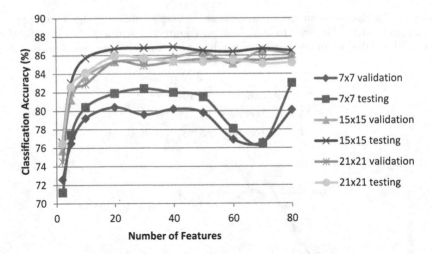

Fig. 5. Number of compressed features verses validation and test accuracy for patch sizes of 7 × 7, 15 × 15 and 21 × 21. For all patch sizes, the number of features can be compressed to as few as 20 before accuracy drops off.Number of compressed features verses validation and test accuracy for patch sizes of 7 × 7, 15 × 15 and 21 × 21. For all patch sizes, the number of features can be compressed to as few as 20 before accuracy drops off.

hours. While the SVM only needs to be trained once for a particular mine site, the shortened training time due to the compressed features allows many more design options to be explored. The time to predict labels is a more important issue, as it is incurred for processing every image. The times taken to predict labels for the test image using 15 × 15 patches are given in Table 2, where the entry for 225 features is for the raw intensity patch and the other entries are for compressed features. The use of compressed features reduces the processing time by over 25×, making the overall approach feasible.

Table 2. Time in seconds for classifying the test image using 15 × 15 patches for different numbers of features. The entry with 225 is the raw patch with no compression. This demonstrates that feature compression is important for good performance, achieving more than 25× speed up for 20 features.

No. Features	20	50	80	225 (raw)
Prediction time (s)	21	30	120	579

4   Conclusion

A method has been proposed to delineate rock fragments with a possible application for the mining industry. The method overcomes some disadvantages of existing techniques by using a machine learning approach, combined with random compressed features, to classify small image patches. For the data set used, the optimum patch size was found to be 15×15. Just 20 compressed features were sufficient to give as good, or better, classification accuracy than the full 225 pixels in a raw image patch. Approximately 85 % of rock fragments are classified correctly. The method compares favorably to the watershed algorithm and has the advantage of not needing markers. The small size of the compressed feature improved training times by a factor of 10 and classification times by more than 25, over using uncompressed image patches. Future work will look at enhancing the feature to improve accuracy, particularly for the case of overlapped rocks, and differentiating fine particles from solid rock.

Acknowledgements. The research is sponsored by the Newcrest Mining Project at CSU and the Compact Grant from the Faculty of Business at CSU.

References

1. Achlioptas, D.: Database-friendly random projections. In: Proceedings of the Twentieth ACM SIGMOD-SIGACT-SIGART Symposium on Principles of Database Systems, PODS 2001, pp. 274–281. ACM, New York (2001)
2. Amankwah, A., Aldrich, C.: Automatic ore image segmentation using mean shift and watershed transform. In: 2011 21st International Conference on Radioelektronika (RADIOELEKTRONIKA), pp. 1–4 (2011)
3. Baraniuk, R., Davenport, M., DeVore, R., Wakin, M.: A simple proof of the restricted isometry property for random matrices. Constructive Approximation **28**(3), 253–263 (2008)
4. Beucher, S., Lantuejoul, C.: Use of watersheds in contour detection. In: International Workshop on Image Processing, Real-time Edge and Motion Detection/Estimation (1979)
5. Bingham, E., Mannila, H.: Random projection in dimensionality reduction: applications to image and text data. In: Proceedings of the Seventh ACM SIGKDD International Conference on Knowledge Discovery and Data Mining, KDD 2001, pp. 245–250. ACM, New York (2001)
6. Bull, G., Gao, J., Antolovich, M.: Image segmentation using random features. In: The 2013 5th International Conference on Graphic and Image Processing (ICGIP 2013) (2013)
7. Calderbank, R., Jafarpour, S., Schapire, R.: Compressed learning: universal sparse dimensionality reduction and learning in the measurement domain. In: Manuscript (2009)
8. Candes, E., Tao, T.: Decoding by linear programming. IEEE Trans. Inf. Theory **51**(12), 4203–4215 (2005)
9. Chang, C.C., Lin, C.J.: LIBSVM: a library for support vector machines. ACM Trans. Intell. Syst. Technol. **2**, 27:1–27:27 (2011). http://www.csie.ntu.edu.tw/cjlin/libsvm
10. Comaniciu, D., Meer, P.: Mean shift analysis and applications. In: The Proceedings of the Seventh IEEE International Conference on Computer Vision, 1999, vol. 2, pp. 1197–1203 (1999)
11. Cortes, C., Vapnik, V.: Support-vector networks. Mach. Learn. **20**, 273–297 (1995)

12. Demenegas, V.: Fragmentation analysis of optimized blasting rounds in the aitik mine: effect of specic charge. Masters thesis, Luleå Tekniska Universitet (2008)
13. Dollar, P., Tu, Z., Tao, H., Belongie, S.: Feature mining for image classication. In: IEEE Conference on Computer Vision and Pattern Recognition, 2007 (CVPR 2007), pp. 1–8 (2007)
14. Donoho, D.: Compressed sensing. IEEE Trans. Inf. Theory 52(4), 1289–1306 (2006)
15. Gao, J., Shi, Q., Caetano, T.S.: Dimensionality reduction via compressive sensing. Pattern Recogn. Lett. 33(9), 1163–1170 (2012)
16. Girdner, K., Kemeny, J., Srikant, A., McGill, R.: The split system for analyzing the size distribution of fragmented rock. In: Franklin, J.A., Katsabanis, T. (eds.) Measurement of Blast Fragmentation - Proceedings of the FRAGBLAST 5 Workshop, pp. 101–108. CRC Press/Balkema, Boca Raton, Fla (1996)
17. Li, P., Hastie, T.J., Church, K.W.: Very sparse random projections. In: Proceedings of the 12th ACM SIGKDD International Conference on Knowledge Discovery and Data Mining, KDD 2006, pp. 287–296. ACM, New York (2006)
18. Liu, L., Fieguth, P.: Texture classication from random features. IEEE Trans. Pattern Anal. Mach. Intell. 34(3), 574–586 (2012)
19. Meyer, F., Beucher, S.: Morphological segmentation. J. Vis. Commun. Image Represent. 1(1), 21–46 (1990)
20. Noy, M.J.: Automated rock fragmentation measurement with close range digital photogrammetry. In: Blanco, J.A.S., Singh, A.K. (eds.) Measurement and Analysis of Blast Fragmentation: Workshop FRAGBLAST 10 - The 10th International Symposium on Rock Fragmentation by Blasting, pp. 13–21. CRC Press/Balkema, Boca Raton, Fla (2013)
21. Siddiqui, F., Shah, S.A., Behan, M.: Measurement of size distribution of blasted rock using digital image processing. Eng. Sci. 20(2), 81–93 (2009)
22. Thurley, M.J.: Fragmentation size measurement using 3D surface imaging. In: Blanco, J.A.S. (ed.) Fragblast 9 : Rock Fragmentation By Blasting, Proceedings of the 9th International Symposium On Rock Fragmentation by Blasting, pp. 229–237. CRC Press/Balkema, Boca Raton, Fla (2009)
23. Thurley, M.J.: Automated, on-line, calibration-free, particle size measurement using 3D prole data. In: Blanco, J.A.S., Singh, A.K. (eds.) Measurement and Analysis of Blast Fragmentation: FRAGBLAST 10 - The 10th International Symposium on Rock Fragmentation by Blasting, pp. 23–32. CRC Press/Balkema, Boca Raton, Fla (2013)
24. Thurley, M.J., Ng, K.C.: Identifying, visualizing, and comparing regions in irregularly spaced 3D surface data. Comput. Vis. Image Underst. 98(2), 239–270 (2005)
25. Viola, P., Jones, M.: Robust real-time object detection. Int. J. Comput. Vision 57(2), 137–154 (2002)
26. Wright, J., Ganesh, A., Rao, S., Peng, Y., Ma, Y.: Robust principal component analysis: exact recovery of corrupted low-rank matrices via convex optimization. In: Bengio, Y., Schuurmans, D., Laerty, J., Williams, C.K.I., Culotta, A. (eds.) Advances in Neural Information Processing Systems, vol. 22, pp. 2080–2088 (2009)
27. Zhang, K., Zhang, L., Yang, M.-H.: Real-time compressive tracking. In: Fitzgibbon, A., Lazebnik, S., Perona, P., Sato, Y., Schmid, C. (eds.) ECCV 2012, Part III. LNCS, vol. 7574, pp. 864–877. Springer, Heidelberg (2012)

Retargeting Framework Based on Monte-carlo Sampling

Roberto Gallea$^{(\boxtimes)}$, Edoardo Ardizzone, and Roberto Pirrone

DICGIM, Universita' degli Studi di Palermo, Viale delle Scienze Ed. 6,
III Piano, 90128 Palermo, Italy
{roberto.gallea,edoardo.ardizzone,roberto.pirrone}@unipa.it

Abstract. Advance in image technology and proliferation of acquisition devices like smartphones, digital cameras, etc., made the display of digital images ubiquitous. Many displays exist in the market, spanning within a large variety of resolutions and shapes. Thus, displaying content optimizing the available number of pixels has become a very important issue in the multimedia community, and the image retargeting problem is being widely faced. In this work, we propose an image retargeting framework based on monte-carlo sampling. We operate the non-homogeneous resizing as the composition of several simple atomic resizing functions. The shape of such atomic operator can be chosen within a set of tested functions or the user could design additional ones. Using independent atomic operators allows parallelizing the retargeting procedure. Additionally, since the algorithm does not require any optimization, it could be executed in real-time, which is a key aspect for on-line visualization of multimedia content.

Keywords: Retargeting · Monte-carlo · Saliency · Image resizing

1 Introduction

The diffusion of display devices coming with different aspect ratios and resolutions, entails using content-aware resizing techniques. Simple cropping is not sufficient due to severe information loss. On the other hand, homogeneous scaling with aspect ratio variation, introduces unwanted distortions in the images. A proper non-homogeneous resizing operator is required in order to preserve image content, introducing deformations just in the low-importance regions of the image.

In this paper, we present a novel image retargeting technique, which is both efficient and effective. Differently from many literature approaches, such a method does not require neither energy minimization nor functional optimization, and relies just on Monte Carlo sampling. Our model estimates the deformation likelihood of each image region, according to the image saliency. Then,

Electronic supplementary material The online version of this chapter (doi:10. 1007/978-3-319-25117-2_18) contains supplementary material, which is available to authorized users.

© Springer International Publishing Switzerland 2015
S. Battiato et al. (Eds.): VISIGRAPP 2014, CCIS 550, pp. 287–299, 2015.
DOI: 10.1007/978-3-319-25117-2_18

by extracting random samples over this probability distribution, less important regions get more deformed, while high-saliency ones are preserved. In order to perform unhomogeneous resizing across the image, a set of atomic resizing operator is proposed. Additionally the system has modular capabilities, and any proper algebraic function could be used for the purpose. Another advantage of using a sample-based approach is that it can be implemented easily using a parallel scheme, thus improving efficiency.

Salient regions can be extracted using several content relevance estimators, such as visual saliency maps [1,2], corner detectors [3], eye-gaze measurement [4], etc. Additionally, both automatic or interactive cues can be given to improve the results: people detectors [5] or face detectors [6] can help in preserving people and faces in the images. Finally, other geometric constraints can be provided by the user to preserve structures explicitly.

2   Related Work

In general, the resizing operators used by image processing applications work by resizing images to a target size by means of homogeneous shrinking or enlarging operators. After early works based on cropping, like [7], more recent approaches use adaptive image resizing. The idea is to preserve important image features by applying a non-linear content driven resize operator. Remarkable works were done using seam carving, [8,9] where 1D seams are removed/added to reduce/increase the image size. Such seams are chosen from low energy regions of the image. However, due to the discrete nature of this method, notches in the objects may appear. In addition, when no more discardable information exists, important details get removed and severe distortions may appear. Warping methods [10] overcome this limitation by squeezing or stretching homogeneous regions, while minimizing the distortion in relevant regions. In [11] regions are scaled by different factors in order to preserve aspect ratio too. Multi-operator approach [12], uses a combination of seam carving, scaling and cropping. Seam carving is very efficient but limited in its use, warping methods are more effective but computationally expensive, almost prohibiting their use in real-time applications with high resolution images or embedded devices with low power profiles. A comprehensive evaluation of several reference literature methods is provided in [13].

3   Image Resizing Approach

In our model an input image $\mathbf{I}$ is considered as a set of n lines (the columns or the rows) $\mathbf{I} = \{l_0, l_1, \ldots, l_{s-1}\}$, where l_i are the initial lines positions and s is the initial image size along the considered dimension. Thus, $l_i = i \ \forall \ i \in \{0, s-1\}$. To resize the image to the new dimension s' we look for the new set $\mathbf{I}' = \{l'_0, l'_1, \ldots, l'_{s-1}\}$ where distances between two consequent lines should be preserved in most informative image regions in order not to introduce distortions as in Eq. (1),

$$(l_i - l_{i-1}) = (l'_i - l'_{i-1}).$$ (1)

Obviously, some distances have to be necessarily changed due to resizing, and some deformation must be introduced. The model is built in order to spread the required deformations across the whole image in a non uniform way that obeys to a probability distribution. This is done applying multiple atomic resize operators centered in image regions sampled from a proper probability mass function. Such function is built according to lines significance. The idea is to apply less deformations in salient regions of the image, while the most deformation affects the unimportant zones.

The whole system is realized by means of a chain, which is schematized in the block diagram in Fig. 1. The input image I is given as input to the system. The saliency estimator generates a saliency map, which is used in turn to build a deformation probability mass function $dpmf$. Such a function is sampled to move the lines of the image non-homogeneously. Finally, the image is reconstructed and the final retargeted image is produced as output.

Fig. 1. Block diagram of the retargeting system. The input image I is the input of the system. The saliency estimator generates a saliency map, which is then used to build a deformation probability mass function $dpmf$; in turn the $dpmf$ is sampled to move the lines of the image non-homogeneously. Finally, the image is reconstructed, and the final retargeted image is produced as output.

3.1   Model Formulation

The proposed method is based on two concepts:

- a resizing operator
- a deformation strategy.

Resizing Operator. The global resizing operator we introduce is considered as the multiple application of several atomic resizing operations. Each atomic resizing operates on the image lines $l_0 \cdots l_{s-1}$, moving each one it by a given quantity defined by the function $r(l_i)$, which is expressed in (fractions of) pixels, and defines the level of detail of the transformation. Relative movement between consequent lines has the effect of deforming the underlying image. Varying the function $r(l_i)$ affects:

- the amount of displacement of each atomic resizing, which is equal to $k_{lod} = max(r(l_i))$
- the number of atomic resizing n_r (see Eq. 2) to produce the required image final size.

These two quantities define the resolution of the global resizing operator. Of course, as the level of detail gets finer, the computational burden gets heavier, due to an increasing number of atomic resizing operations. In the following paragraphs, the selection of the k_{lod} parameter will be discussed in detail.

$$n_r = s'/k_{lod}. \tag{2}$$

Deformation Strategy. In order to apply the described resizing operator, a line selection strategy needs to be designed to determine which line should be moved at each resizing step. The required procedure has to exhibit a dual behavior: firstly, it should select less important lines from a visual content importance perspective, as the candidates for resizing since distortions should be preferentially introduced in low-importance or homogeneous regions. On the other hand, deformations should be distributed across the whole image in order not to remove whole image regions, thus introducing severe artifacts. In order to attain such behaviors, we define a *deformation probability mass function - dpmf* $p_d(x)$ over the image **I**. Such a *dpmf* indicates the likelihood that a single resizing operation would affect the image line l_x.

Intuitively, such a probability should be related to the image visual content importance. In particular, line relevance $R(i)$ is extracted from the dual form of visual saliency, i.e. visual inconspicuousness. Visual saliency $S(i, j)$ of each pixel is extracted using either Itti's saliency detector [1] or signature saliency [2]. Such values are then projected along the considered resizing axis using the maximum operator, and are complemented as in Eq. (3).

$$R(i) = 1 - max_j S(i, j). \tag{3}$$

Finally, the values are normalized w.r.t. their summation, as in Eq. (4) to recover $p_d(i)$; an example is reported in Fig. 2.

$$p_d(i) = R(i)/\sum_{j=0}^{s-1} R(j). \tag{4}$$

Here high-value points correspond to image regions with high-probability of being deformed, while low-value points correspond to image regions that should be preferentially preserved. Note that visual saliency can also be improved either interactively by adding constraints, or automatically using people [5] or face detectors [6], and modifying $R(i)$ to have low values in presence of constraints or people/faces. Note that this operation must be performed prior to normalization reported in Eq. (4) to preserve the $p_d(i)$ integral to sum to 1, thus being a valid *dpmf*.

In order to obtain the actual retargeting, we run a Monte Carlo process. As defined in Eq. (2), n_r samples are drawn from the $p_d(x)$ distribution and each extracted corresponding line is used as the application point l_c of the atomic resize function $r(l_i)$. The result is that each line l_i in the image gets a chance to

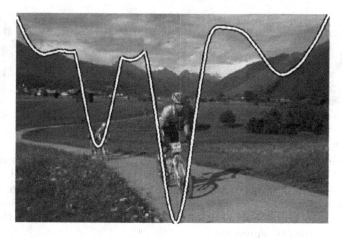

Fig. 2. Plot of the *deformation probability mass function $p_d(i)$* related to the example image. The function represents the probability that a given line l_i will be subject to an atomic resizing step during the retargeting operation.

be moved, proportionally to its incospicuousness value. Statistically, the deformations are spread across the whole image, limiting the presence of artifacts, while still preserving important regions.

After recovering new lines position l_i', the resulting image needs to be reconstructed. This process requires an interpolation procedure, since l_i' values are generally real values. Any interpolation scheme could be used for this purpose. Choosing the best interpolating function is out of the scope of this paper, so no further investigations were done in this direction. However simple linear interpolation gives satisfying results, so it has been used for generating all of the results in this work. For illustrating the whole process, Fig. 4 reports the sampling of $p_d(i)$ for the image of Fig. 2 for a width scaling ratio $s_w = 0.5$ and $k_{lod} = 0.1$. In the picture y values correspond to how many times the line l_i was drawn from the *dpmf*. In the plot is evident how lines belonging to salient regions are drawn rarely or not drawn at all, leaving the underlying content undeformed.

3.2 $r(l_i)$ Function Design

Basically, any algebraic function could be used as the atomic retargeting operator r_i. However, in order to provide meaningful results, it should be designed following some guidelines:

- Since the value recovered from $r(l_i)$ represents the amount of displacement of the line l_i, in order to avoid artifacts, the function should be monotonically increasing.
- The function should provide values comprised in $[0, k_{lod}]$, i.e. the values should be positive and within the chosen level of detail.
- Impose $k_{lod} \leq 1$ to avoid overlap between lines while resizing.

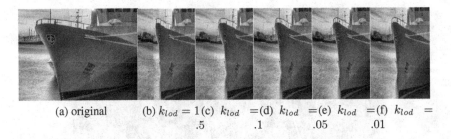

(a) original   (b) $k_{lod} = 1$ (c) k_{lod} =(d) k_{lod} =(e) k_{lod} =(f) k_{lod} =
.5   .1   .05   .01

Fig. 3. Detail of an output image using different values for k_{lod} and scaling ratio $s_w = 0.5$. From left to right: (a) original image, (b) $k_{lod} = 1$, (c) $k_{lod} = 0.5$, (d) $k_{lod} = 0.1$, (e) $k_{lod} = 0.05$, (f) $k_{lod} = 0.01$. Even though the effect is more noticeable when the images are larger, as it can be seen in this image portion, the finest results are provided using a high level of detail. However, values of k_{lod} smaller than 0.1 do not provide remarkable improvements.

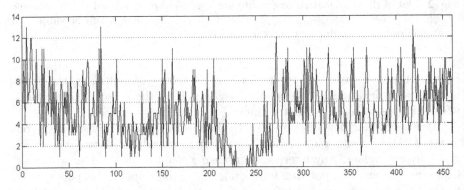

Fig. 4. Plot of the sampling of the $dpmf$ for the picture shown in Fig. 2. Each column corresponds to the number of times each column was drawn from the $dpmf$.

Keeping in mind these guidelines, several atomic retargeting operators were tested, which are: step function (5), line segment function (6), sigmoid function (7), s-shaped function (sshaped).

$$r(l_i) = \begin{cases} 0 & if \quad l_i < l_c \\ k_{lod} & if \quad l_i \geq l_c \end{cases}, \tag{5}$$

$$r(l_i) = \begin{cases} 0 & if \quad l_i < l_c - d \\ k_{lod}\frac{l_i - (l_c - d)}{d} & if \quad l_c - d \leq l_i < l_c \\ k_{lod} & if \quad l_i \geq l_c \end{cases}, \tag{6}$$

$$r(l_i) = \frac{k_{lod}}{1 + e^{-d[l_i - (l_c - d/2)]}} \tag{7}$$

$$r(l_i) = \begin{cases} 0 & if \quad l_i < l_c - d \\ k_{lod}2\left(\frac{l_i-(l_c-d)}{d}\right)^2 & if \quad l_c - d \leq l_i \leq \frac{2l_c-d}{2} \\ k_{lod}(1 - 2\left(\frac{l_i-l_c}{d}\right)^2) & if \quad \frac{2l_c-d}{2} \leq l_i \leq l_c \\ k_{lod} & if \quad l_i \geq c \end{cases} \qquad (8)$$

where l_c is the function application point and d is a parameter defining the region where the function increases (i.e. where the image get deformed). The used function and the meaning of the parameters are shown in Fig. 5.

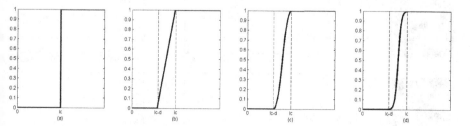

Fig. 5. Shapes of the atomic resizing functions. (a) Step function, (b) line segment, (c) sigmoid function, (d) s-shaped function. L_c represents the application point of the function, while d defines the region of increment of the function.

Regarding the quality of the produced images, the differences are mainly related to the smoothness of the global displacement functions. Figure 6 depicts the results of using the four different atomic resizing functions proposed. In Fig. 6a the final retargeted image is shown, while in Fig. 6b-c are plotted the global displacement function and the sampled atomic resizings respectively. Even though the results are not deterministring, and some differences can arise between different samplings, the trend for the four different global displacement curves is similar. The smoother curve is the one obtained using sigmoid function, at a cost of a slighter higher computational effort. Note that these results were obtained with $k_{lod=1}$.

3.3 k_{lod} Parameter Selection

The whole procedure is automatic, just the parameter k_{lod} requires to be tuned. Since it influences the quality of the result, it should be as smaller as possible. However, the computation time is in inverse proportion to the parameter value, so it should be determined as the best trade-off between quality and efficiency. For a visual evaluation purpose, we report the results using different values for k_{lod} (1, 0.5, 0.1, 0.05 and 0.01), see Fig. 3. The results show that using a coarse level of detail causes artifacts. However, using too fine level of detail is not useful, since the resulting image quality does not get remarkable improvements. However, visual inspection is not sufficient to determine how to choose k_{lod}. More

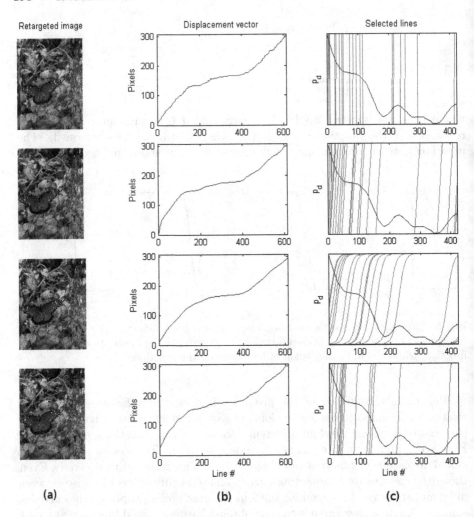

Fig. 6. Results using different atomic resizing functions. From row 1 to 4 are shown respectively: step function, line segment function, sigmoid function and s-shaped function. In (a) the final retarget image is shown. (b) is the plot of the global displacement curve, and (c) shows the sampled atomic resizing over the saliency curve of the lines in the image.

objective cues are derived by measuring the variations of two image difference indexes: Root Mean Squared Error - $RMSE$ and Structural Similarity - $SSIM$ [14]. Measures have been computed between the image produced using the highest level of detail (approximated using a very low value of $k_{lod} = 0.001$) and the one resulting using a k_{lod} value varying in the interval $[0.001, 1]$. The results of this experimentation are shown in Fig. 7. In Fig. 7(a) the results are better for lower values of $RMSE$. In Fig. 7(b) the results are better for higher values of $SSIM$. The plots exhibit jumps which are due to interpolation artifacts that arise during image reconstruction. Such artifacts get smaller as k_{lod} decreases. When k_{lod} approaches values around 0.1, jumps disappear, the trend becomes asymptotic and the quality of the result has very low variations. As a consequence $k_{lod} = 0.1$ is assumed to be used for the referred results in the next sections.

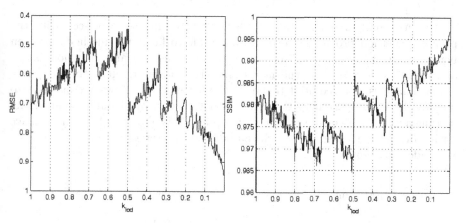

Fig. 7. Plot of $RMSE$ (a) and $SSIM$ (b) against k_{lod} value with scaling ratio $s_w = 0.5$. In (a) the results are better for lower values of $RMSE$. In (b) the results are better for higher values of $SSIM$. Image quality plots exhibit jumps due to interpolation artifacts that arise during image reconstruction. However, such artifacts get smaller as k_{lod} decreases. When k_{lod} approaches values around 0.1, jumps disappear, the trend becomes asymptotic and the quality of the result has very low variations.

4   Experimental Results and Discussion

The described method was implemented on a PC with Quad CPU 2.30 GHz. The system can benefit from parallel computation by leveraging gpGPU capabilities being implemented using Nvidia CUDA API extensions [15].

Comparison. To evaluate the results of our system, we compared it with other literature retargeting systems. For space reasons, this paper references four methods: Multi-operator [12], non-homogeneous warping [10], seam carving [9] and scale-and-stretch [11]. However, several other methods were compared and the reader is referred to the supplemental material provided with this paper. The evaluation was assessed using the datasets and measures provided by the *RetargetMe* comparative study [13]. Examples of comparisons are shown in Fig. 8.

Additionally to qualitative images inspection, an objective evaluation was also taken. Two comparative measures were used for this purpose: Earth Mover's Distance (EMD) [16] and SIFTflow [17]. These are two commonly used similarity metrics, which do not require the two datasets to be the same size, a binding property for image retargeting. Both measures use a dense SIFT descriptor [18], which captures structural properties of the image robustly, while EMD also uses a state of the art color descriptor (ciede2000). The two measures both endorse their solutions to small and smooth local displacements, reflecting the way human vision system tolerates deformations and the operations applied by retargeting operators.

Results, summarized in Table 1, show that the images produced with the proposed method provide measures comparable to literature methods, or even better. Most of the existing literature methods, tend to warp the whole image and make them fit it into the new frame size. However, often the periphery of the image is not required to be kept. Our method allows intrinsically to discard the whole periphery data, if it is not salient, by strongly compressing it, thus achieving a certain extent of cropping. This allows to keep more space for important image regions, which can be better preserved without introducing heavy deformations.

Table 1. EMD and SIFTflow measures for images of *RetargetMe* framework.

Measure	EMD	SIFTflow
Monte-carlo	$8.01 \pm 3.23 \cdot 10^3$	$3.98 \pm 2.02 \cdot 10^5$
Multi-operator	$8.30 \pm 3.58 \cdot 10^3$	$3.94 \pm 1.99 \cdot 10^5$
Non-homogeneous	$8.68 \pm 3.73 \cdot 10^3$	$4.12 \pm 2.15 \cdot 10^5$
Seam carving	$8.69 \pm 3.60 \cdot 10^3$	$4.09 \pm 2.38 \cdot 10^5$
Scale and stretch	$8.95 \pm 3.82 \cdot 10^3$	$5.37 \pm 2.69 \cdot 10^5$

4.1 Complexity Considerations

Looking at the proposed retargeting operator from a complexity perspective, is possible to take both memory and computational considerations.

The memory amount required to store all the data needed to retarget an image composed of s into one composed of s' lines is the following:

- s real values to store the positions of the lines l_i,
- s real values to store the $dpmf$,
- $s' \cdot k_{lod}$ real values to store the samples extracted from the $dpmf$,

As a consequence, the proposed method needs a total of $2s \cdot s' \cdot k_{lod}$ real values, keeping the memory complexity polynomial.

From a computational point of view, the main burden is related to the saliency extraction which is common in all of the retargeting methods, so it is not considered. For the same reason, image reconstruction is not taken into account. The rest of the process is accomplished by the following operations:

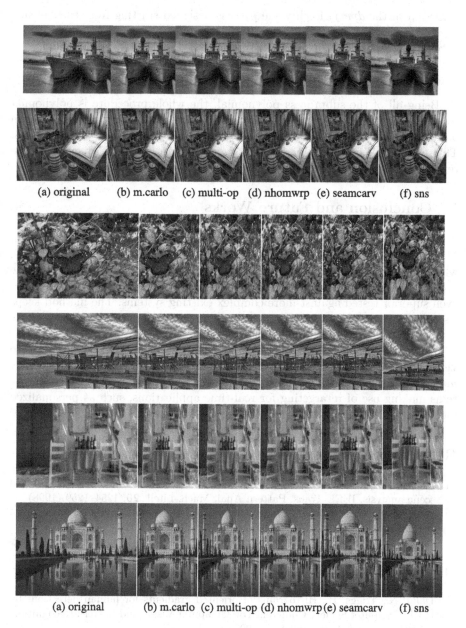

(a) original   (b) m.carlo   (c) multi-op   (d) nhomwrp   (e) seamcarv   (f) sns

(a) original   (b) m.carlo (c) multi-op (d) nhomwrp(e) seamcarv   (f) sns

Fig. 8. Comparison results for some (a) test images: methods reported are (b) our Monte Carlo method, (c) Multi-operator [12], (d) non-homogeneous warping [10], (e) seam-carving [9] and (f) scale-and-stretch [11]. The first two rows are compressed using $s_w = 0.75$, while the last four rows are compressed with $s_w = 0.5$. Note how in the *butterfly* and *deck* images, low-saliency periphery content has been cropped by extreme line compression, allowing more space for important image data.

- Design of the $dpmf$. Each value $p_d(i)$ is designed starting from the saliency $S(i,j)$ using the max($\cdot$) operator $\rightarrow$ polynomial,
- Sampling $p_d(i)$. This operation is repeated $s' \cdot k_{lod}$ times $\rightarrow$ polynomial,
- Updating of the lines position l_i according to the extracted samples $\rightarrow$ polynomial.

Being all of the subprocess polynomial, the whole procedure is polynomial too. In addition, all of the previous operations can be easily implemented in parallel, since little or no dependencies exists both between data and processes. This allow very fast one-shot retargeting of images, opposed to many of the reference literature methods relying onto iterative optimization.

5   Conclusion and Future Works

A novel efficient method for image retargeting was presented. It is based on Monte Carlo sampling of the deformation probability mass function of the image, which is defined using the image saliency map. This allows its use for real-time applications. Experimental results show that its performance are comparable or even superior tested against more complex existing systems. The method keeps its complexity very low both from a memory and computational perspective, also leveraging the parallelization of its processes.

Further work will involve overall system improvements and its extension to video resizing. This issue requires the introduction of a time-coherent saliency map and further constraints. Additionally, the model will be embedded in systems making use of retargeting for real-time applications, such as personalized media content distribution on mobile devices or the web.

References

1. Itti, L., Koch, C., Niebur, E.: A model of saliency-based visual attention for rapid scene analysis. IEEE Trans. Pattern Anal. Mach. Intell. **20**, 1254–1259 (1998)
2. Hou, X., Harel, J., Koch, C.: Image signature: Highlighting sparse salient regions. IEEE Trans. Pattern Anal. Mach. Intell. **34**, 194–201 (2012)
3. Harris, C., Stephens, M.: A combined corner and edge detection. In: Proceedings of the Fourth Alvey Vision Conference, pp. 147–151 (1988)
4. Santella, A., Agrawala, M., Decarlo, D., Salesin, D., Cohen, M.: Gaze-based interaction for semi-automatic photo cropping. In. In CHI 2006, pp. 771–780 (2006)
5. Dalal, N., Triggs, B.: Histograms of oriented gradients for human detection. In: IEEE Computer Society Conference on Computer Vision and Pattern Recognition, CVPR 2005, vol. 1, pp. 886–893 (2005)
6. Viola, P., Jones, M.: Robust real-time object detection. Int. J. Comput. Vis. (2001)
7. Suh, B., Ling, H., Bederson, B.B., Jacobs, D.W.: Automatic thumbnail cropping and its effectiveness. In: UIST 2003: Proceedings of the 16th Annual ACM Symposium on User Interface Software and Technology, pp. 95–104. ACM, New York (2003)
8. Avidan, S., Shamir, A.: Seam carving for content-aware image resizing. ACM Trans. Graph. **26**, 10 (2007)

9. Rubinstein, M., Shamir, A., Avidan, S.: Improved seam carving for video retargeting. ACM Trans. Graph. **27**, 1–9 (2008)
10. Wolf, L., Guttmann, M., Cohen-Or, D.: Non-homogeneous content-driven video-retargeting. In: Proceedings of the Eleventh IEEE International Conference on Computer Vision (ICCV-07) (2007)
11. Wang, Y.S., Tai, C.L., Sorkine, O., Lee, T.Y.: Optimized scale-and-stretch for image resizing. In: ACM Transactions on Graphics, Proceedings of ACM SIGGRAPH ASIA, vol. 27 (2008)
12. Rubinstein, M., Shamir, A., Avidan, S.: Multi-operator media retargeting. In: ACM Transactions on Graphics, Proceedings SIGGRAPH 2009, vol. 28, pp. 1–11 (2009)
13. Rubinstein, M., Gutierrez, D., Sorkine, O., Shamir, A.: A comparative study of image retargeting. ACM Transactions on Graphics, Proceedings SIGGRAPH Asia, vol. 29 (2010)
14. Wang, Z., Bovik, A.C., Sheikh, H.R., Member, S., Simoncelli, E.P.: Image quality assessment: from error measurement to structural similarity. IEEE Trans. Image Process. **13**, 600–612 (2004)
15. NVIDIA CUDA Compute Unified Device Architecture - Programming Guide (2007)
16. Pele, O., Werman, M.: Fast and robust earth mover's distances. In: ICCV (2009)
17. Liu, C., Yuen, J., Torralba, A., Sivic, J., Freeman, W.T.: SIFT Flow: Dense Correspondence across Different Scenes. In: Forsyth, D., Torr, P., Zisserman, A. (eds.) ECCV 2008, Part III. LNCS, vol. 5304, pp. 28–42. Springer, Heidelberg (2008)
18. Lowe, D.G.: Distinctive image features from scale-invariant keypoints. Int. J. Comput. Vis. **60**, 91–110 (2004)

Author Index

Printed in the United States
By Bookmasters